BS 7671: REQUIREMENTS FOR ELECTRICAL INSTALLATIONS

IEE WIRING REGULATIONS
SEVENTEENTH EDITION

Study Notes

constructionskills

E1

Published by ConstructionSkills, Bircham Newton, King's Lynn, Norfolk, PE31 6RH

© **Construction Industry Training Board 1982, 1992, 2008**

The Construction Industry Training Board otherwise known as CITB-ConstructionSkills and ConstructionSkills is a registered charity (Charity Number: 264289)

First published 1982
New edition published 1992
Revised 1994
Revised 1995
Revised 1996
Revised 1997
Revised 1998 (twice)
Revised 2000
Revised 2001
Revised 2002
Revised June 2002
Revised June 2003
Revised June 2004
Revised June 2005
Revised June 2006
New edition published February 2008
Reprinted with amendments June 2008

ISBN: 978-1-85751-278-6

ConstructionSkills has made every effort to ensure that the information contained within this publication is accurate. Its content should be used as guidance material and not as a replacement for current Regulations or existing standards.

All rights reserved. No part of this publication may be reproduced, stored in a retrieval system or transmitted in any form or by any means, electronic, mechanical, photocopying, recording or otherwise, without the prior permission in writing from ConstructionSkills.

CONTENTS

Acknowledgements

Preface

Introduction

1 Scope, object and fundamental principles

Scope
Object and effects
Fundamental principles

2 Definitions

3 Assessment of general characteristics

Purposes, supplies and structure
Classification of external influences
Compatibility
Maintainability
Safety services
Continuity of service
Technical data sheets:
    3A   Electrical supply systems
    3B   Accessories and diversity
    3C   Standard circuits

4 Protection for safety

Protection against electric shock
Protection against thermal effects
Protection against overcurrent
Protection against voltage disturbances and electromagnetic disturbances
Technical data sheets:
    4A   Residual Current Device (RCD)
    4B   Consumer units
    4C   Fuses
    4D   Circuit breakers
    4E   Operating characteristics of overcurrent protective devices

5 Selection and erection of equipment

Common rules
Selection and erection of wiring systems
Protection, isolation, switching and monitoring
Earthing arrangements and protective conductors
Other equipment
Safety services
Technical data sheets:
    5A   Index of protection (IP) code
    5B   Types and methods of installing earth electrodes
    5C   Earth monitoring equipment
    5D   Protective multiple earthing (PME)

6 Inspection and testing

Initial verification
Periodic inspection and testing
Certification and reporting
Technical data sheets:
- 6A Safely isolate the supply
- 6B NAPIT certificates and reports

7 Special installations or locations

Introduction
Locations containing a bath or shower
Swimming pools and other basins
Rooms and cabins containing sauna heaters
Construction and demolition site installations
Agricultural and horticultural installations
Conducting locations with restricted movement
Electrical installations in caravan/camping parks and similar locations
Marinas and similar locations
Exhibitions, shows and stands
Solar photovoltaic (PV) power supply systems
Mobile or transportable units
Electrical installations in caravans and motor caravans
Temporary electrical installations for structures, amusement devices and booths at fairgrounds, amusement parks and circuses
Floor and ceiling heating systems
Technical data sheet:
- 7A Electrical supplies on construction and demolition sites

8 Cable selection

Current-carrying capacities of cables

9 Sizing conduit and trunking systems

Other mechanical stresses
Projects
Model answers

10 Major changes of BS 7671: 17th Edition IEE Wiring Regulations 2008

Appendix 1 Step by step procedure for the selection of a final circuit cable and protective device

Appendix 2 Cable selection procedure

Index

ACKNOWLEDGEMENTS

ConstructionSkills wishes to express its thanks to the following organisations for their co-operation in allowing extracts and illustrations from their various publications to be reproduced in these Study Notes:

- The Institution of Electrical Engineers
- On-site Guide to the 16th Edition Wiring Regulations
- NAPIT (National Association of Professional Inspectors and Testers)
- The British Standards Institution

We also wish to thank the following companies for permission to illustrate their products:

- Crabtree Electrical Industries Ltd
- Evershed and Vignoles Ltd
- GEC Fusegear Ltd
- WJ Furse & Co. Ltd
- Davis Trunking Ltd
- BICC Components Ltd
- BICC Pyrotenax Ltd
- Clare Instruments Ltd

Note: These study notes have been prepared using existing information and table from the On-site Guide to the 16th Edition of BS 7671, as the updated version to the 17th Edition of BS 7671 was not available at the time of printing.

PREFACE

This publication seeks to reconcile design requirements for electrical installations, with actual work methods employed by practising installation electricians.

ConstructionSkills wishes to express its thanks to WR Allan C.Eng. MIET technical adviser.

Note: These study notes contain abbreviated extracts and paraphrases of BS 7671:2008. It is emphasised that these interpretations of the British Standard have been devised for the purpose of training and should not be regarded as authoritative in any other context. When necessary, the British Standard should be referred to directly.

Technical data sheets, where appropriate, have been included in the various sections of this publication, where detailed information on equipment or systems may be required to fully understand the Regulations.

INTRODUCTION

Plan and style of Regulations

The 17th Edition is based on international Regulations produced initially by the International Electrotechnical Commission (IEC) on a world-wide basis. The standards are then voted on by CENELEC for adoption by European countries. When agreed, these standards are published as CENELEC Harmonisation Documents (HDs) with the obligation that member countries include the documents, or at least the technical intent of them, in their national standard. The Preface to BS 7671 shows the HDs which have been agreed so far in CENELEC, the technical intent of which has been included in the UK national standard, BS 7671.

Plan

The plan of BS 7671:2008 has been agreed internationally for the arrangement of safety rules for electrical installations.

The Regulations are comprised of seven parts with 15 appendices. This structure is illustrated overleaf.

Numbering

The Regulation numbering follows the pattern and corresponding references of IEC 60364.

Each part of the Regulations is numbered consecutively, being identified by the first number of each group of digits. The parts are divided into chapters, identified by the second digit and each chapter is split into sections, identified by the third digit. After the first group of three digits, the digits separated by full stops identify the Regulation itself.

Example 1

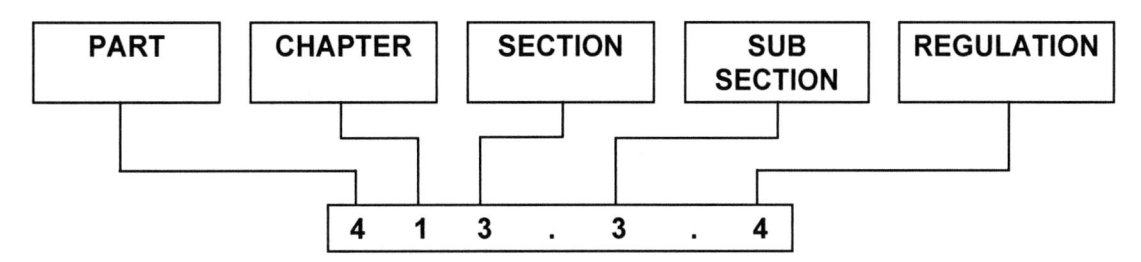

PART	CHAPTER	SECTION	SUB SECTION	REGULATION
4	1	3	3	4

Example 2

Part 4	– Protection for safety
Chapter 41	– Protection against electric shock
Section 413	– Protective measures: Electrical separation
Subsection 413.3	– Requirements for fault protection
Regulation 413.3.4	– Flexible cable and cords

Regulation number: 4 1 3 . 3 . 4

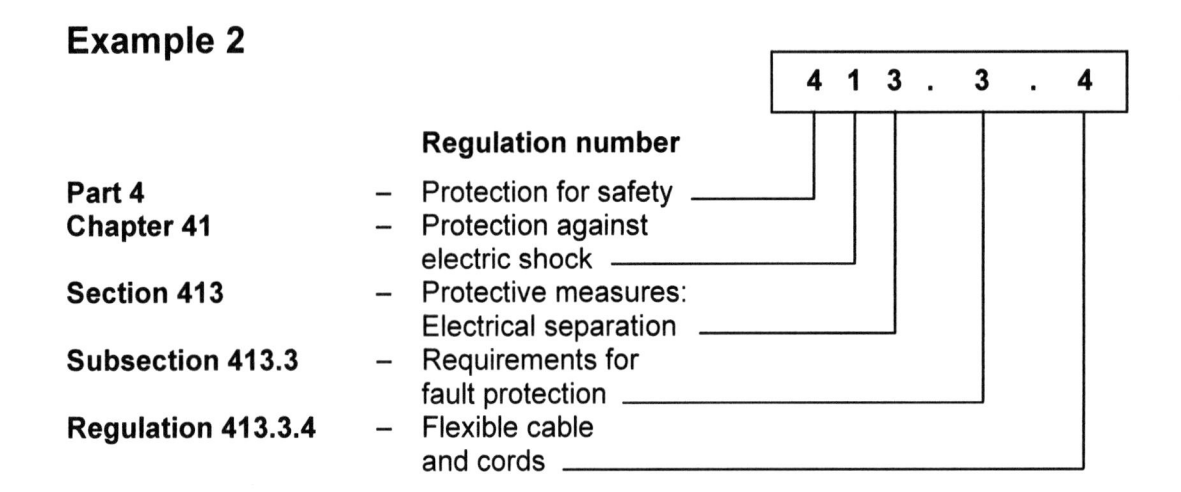

Each Part has a contents page and each chapter starts with an introduction, general scope or other statement explaining what the chapter covers or contains.

Part 7 is different as the number appearing after the section number refers to the corresponding chapter, section or Regulation within Parts 1 to 6. For example, Regulation 701.415.2 concerns supplementary bonding and is made up as follows:

- Section 701 – Locations containing a bath or a shower
- Regulation 415.2 – Additional protection: Supplementary equipotential bonding.

Cross-referencing related sections may be simplified:

- through the use of the index
- by reference to the diagram on the following page.

Relevance to statutory Regulations

The following statutory Regulations recognise BS 7671 as a code of good practice:

- Health and Safety at Work etc. Act 1974 and HSE guidance
- Electricity at Work Regulations 1989
- Electricity Safety, Quality and Continuity Regulations 2002.

Plan of 17th Edition

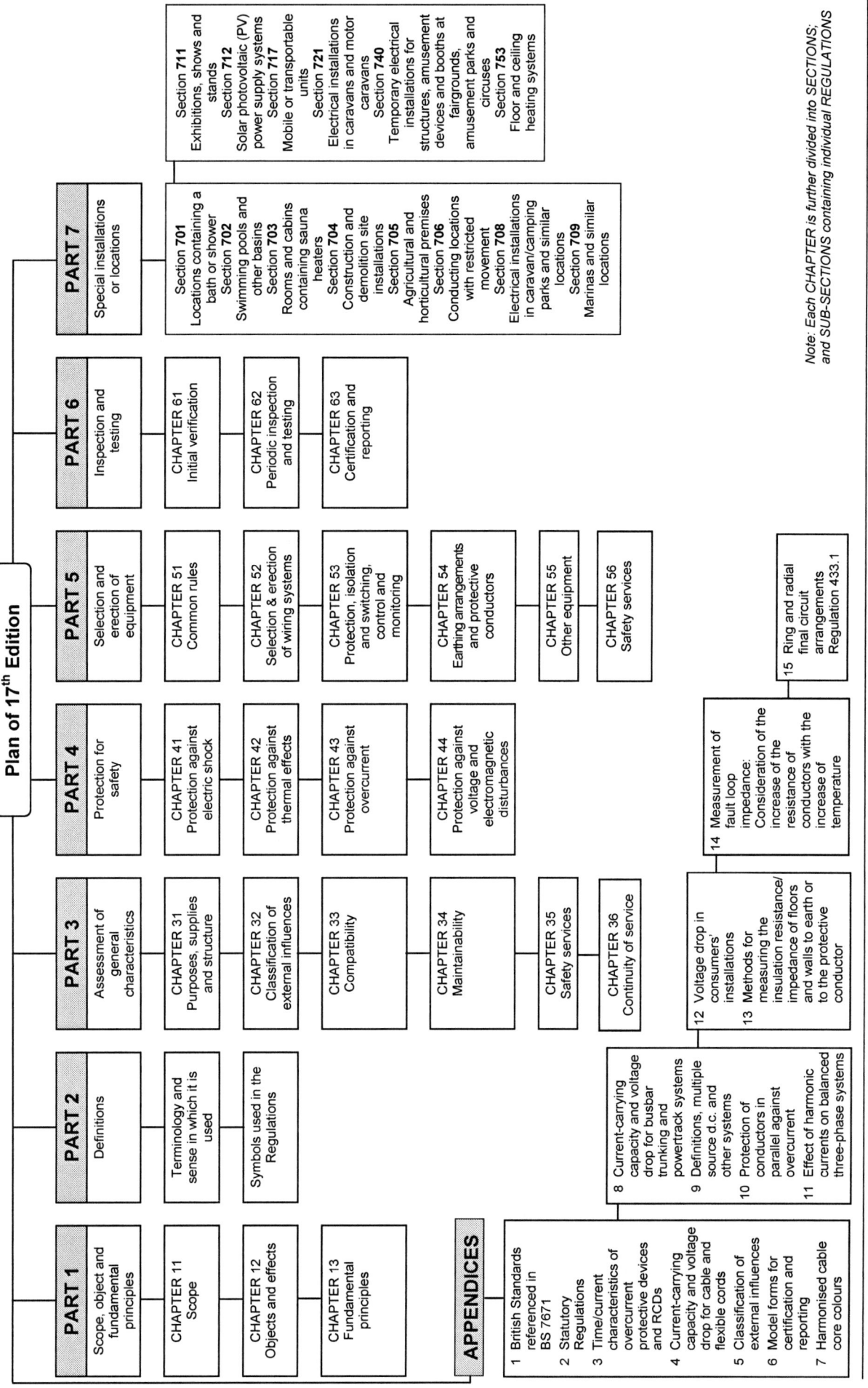

PART 1 — Scope, object and fundamental principles
- CHAPTER 11 Scope
- CHAPTER 12 Objects and effects
- CHAPTER 13 Fundamental principles

PART 2 — Definitions
- Terminology and sense in which it is used
- Symbols used in the Regulations

PART 3 — Assessment of general characteristics
- CHAPTER 31 Purposes, supplies and structure
- CHAPTER 32 Classification of external influences
- CHAPTER 33 Compatibility
- CHAPTER 34 Maintainability
- CHAPTER 35 Safety services
- CHAPTER 36 Continuity of service

PART 4 — Protection for safety
- CHAPTER 41 Protection against electric shock
- CHAPTER 42 Protection against thermal effects
- CHAPTER 43 Protection against overcurrent
- CHAPTER 44 Protection against voltage and electromagnetic disturbances

PART 5 — Selection and erection of equipment
- CHAPTER 51 Common rules
- CHAPTER 52 Selection & erection of wiring systems
- CHAPTER 53 Protection, isolation and switching, control and monitoring
- CHAPTER 54 Earthing arrangements and protective conductors
- CHAPTER 55 Other equipment
- CHAPTER 56 Safety services

PART 6 — Inspection and testing
- CHAPTER 61 Initial verification
- CHAPTER 62 Periodic inspection and testing
- CHAPTER 63 Certification and reporting

PART 7 — Special installations or locations
- Section 701 Locations containing a bath or shower
- Section 702 Swimming pools and other basins
- Section 703 Rooms and cabins containing sauna heaters
- Section 704 Construction and demolition site installations
- Section 705 Agricultural and horticultural premises
- Section 706 Conducting locations with restricted movement
- Section 708 Electrical installations in caravan/camping parks and similar locations
- Section 709 Marinas and similar locations
- Section 711 Exhibitions, shows and stands
- Section 712 Solar photovoltaic (PV) power supply systems
- Section 717 Mobile or transportable units
- Section 721 Electrical installations in caravans and motor caravans
- Section 740 Temporary electrical installations for structures, amusement devices and booths at fairgrounds, amusement parks and circuses
- Section 753 Floor and ceiling heating systems

APPENDICES
1. British Standards referenced in BS 7671
2. Statutory Regulations
3. Time/current characteristics of overcurrent protective devices and RCDs
4. Current-carrying capacity and voltage drop for cable and flexible cords
5. Classification of external influences
6. Model forms for certification and reporting
7. Harmonised cable core colours
8. Current-carrying capacity and voltage drop for busbar trunking and powertrack systems
9. Definitions, multiple source d.c. and other systems
10. Protection of conductors in parallel against overcurrent
11. Effect of harmonic currents on balanced three-phase systems
12. Voltage drop in consumers' installations
13. Methods for measuring the insulation resistance/impedance of floors and walls to earth or to the protective conductor
14. Measurement of fault loop impedance: Consideration of the increase of the resistance of conductors with the increase of temperature
15. Ring and radial final circuit arrangements Regulation 433.1

Note: Each CHAPTER is further divided into SECTIONS; and SUB-SECTIONS containing individual REGULATIONS

© Construction Industry Training Board

E1: BS 7671 (February 2008)

SCOPE, OBJECT AND FUNDAMENTAL PRINCIPLES

BS 7671 Part 1

Scope *(Chapter 11)*

The Regulations apply to electrical installations generally, such as those found in:

- residential premises
- industrial and commercial premises
- public premises
- agricultural and horticultural premises
- prefabricated buildings
- caravans, caravan parks and similar sites
- construction sites, exhibitions, shows, fairgrounds and other installations for temporary purposes, e.g. professional stage and broadcasting applications
- marinas
- external lighting and similar installations
- mobile or transportable units
- photovoltaic systems
- low voltage generating sets
- highway equipment and street furniture.

Requirements are included for:

- circuit supplies at nominal voltages up to and including 1,000 V a.c. or 1,500 V d.c. (preferred a.c. frequencies are 50 Hz, 60 Hz and 400 Hz. The use of other frequencies for special purposes is permitted)
- circuits, other than the internal wiring of equipment operating at greater than 1,000 V and derived from an installation having a voltage not exceeding 1,000 V a.c. (such as discharge lighting)
- any wiring systems and cables not covered by an appliance standard
- all consumer installations external to buildings
- fixed wiring for IT and communication equipment, signalling, control, etc. (not including the internal wiring)
- alterations and additions to installations and the subsequently affected parts of existing installations.

Exclusions from scope *(110.2)*

The Regulations **do not** apply to:

- the distributor's equipment (distributors are defined as providers of electricity to consumers using electrical lines and equipment that they own or operate)
- railway traction equipment, rolling stock and railway signalling equipment
- motor vehicles (except those to which the Regulations concerning caravans or mobile units apply)
- ships
- offshore installations
- aircraft
- mines and quarries
- radio interference suppression equipment, except where it affects the safety of an installation
- lightning protection of buildings and structures, except where bonded to the electrical installation covered by BS EN 62305
- lift installations covered by BS 5655 and BS EN 81-1
- machines and electrical equipment covered by BS EN 60204.

Equipment

The Regulations do not apply to the construction of electrical equipment, but only to its selection and application in an installation.

Voltage ranges (Part 2)

The Regulations cover installations with the following operating voltages:

- **Extra-low voltage**: 0 V to 50 V a.c. or 120 V ripple-free d.c., whether between conductors or to earth
- **Low voltage**: Exceeding extra-low voltage to 1,000 V a.c. or 1,500 V d.c. between conductors, or 600 V a.c. or 900 V d.c. between conductors and earth.

Voltage bands (Part 2)

Band I

- Installations where under certain conditions the value of voltage provides protection against electric shock.
- Installations that for operational reasons have their voltage limited, e.g. alarms, controls, telecommunications, signalling, bells, etc.
- Extra-low voltage.

Band II

- The supplies to domestic, commercial and industrial installations.
- Low voltage (Band II voltages do not exceed 1,000 V rms a.c. or 1,500 V d.c.).

Relationship with statutory Regulations *(114)*

BS 7671: Requirements for Electrical Installation are non-statutory, although they may be used as evidence in a court of law to prove compliance with a statutory Regulation, such as those included in Appendix 2 of BS 7671:2008.

Installations in premises subject to licensing

When installations are to be carried out in premises which a licensing or other authority have a statutory control, the requirements of the authority shall be complied with in the design and the installation.

Object and effects *(Chapter 12)*

BS 7671 contains the rules for the design and erection of electrical installations that ensures their safe operation and proper functioning for the intended use.

Chapter 13 contains the fundamental principles. It does not include detailed technical measurements, which may be subject to change due to advances in technology.

The remaining parts of BS 7671 contain the technical requirements to enable compliance with Chapter 13:

- Part 3   Assessment of general characteristics
- Part 4   Protection for safety
- Part 5   Selection and erection of equipment
- Part 6   Inspection and testing
- Part 7   Special installations or locations.

Intentional departures from these parts require the special consideration by the installation designer and must be recorded on the Electrical Installation Certificate.

New materials and inventions

Where the installation of new materials or inventions creates a departure from the Regulations, the level of safety must not be less than that expected by compliance with the Regulations. This use is to be recorded on the Electrical Installation Certificate.

Fundamental principles *(Chapter 13)*

Protection for safety *(131.1)*

This section contains the general requirements for the safety of:

- persons
- livestock
- property

against the risks which could arise in the normal use of electrical installations.

Causes of injury could include:

- shock
- excessive temperatures likely to cause burns or fires, etc.
- mechanical movement of electrically actuated equipment
- explosion.

Protection against electric shock *(131.2)*

Persons and livestock must be protected, so far as is reasonably practicable, against the dangers arising from contact with live parts.

Basic protection *(131.2.1)*

This was previously protection against direct contact and can be achieved by:

- preventing current flowing through persons or livestock
- limiting the value of current that can flow through a body to a non-hazardous value.

Fault protection *(131.2.2)*

This was previously protection against indirect contact and can be achieved by:

- preventing current flowing through persons or livestock
- limiting the value of current that can flow through a body to a non-hazardous value
- limiting the time a fault current can pass through a body to a non-hazardous duration.

The application of equipotential bonding is an important safety principle.

Protection against thermal effects *(131.3)*

Installations must be designed so that the risks of fire due to high temperature or arcing are reduced, so far as is reasonably practicable.

During the normal use of equipment, the risk of burns to persons or livestock must be minimised.

Persons, fixed equipment or fixed materials next to electrical equipment must be protected from the harmful effects of heat or thermal radiation emitted by the electrical equipment, especially:

- combustion, ignition or degradation of materials
- burns
- unsafe functioning of equipment.

Electrical equipment must not create a fire hazard to adjacent materials.

Overcurrent protection *(131.4)*

Persons and livestock are to be protected from injury, and property from damage, due to high temperatures and electromechanical stresses caused by overcurrents in live conductors.

Fault current protection *(131.5)*

All conductors other than live conductors and any other parts intended to carry fault currents are to be capable of carrying the currents without reaching excessive temperatures.

Protection against voltage disturbances and measures against electromagnetic influences *(131.6)*

Persons and livestock are to be protected from injury, and property from damage, due to faults between live parts of circuits supplied at different voltages and any overvoltages that may arise due to switching or atmospheric conditions and as a consequence of undervoltage and any subsequent voltage recovery.

Installations shall be designed to take into account anticipated electronic emissions by the installation or equipment, to ensure an adequate level of immunity against electromagnetic disturbances to enable them to function correctly.

Protection against interruption of the power supply *(131.7)*

Installation or equipment shall be protected from damage due to any interruption of the power supply.

Additions and alterations *(131.8)*

Before an addition or alteration is made to an existing installation it must be verified that the rating and condition of the existing equipment (including the distributor's equipment) is adequate for any anticipated increase in load and that the earthing and bonding arrangements are adequate.

Design *(132)*

An electrical installation must provide for the:

- protection of persons, livestock and property
- proper functioning of the electrical installation for the intended use.

Supply characteristics *(132.2)*

The supply characteristics are to be determined by calculation, measurement, enquiry or inspection.

1. Nature of current (a.c. and/or d.c.).
2. Purpose and number of conductors.

   For a.c:
   - Line conductor(s).
   - Neutral conductor.
   - Protective conductor.
   - PEN conductor.

   For d.c:
   - Equivalent to those conductors opposite.
   - Outer/middle/earthed live conductors.
   - Protective conductor.
   - PEN conductor.

3. Values and tolerances:
   - Nominal voltage and tolerances.
   - Nominal frequency and tolerances.
   - Maximum current allowable.
   - Prospective fault current (PFC).
   - Earth fault loop impedance.
4. Protective measures in the supply (e.g. earthed neutral).
5. The distributor's particular requirements.

Nature of demand *(132.3)*

The circuits required for lighting, power, control, signalling, communication and IT equipment, etc. are to be determined from:

- location of points of power demand
- loadings
- daily and annual power demand variations
- special conditions (e.g. harmonics)
- signalling, control, IT and communication requirements
- anticipated future demand.

Electric supplies for safety services or standby supply systems *(132.4)*

Where these supplies are required, the supply characteristics and the number of circuits to be supplied must be determined.

Environmental conditions *(132.5)*

Equipment must be constructed or protected so as to prevent danger from exposure to the weather, corrosive atmospheres or other adverse conditions. Equipment in areas with a high risk of fire or explosion must be constructed or protected and other such special precautions taken to prevent danger.

Cross-sectional area (csa) of conductors *(132.6)*

Conductor size is to be decided according to:

- admissible maximum temperature
- volt drop
- electromechanical stresses occurring due to short circuit or earth fault currents
- other likely mechanical stresses on the conductors
- the maximum impedance values for operation of the short circuit or earth fault protection
- the method of installation
- harmonics
- thermal insulation.

Type of wiring and method of installation *(132.7)*

This is to be determined with consideration of:

- nature of location
- structure supporting the wiring
- accessibility of the wiring to persons and livestock
- voltage
- electromechanical stresses due to short circuit and earth fault currents
- electromagnetic interference
- any other stresses to which the wiring could be exposed during the erection or use of the installation (e.g. thermal, mechanical and fire-related stresses).

Protective equipment *(132.8)*

Protective equipment has to be selected for:

- overcurrent (overload and short circuit)
- earth fault current
- overvoltage
- undervoltage and no voltage.

Protective devices must operate at values of current, voltage and time with respect to the circuit characteristics and the possibilities of danger.

Emergency control *(132.9)*

A means of emergency control is to be provided where there is a risk of danger. This must interrupt the supply rapidly and effectively, and the device used must be easily recognised.

Disconnecting devices *(132.10)*

A means of disconnecting the supply must be provided to enable maintenance, testing, fault detection and repair to be carried out on installations, circuits or individual items of equipment.

Prevention of mutual detrimental influence *(132.11)*

No mutual detrimental influences are to occur between individual electrical installations and non-electrical installations. Electromagnetic interference is also to be taken into account.

Accessibility of electrical equipment *(132.12)*

There must be:

- enough space for the original installation and any future replacement of individual items of electrical equipment
- accessibility to allow for operation, inspection, testing, fault detection, maintenance and repair.

Documentation for the electrical installation *(132.13)*

Appropriate documentation shall be provided for every electrical installation.

Protective devices and switches *(132.14)*

- Single pole protective devices or switches to be in the **line conductor only**.
- No switch, circuit breaker (except where linked) or fuse to be in an earthed neutral conductor. **Linked switches or circuit breakers must break all the associated line conductors**.

Isolation and switching *(132.15)*

There must be effective means, suitably placed for ready operation, to enable all voltage to be cut off from every installation to prevent or remove danger.

To prevent danger, every fixed electric motor must have an efficient means of switching off that is readily accessible, easily operated and suitably located.

Selection of electrical equipment *(133)*

Every item of equipment must comply with the appropriate British Standard. In all other cases, reference must be made to the appropriate IEC standard or to the appropriate national standard of another country.

Equipment must be suitable for the nominal voltage and any overvoltage possible. For certain equipment, the lowest voltage that could occur needs to be taken into account.

All items of equipment must be suitable for their design current and any currents likely to flow in abnormal conditions, including the time period for operation of protective devices.

Electrical equipment must be suitable for the frequency likely to occur in the circuit and any equipment selected on the basis of its power characteristics must be suitable for duty demanded of the equipment.

Equipment must be able to safely withstand the stresses, environmental conditions and characteristics of its location. Any equipment that is not suitable for its location can only be used if adequate further protection is provided.

Electrical equipment must not cause any harmful effects on other equipment or disrupt the supply during normal use.

The effect of switching operations is to be taken into account.

Erection, verification and periodic inspection and testing *(134)*

Good workmanship and the correct materials must be used.

The equipment must not be impaired by the erection process.

Conductors must be identified and joints and connections must be electrically and mechanically suitable for use.

Equipment design temperatures must not be exceeded and equipment that could cause high temperatures or arcing must be placed or guarded to minimise any risks of ignition. If the temperature of an item of electrical equipment could cause burns to persons or livestock, that too must be located or guarded to prevent any accidental contact.

Suitable warning signs and/or notices where necessary for safety, must be provided.

Installations, additions or alterations must be inspected and tested on completion by competent persons to establish whether or not compliance with BS 7671 has been achieved. Appropriate certification must be issued.

The person carrying out the inspection and test is required to make recommendations for future periodic inspection and testing, as specified in Part 6 of BS 7671.

BS 7671:2008 recommends that periodic inspection and testing is carried out on every electrical installation.

DEFINITIONS

**BS 7671
Part 2**

The following is a list of new definitions. Reference should be made to BS 7671:2008 for the complete list and also for the actual definitions.

Agricultural and horticultural premises.

Amusement device.

Arrangements for livestock keeping.

Backup protection.

Basic protection.

Basin of fountain.

Booth.

Busbar trunking system.

Caravan/camping park.

Caravan pitch.

Caravan pitch electrical supply equipment.

Central power supply system.

Cold tail.

Competent person.

Complementary floor heating.

Conducting location with restricted movement.

Conventional impulse withstand voltage.

Direct heating system.

Disconnector.

Earth electrode.

Electric circuit for safety services.

Electric source for safety services.

Electric supply system for safety services.

Escape route.

Exhibition.

Exposed-conductive-part.

Fairground.

Heating cable.

Heating-free area.

Heating unit.

High-density livestock rearing.

Houseboat.

Inspection.

Luminaire.

Maintenance.

Marina.

Monitoring.

Motor caravan.

Origin of the temporary electrical installation.

Pleasure craft.

Powertrack.

Powertrack system (PT system).

Prospective fault current (Ipf).

Protective bonding conductor.

Protective conductor (PE).

Protective earthing.

Protective equipotential bonding.

Protective separation.

PV.

PV a.c. module.

PV array.

PV array cable.

PV array junction box.

PV cell.

PV d.c. main cable.

PV generator.

PV generator junction box.

PV installation.

PV converter.

PV module.

PV string.

PV string cable.

PV supply cable.

Rated current.

Reporting.

Residences and other locations belonging to agricultural and horticultural premises.

Residential park home.

Residual current operated circuit breaker with integral overcurrent protection (RCBO).

Residual current operated circuit breaker without integral overcurrent protection (RCCB).

Response time.

Short-circuit current under standard test conditions $I_{sc\ STC}$.

Show.

Simple separation.

Stand.

Standard test conditions (STC).

Standby electrical source.

Standby electrical supply system.

Static convertor.

Street furniture.

Temporary electrical installation.

Temporary structure.

Testing.

Thermal storage floor heating system.

Verification.

Three definitions which, although they were included in the previous edition of BS 7671, are especially important in terms of BS 7671:2008 are:

Skilled person.
A person with technical knowledge or sufficient experience to enable him/her to avoid dangers which electricity may create

Instructed person.
A person adequately advised or supervised by skilled persons to enable him/her to avoid dangers which electricity may create

Ordinary person.
A person who is neither a skilled person nor an instructed person.

ASSESSMENT OF GENERAL CHARACTERISTICS

BS 7671 Part 3

This part of the Regulations deals with the need to assess the general characteristics of the energy source or supply and the installation itself.

The following characteristics are to be assessed:

- The purpose for which the electrical installation is to be used, its general structure and supplies (Chapter 31).
- The external influences to which it is exposed (Chapter 32).
- The compatibility of its equipment (Chapter 33).
- Its maintainability (Chapter 34).
- Safety services (Chapter 35).
- Assessment for continuity of service (Chapter 36).

Purposes, supplies and structure *(Chapter 31)*

Maximum demand and diversity *(311)*

The maximum demand of the electrical installation, expressed as a current value, must be assessed. Diversity may be taken into account when determining the maximum demand.

Arrangement of live conductors and type of earthing *(312)*

Number and type of live conductors

The number and type of live conductors, e.g. single-phase two-wire or three-phase four-wire (a.c.) for the source of energy and for the circuits to be used in the installation need to be assessed. The distributor must be consulted where necessary.

Type of earthing arrangement

The type of earthing arrangement to be used must also be determined.

The choice of arrangements may be influenced by the characteristics of the energy source and any facilities for earthing. See Technical Data Sheet 3A.

Supplies *(313)*

The following characteristics of supply sources and their range where appropriate shall be determined by measurement, calculation, enquiry or inspection. This information will be required when completing installation certificates and periodic reports.

- Nominal voltage(s) and its characteristics.
- Current and frequency.
- Prospective fault current at the installation origin (PFC).
- External earth loop impedance (Z_e).
- Installation requirements, suitability and maximum demand.
- Overcurrent protective device(s), type and rating at the origin of the installation.

Supplies for safety services and standby systems *(313.2)*

Where safety services are required (such as those required by the fire authorities for fire detection, alarm systems and emergency lighting), and where the provision of standby supplies needs specifying, the characteristics of the source(s) of supply for safety services and/or standby systems shall be separately assessed. These supplies shall have adequate capacity, reliability and rating, with an appropriate changeover time for the operation specified. There is no particular requirement for standby systems in BS 7671:2008.

Division of installation *(314)*

Every installation shall be divided into circuits in order to:

- avoid damage and minimise inconvenience in the event of a fault
- enable safe inspection, testing and maintenance
- be aware of any danger that could arise in the event of failure of a single circuit, such as a lighting circuit which may give rise to trips or falls
- reduce unwanted tripping of RCDs, resulting from equipment producing excessive protective conductor currents in normal operation
- reduce the effects of electromagnetic interferences (EMI)
- prevent the indirect energising of a circuit which has been isolated.

Separate circuit(s) shall be provided for each part of an installation that has to be separately controlled so that those circuits remain energised in the event of failure of other circuits in the installation. For example, an emergency stop circuit controlling the power supply in a workshop, when operated, cuts off the power to the machines but not the lighting, which has to be separately maintained to prevent danger.

The number of final circuits required in an installation and the number of points supplied by a final circuit shall be arranged to comply with the requirements for overcurrent protection (Chapter 43), isolation and switching (Section 537), and current-carrying capacity of conductors (Chapter 52).

Each final circuit must be connected to a separate way in a distribution board, and the wiring of each final circuit must be electrically separate from every other final circuit.

Classification of external influences *(Chapter 32)*

Reference should be made to Chapter 51 and Appendix 5 of BS 7671:2008.

Compatibility *(Chapter 33)*

Compatibility of characteristics *(331)*

An assessment must be made of any characteristics of equipment likely to have harmful effects upon other electrical equipment or other services, or likely to impair the supply.

The following characteristics have (for example) been identified:

- transient overvoltage
- undervoltage
- rapidly fluctuating loads
- unbalanced loads
- starting currents
- harmonic currents (e.g. fluorescent lighting loads and thyristor drives)
- d.c. feedback
- high-frequency oscillations
- earth leakage currents
- protective conductor currents not due to a fault
- any need for additional connections to earth (e.g. for equipment needing a connection with earth independent of the means of earthing of the installation, for avoidance of interference with its operation)
- power factor
- for an external energy source, the distributor should be consulted regarding any equipment within the installation that could have significant influence on the supply.

Electromagnetic compatibility *(332)*

Planners and designers of electrical installations shall take measures to reduce the effects of induced voltage disturbances and electromagnetic interferences (EMI). All electrical equipment shall meet the appropriate EMC requirements and standards. (See Chapter 44 of BS 7671:2008.)

Maintainability *(Chapter 34)*

Maintainability is also a very important factor to consider when deciding on the design of an installation.

An assessment must be made of the frequency and quality of maintenance that the installation can reasonably be expected to receive during its intended life. This will include (where practicable) consultation with the person or body responsible for the operation and maintenance of the installation.

Only then can the Regulations be applied so that:

- any periodic inspection, testing, maintenance and repairs likely to be necessary during the intended life can be readily and safely carried out
- the protective measures for safety remain effective
- the reliability of equipment is appropriate to the intended life.

Safety services *(Chapter 35)*

Safety services, which include emergency escape lighting, fire alarm systems, smoke and heat extraction equipment, fire rescue service lifts and installations for fire pumps, are often regulated by statutory authorities, such as the fire service, and their requirements have to be observed.

The following are recognised safety services:

- storage batteries
- primary cells
- independent generator sets
- a supply network feeder which is independent of the normal feeder.

Continuity of service *(Chapter 36)*

Every circuit shall be assessed for any need for continuity of service necessary during its life. The following characteristics should be considered:

- system earthing
- the selection of a protective device to achieve discrimination
- number of circuits
- multiple power supplies
- monitoring devices.

Technical data

In order to relate the following parts of the Regulations to the different types of supply systems, electrical accessories and standard circuits, the following technical data sheets have been produced:

- electrical supply systems (3A)
- accessories and diversity (3B)
- standard circuits (3C).

TECHNICAL DATA SHEET   3A

Electrical supply systems

Types of system

An electrical system consists of a single source of electrical energy and an installation. Types of system are identified as follows, depending upon the relationship of the source, and of exposed-conductive-parts of the installations, to earth:

TN system. A system having one or more points of the source of energy directly earthed, the exposed-conductive-parts of the installation being connected to that point by protective conductors.

TN-C system. In which neutral and protective functions are combined in a single conductor through the system.

TN-S system. A system having separate neutral and protective conductors throughout.

TN-C-S system. A system in which neutral and protective functions are combined in a single conductor in part of the system.

TT system. A system having one point of the source of energy directly earthed, the exposed-conductive-parts of the installation being connected to earth electrodes electrically independent of the earth electrodes of the source.

IT system. A system having no direct connection between live parts and earth, the exposed-conductive-parts of the electrical installation being earthed.

Low voltage

Exceeding ELV but not exceeding 1,000 V a.c. or 1,500 V d.c. between conductors or 600 V a.c. (rms) or 900 V d.c. between conductors and earth.

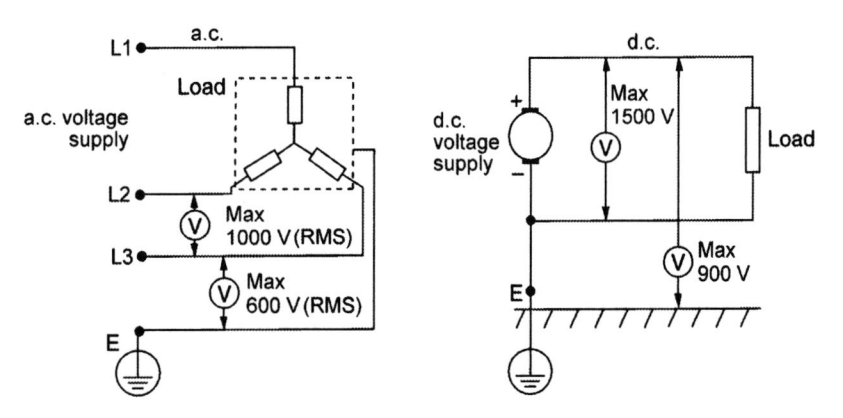

Low voltage systems

Classification of systems

Systems are classified with the following letter designations.

SUPPLY earthing arrangements are indicated by the first letter.

- **T** – one or more points of the supply are directly connected to earth.
- **I** – supply system not earthed, or one point is earthed through a fault limiting impedance.

INSTALLATION earthing arrangements are indicated by the second letter.

- **T** – exposed-conductive-parts connected directly to earth.
- **N** – exposed-conductive-parts connected directly to the earthed point of the source of the electrical supply. (The point where neutral normally originates.)

The RELATIONSHIP between the NEUTRAL and PROTECTIVE CONDUCTORS is indicated, where appropriate, by the third and fourth letters.

- **S** – separate neutral and protective conductors.
- **C** – neutral and protective conductors combined in a single conductor.

System earthing arrangements

TN-S systems

This is likely to be the type of system used where the distributor's installation is fed from underground cables with metal sheaths and armour. In TN-S systems, the consumer's earthing terminal is connected by the distributor to the distributor's protective conductor (i.e. the metal sheath and armour of the underground cable network) which provides a continuous path back to the star point of the supply transformer, which is effectively connected to earth.

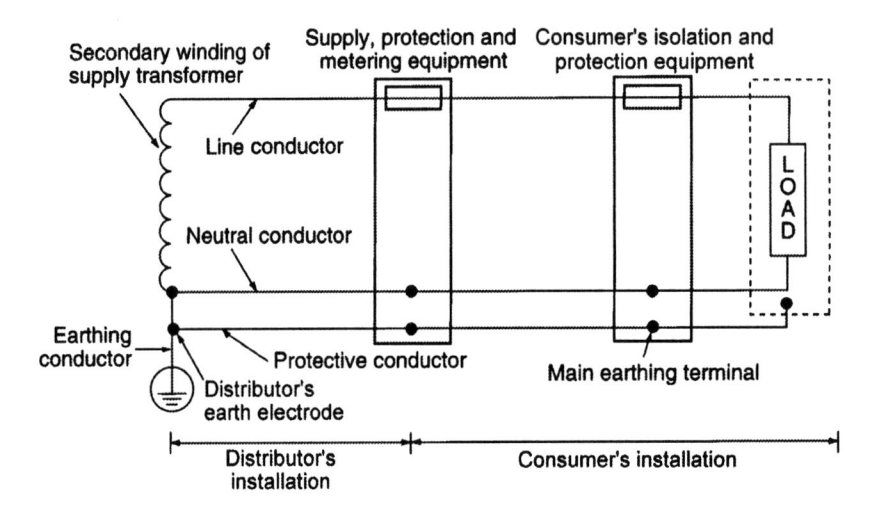

TT systems

This is likely to be the system used where the distributor's installation is fed from overhead cables. With such systems, no earth terminal is provided by the distributor. An earth electrode for connecting the circuit protective conductors to earth has to be provided by the consumer. An effective earth connection can be difficult to obtain when using an independent earth electrode and, because of this, a residual current device (RCD) must be installed in addition to any overcurrent protective devices.

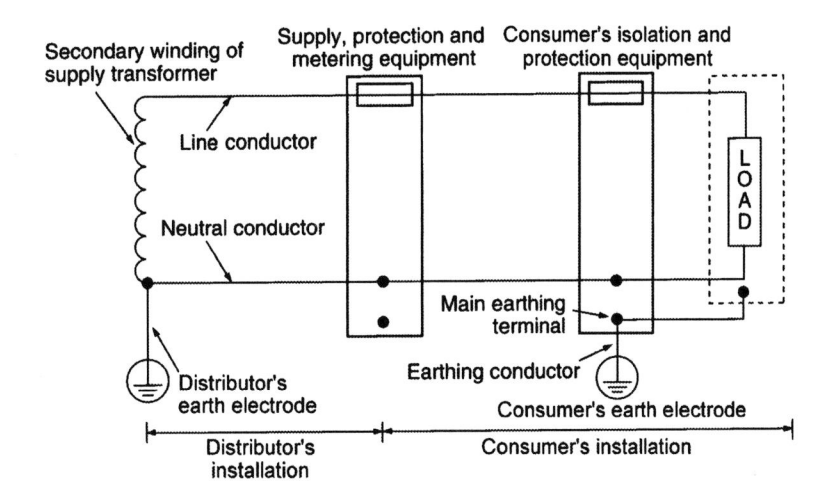

TN-C-S systems

The usual system provided to the majority of new installations is called a TN-C-S system. It is also referred to as a protective multiple earth system (PME). In this system, the distributor uses a PEN conductor, which combines the protective and neutral functions in a single conductor.

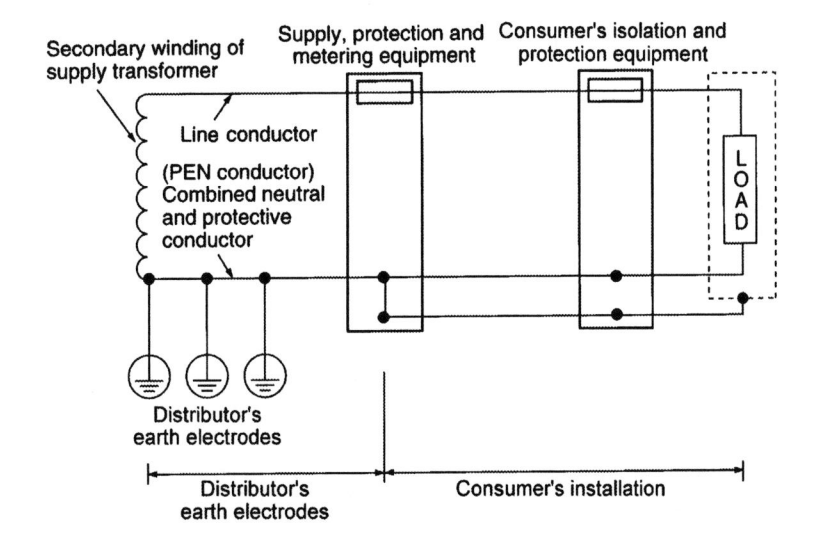

TN-C systems

Where a combined neutral and earth conductor (PEN conductor) is used in both the supply system and the consumer's installation, this is referred to as a TN-C system.

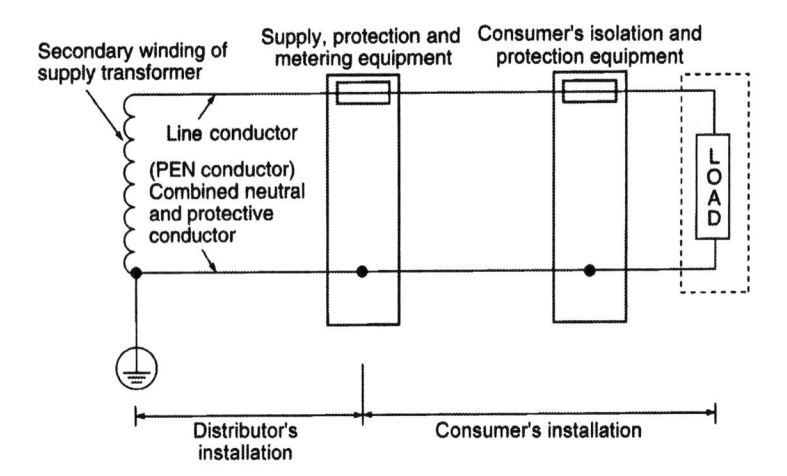

Regulation 8 (4) of the Electricity Safety, Quality and Continuity Regulations 2002 prohibits a consumer from combining the neutral and protective functions in a single conductor within the consumer's installation. Therefore, the TN-C system must not be used for public supplies in the UK.

IT systems

Where the supply system has either no earth or is deliberately earthed through a high impedance, this is known as an IT system.

With this type of system there is no shock or fire risk involved when an earth fault occurs. The protection is afforded by means of devices which monitor the insulation and give an audible or visual signal or disconnect the supply when a fault occurs.

Note: An IT system must not be used for public supply networks in the UK. Thus, IT systems are generally limited to installations which involve a continuous process where the disconnection could result in a hazard and where the installation is not connected directly to the distributor's network.

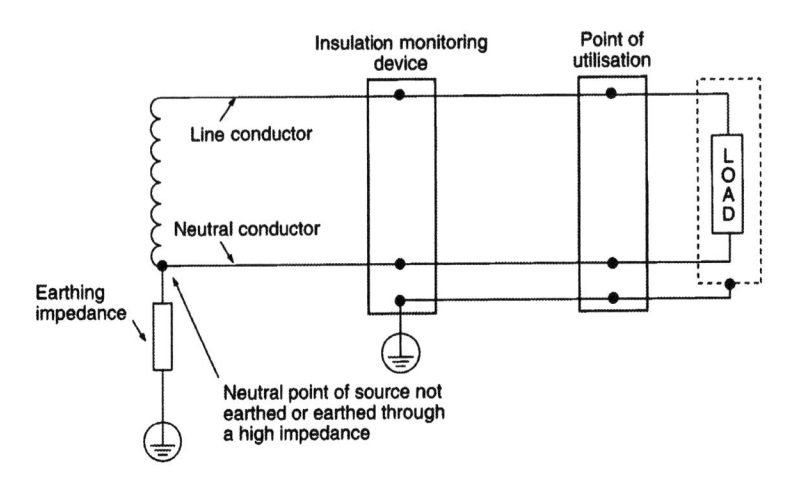

Earthing arrangements and terminations

The diagrams below show what the three most common earthing arrangements look like in practice.

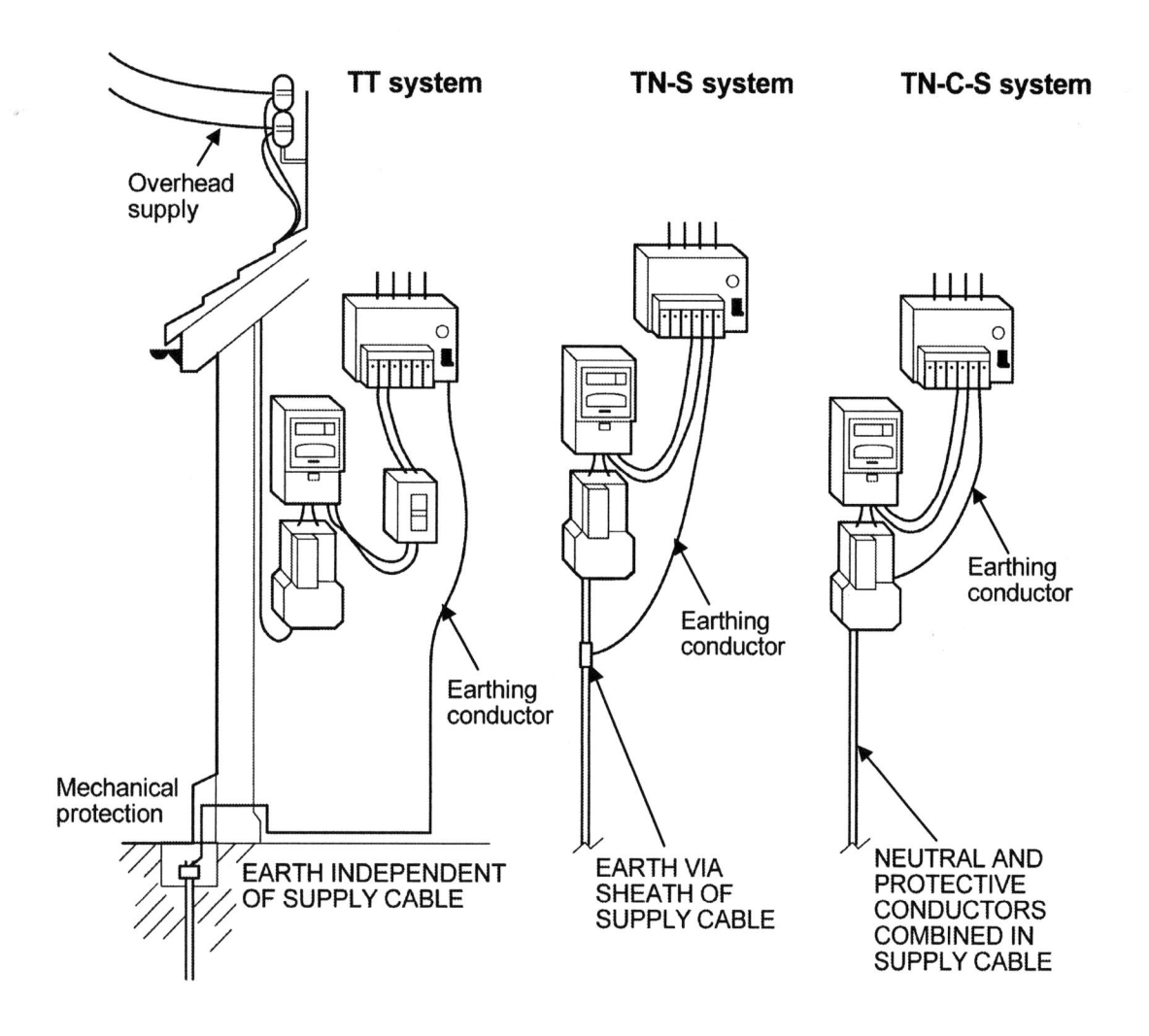

TT system — Overhead supply, Mechanical protection, Earthing conductor, EARTH INDEPENDENT OF SUPPLY CABLE

TN-S system — Earthing conductor, EARTH VIA SHEATH OF SUPPLY CABLE

TN-C-S system — Earthing conductor, NEUTRAL AND PROTECTIVE CONDUCTORS COMBINED IN SUPPLY CABLE

General

In the United Kingdom, distributors have to comply with the Electricity Safety, Quality and Continuity Regulations 2002.

Full discussions with the relevant distributor are essential when planning or installing a customer's installation in order to determine characteristics of the supply, such as the value of the earth loop impedance of that part of the system that is external to the installation (Z_e) and the prospective short-circuit current at the origin of the installation (I_p).

Voltage ranges

Extra-low voltage (ELV)

0 V to 50 V a.c. or 0 V to 120 V ripple-free d.c. (whether between conductors or to earth).

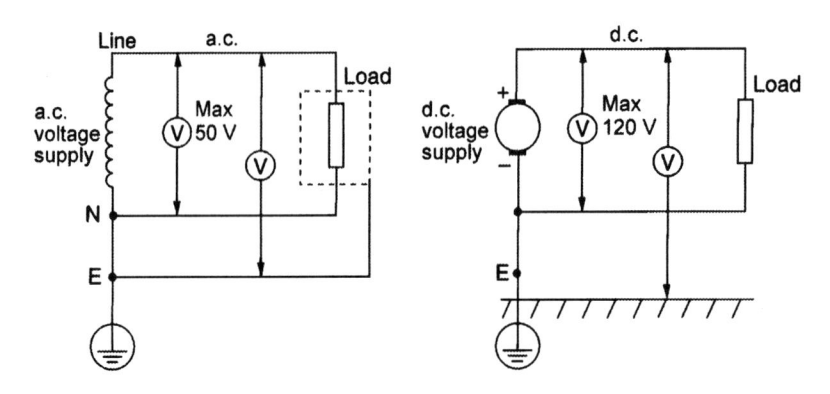

Extra-low voltage systems

SELV (separated extra-low voltage)

An extra-low voltage system which is electrically separated from earth and from other systems in such a way that a single fault cannot give rise to the risk of electric shock.

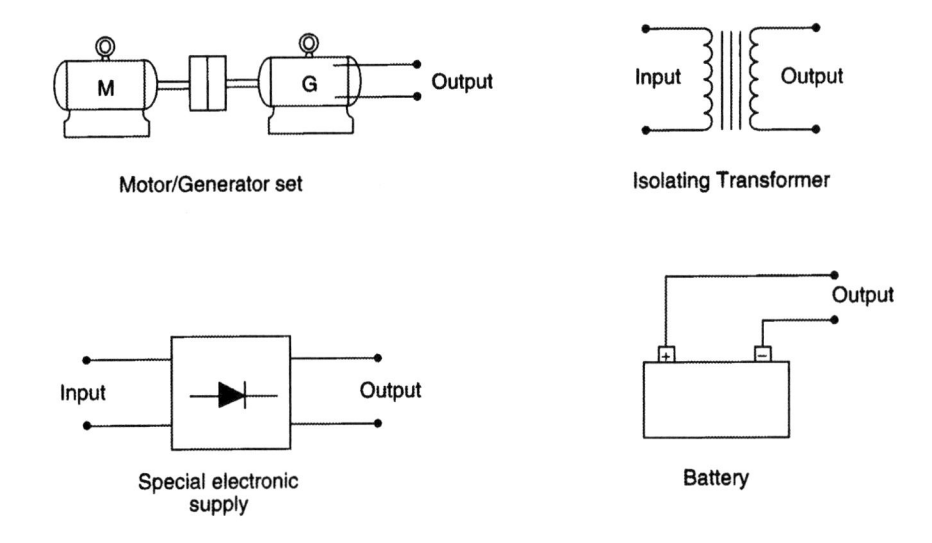

Motor/Generator set

Isolating Transformer

Special electronic supply

Battery

SELV supplies

PELV (protective extra-low voltage)

Any extra-low voltage system which is not electrically separated from earth but satisfies all the other requirements of SELV.

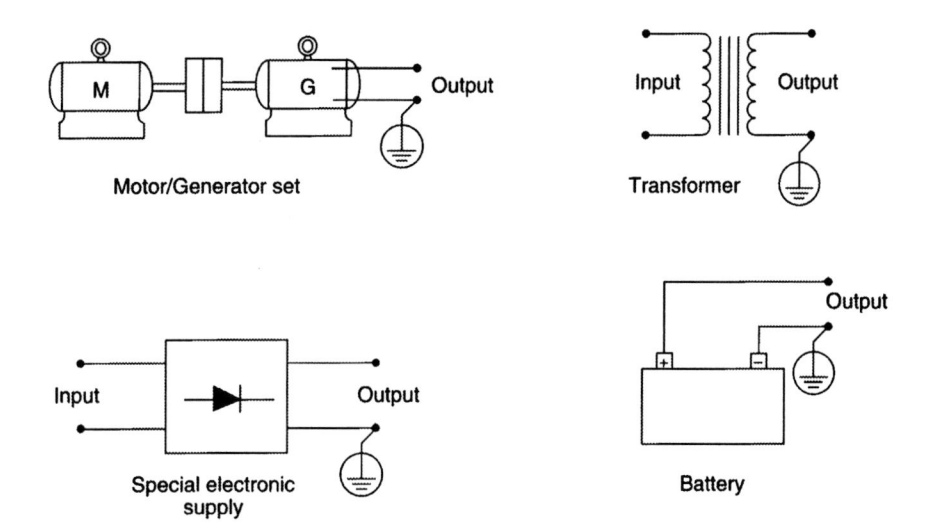

Motor/Generator set

Transformer

Special electronic supply

Battery

FELV (functional extra-low voltage)

An extra-low voltage system in which not all of the protective measures required for a SELV or PELV have been applied, e.g. a supply obtained from an auto transformer.

TECHNICAL DATA SHEET 3B

Accessories and diversity

Accessories

Plugs and socket-outlets *(553.1)*

The plugs and socket-outlets which are recognised as being suitable for low voltage circuits are listed in Table 55.1 of BS 7671 and are shown below.

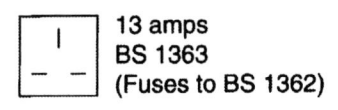

- 13 amps
  BS 1363
  (Fuses to BS 1362)

- 5, 15, 30 amps
  BS 196

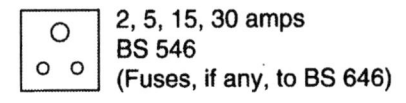

- 2, 5, 15, 30 amps
  BS 546
  (Fuses, if any, to BS 646)

- 16, 32, 63, 125 amps
  BS EN 60309-2

These plugs and socket-outlets are designed so that it is not possible to engage any pin of the plug into a live contact of a socket-outlet whilst any other pin of the plug is exposed (not a requirement for SELV circuits) and the plugs are not capable of being inserted into sockets of systems other than their own.

With the exception of SELV or special circuits having characteristics where danger may arise, all socket-outlets must be of the non-reversible type, with a point for the connection of a protective conductor.

Plugs and socket-outlets, other than those shown on the previous page, may be used on single-phase a.c. or two-wire d.c. circuits operating at voltages not exceeding 250 V for the connection of:

    Electric clocks – clock connector unit incorporating a fuse BS 646 or 1362, not exceeding 3 amperes.

    Electric shaver – BS EN 60742 shaver supply unit for use in bath or shower rooms. In other locations, a socket to BS 4573 can be used.

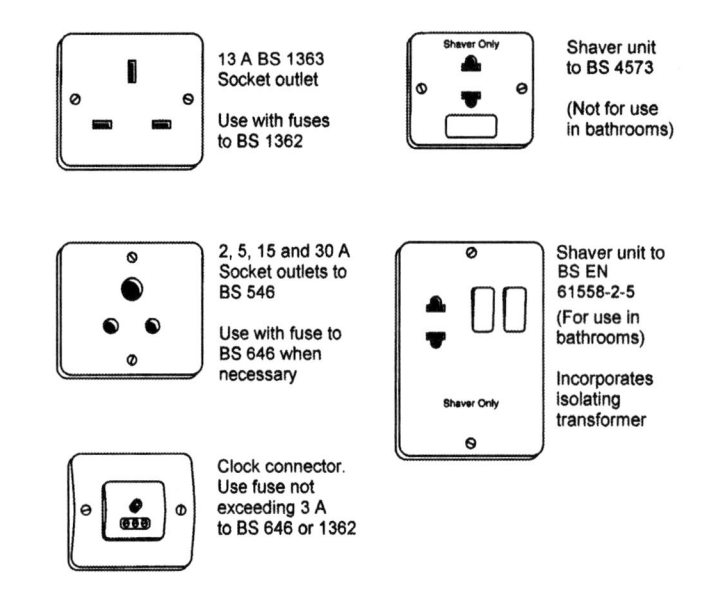

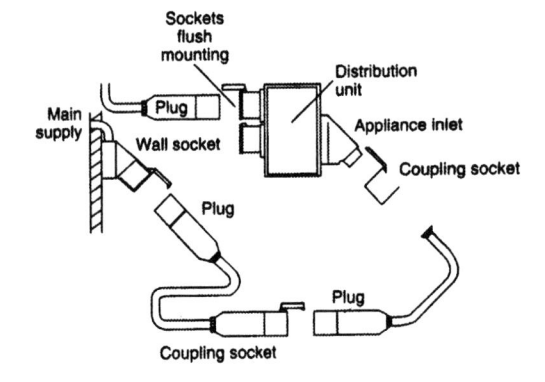

13 A BS 1363 Socket outlet

Use with fuses to BS 1362

Shaver unit to BS 4573

(Not for use in bathrooms)

2, 5, 15 and 30 A Socket outlets to BS 546

Use with fuse to BS 646 when necessary

Shaver unit to BS EN 61558-2-5 (For use in bathrooms)

Incorporates isolating transformer

Clock connector. Use fuse not exceeding 3 A to BS 646 or 1362

Rating (amps)	Colour code
BS 646	
1	Green
2	Yellow
3	Black
5	Red
BS 1362	
3	Red
13	Brown
All other ratings	Black

On construction sites (but not necessarily in site offices, toilets, etc.) only plugs, socket-outlets and couplers to BS EN 60309-2 must be used.

Where socket-outlets are mounted vertically, they must be fixed at a height above floor level or working surface so that the socket-outlet, its associated plug and its flexible cord are not subjected to mechanical damage during insertion, use or withdrawal of the plug.

Where portable equipment is likely to be used, an adequate number of socket-outlets must be provided so that the equipment can be supplied from an adjacent and conveniently accessible socket-outlet.

Cable couplers *(553.2)*

Cable couplers may be used in conjunction with the following types of plug and sockets (but these are not to be used on SELV circuits or Class II circuits):

BS 196	BS EN 60309-2
BS 4491	BS 6991.

Cable couplers must be connected at the end of the cable remote from the supply. They must be non-reversible and have provision for the connection of the protective conductor.

Lampholders

Lampholders must not be connected to any circuit where the rated current of the overcurrent protective device exceeds those given below (reference should be made to BS 7671, Table 55B).

Overcurrent protection of lampholders		
Type of lampholder		**Maximum rating of overcurrent protective device (Amps)**
Small bayonet cap	B15	16
Bayonet cap	B22	16
Small Edison screw	E14	16
Edison screw	E27	16
Giant Edison screw	E40	16

This requirement does not apply where the lampholders and their wiring are enclosed in earthed metal or insulating material (ignitability characteristic 'P' as specified in BS 476 Part 5) or where separate overcurrent protection is provided.

Lampholders for filament lamps must not be used on circuits operating at voltages in excess of 250 V.

For circuits of TN and TT systems (except for E14 and E27 lampholders complying with BS EN 60238), the outer contact of Edison screw lampholders and single centre bayonet cap-type lampholders must be connected to the neutral conductor. The same requirement applies to track-mounted lighting.

Lighting points *(559.6)*

For each fixed lighting point, one of the following must be used:

- ceiling rose to BS 67
- luminaire supporting coupler to BS 6972 or BS 7001
- batten lampholder or a pendant set to BS EN 60598
- luminaire designed to BS EN 60598
- suitable socket-outlet to BS 1363-2, BS 546 or BS EN 60309-2
- plug-in lighting distribution unit to BS 5733
- a connection unit to BS 1363-4
- box to relevant part of BS 4662 or the relevant part of the BS EN 60670 series
- device for connecting luminaire (DCL) outlet to IEC 61995-1.

When luminaire supporting couplers (LSC) are required, they must not be used for the connection of any other equipment as they are designed specifically for mechanical support and the electrical connection of luminaire. If the LSC has a protective conductor contact, it cannot be used on a SELV system.

Lighting accessories or luminaires must be controlled by a switch or switches to BS 3676, BS EN 60669-1 and/or BS EN 606692-2-1 or by a suitable automatic control, and must be appropriate, where necessary, for the control of discharge lighting circuits.

Note: As a rule, for discharge lighting circuits where exact information is not available, the demand in volt-amperes is taken as the rated lamp watts multiplied by not less than 1.8 (reference should be made to Table 1A of the IEE On-site Guide). See example on page 3B/6.

Ceiling roses for filament lamps must not be installed in any circuit operating at voltages in excess of 250 V and must not be used for the attachment of more than one flexible cord, unless specially designed for multiple pendants.

Diversity

Consider a domestic installation. It is extremely unlikely that all appliances and equipment will be in full use at any one time. For example, in normal circumstances, a householder would be unlikely to switch on all the appliances – kettle, fires, water heaters, iron, toaster and cooker – at the same time, and it would be uneconomical to provide cables and switchgear of a capacity for the maximum possible load. The loads they will carry are likely to be less than the maximum. It is this factor which is referred to as 'diversity'. By making allowances for diversity, the size and cost of conductors, protective devices and switchgear can be reduced.

To calculate the diversity factor $\left[\frac{\text{minimum actual load}}{\text{installed load}} \right]$ for every type of electrical installation, specialist knowledge and experience is required.

The following information is based on that of Appendix 1 of the IEE On-site Guide.

The common methods of obtaining the current of a circuit is to add together the current demand of all points of utilisation and equipment in a circuit.

Assumed current demand for points of utilisation and equipment are given in Table 1A of the IEE On-site Guide.

Example – Household cooking appliances

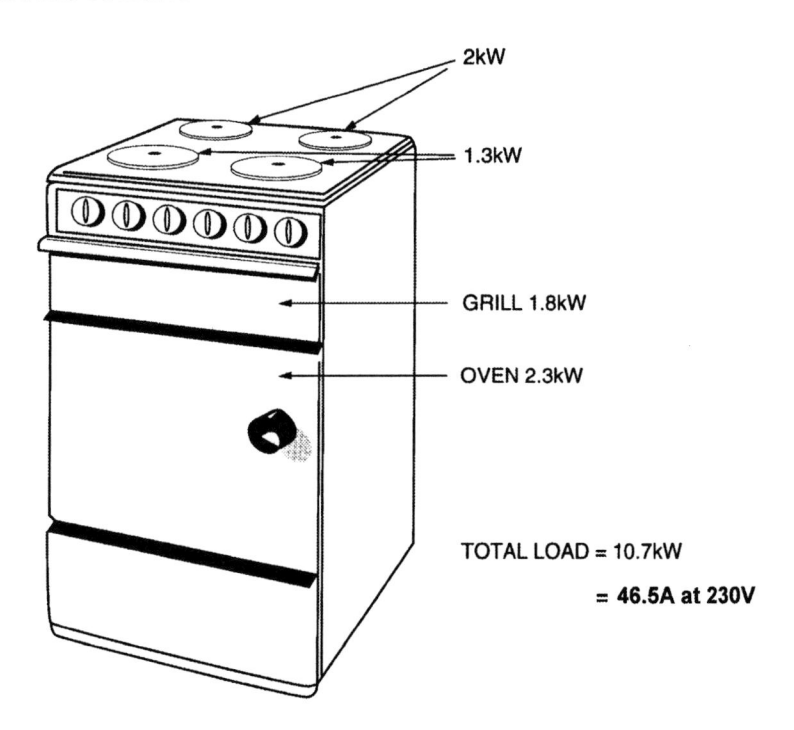

Assumed current demand

- 2kW
- 1.3kW
- GRILL 1.8kW
- OVEN 2.3kW

TOTAL LOAD = 10.7kW

= **46.5A at 230V**

If we consider the electric cooker with a maximum loading of 46.5 amperes, as illustrated, the assessed current demand would be as follows:

The first 10 amperes of the total rated current of the cooker, plus 30% of the remainder of the total rated current of the cooker, plus 5 amperes if a socket-outlet is incorporated in the control unit.

Total current rating		46.5 A
First 10 amperes	=	10
30% of remaining 36.5	=	10.95
Socket-outlet	=	5
Assessed current demand		25.95 A

Discharge lighting

Final circuits supplying discharge lighting must be capable of carrying the total steady current of the lamp and associated control gear. Where exact information is not available, provided the power factor of the circuit is not less than 0.85, the current demand of a discharge lamp can be calculated from the wattage of the lamp, multiplied by not less than 1.8.

Therefore, steady current of a discharge lamp =

$$\frac{\text{lamp power (watts)} \times 1.8}{\text{supply voltage}}$$

Example

A circuit supplies five, 230 V single-phase fluorescent luminaires each rated at 65 watts. The current demand would be:

$$I_b = \frac{5 \times 65 \times 1.8}{230} = 2.54 \text{ A}$$

Methods of applying diversity

Method 1

The current demand of a circuit supplying a number of final circuits can be obtained by adding the current demands of all the equipment supplied by each final circuit of the system and applying the allowances for diversity given in Table 1B of the IEE On-site Guide. For a circuit with socket-outlets, the rated current of the protective device is the current demand of the circuit.

Method 2

The alternative method of assessing the current demand of a circuit supplying a number of final circuits is to calculate the diversified current demand for each circuit and then apply a further allowance for diversity, on the assumption that not all circuits will be in use at the same time.

Note: With this method, the allowances given in Table 1B of the IEE On-site Guide are NOT to be used. The values used should be chosen by the designer of the installation.

Method 3

The current demand of a circuit is determined by a suitably qualified electrical engineer.

Example

Consider a small guest house with 10 bedrooms, 3 bathrooms, lounge, dining room, kitchen and utility room with the following loads connected to 230 V single-phase circuits balanced over a three-phase supply.

Lighting:	3 circuits tungsten lighting. Total 2,860 watts
Power:	3 x 30 A ring circuits to 13 A socket-outlets
Water heating:	1 x 7 kW shower 2 x 3 kW immersion heater thermostatically controlled
Cooking appliances:	1 x 2 kW cooker 1 x 10.7 kW cooker

Calculations and answer on next page

Calculations and answer to example

		Current demand (Amperes)	Table 1B (diversity factor)	Current demand allowing for diversity (Amperes)
Lighting	$\dfrac{2{,}860}{230}$	12.4	75%	9.3
Power	(i) (ii) (iii)	30 30 30	100% 50% 50%	60
Water heaters (inst)	$\dfrac{7{,}000}{230}$	30.4	100%	30.4
Water heaters (thermo)	$\dfrac{6{,}000}{230}$	26	100%	26
Cookers	(i) $\dfrac{10{,}700}{230}$	46.5	100%	46.5
	(ii) $\dfrac{2{,}000}{230}$	8.7	80%	7

Total current demand (allowing diversity)   179.2

Load – spread over 3 phases = $\dfrac{179.2}{3}$

= 59.7 A

= 60 A per phase

TECHNICAL DATA SHEET 3C

Standard circuits *(Appendix 8, IEE On-site Guide)*

Types of final circuit using BS 1363 socket-outlets and fused connection units are:

- ring circuits
- radial circuits.

Final circuits using BS 1363 socket-outlets and connection units (Table 8A of the IEE On-site Guide) are shown below.

Circuit	Minimum conductor size (PVC insulated, copper conductors)	Copper conductor MI cable	Type and rating of overcurrent device	Maximum floor area
Ring	2.5 mm²	1.5 mm²	30 A or 32 A	100 m²
Radial	4 mm²	2.5 mm²	30 A or 32 A	75 m²
Radial	2.5 mm²	1.5 mm²	20 A	50 m²

The table is applicable for circuits protected by:

- fuses to BS 3036, BS 1361 and BS 88
- circuit breakers types B and C to BS EN 60898
- RCBOs to BS EN 61009-1
- circuit breakers to BS 60947-2.

The table gives the minimum conductor size, therefore the actual conductor size must still be calculated or selected in the usual manner. Reference should be made to Appendix 8 of the IEE On-site Guide.

Note: Conductor size may be reduced for fused spurs. Fused spurs are covered on page 3C/4.

Ring circuits

The requirements for ring circuits are:

- if the floor area served by a single 30 A or 32 A ring circuit is less than 100 m², an unlimited number of outlets may be provided. (Each socket-outlet of a twin or multiple socket is to be regarded as one outlet)
- consideration must be given to the loading of the circuit, especially kitchens which may require a separate circuit

- when more than one ring circuit is installed in the same premises, the outlets installed must be reasonably shared amongst the ring circuits so that the assessed load is balanced.

Regulation 433.1.4 also requires that:

- the current-carrying capacity (I_z) of the circuit conductors is not less than 20 A
- under its intended conditions of use, the load current in any part of the ring is unlikely to exceed the current-carrying capacity of the cable for long periods.

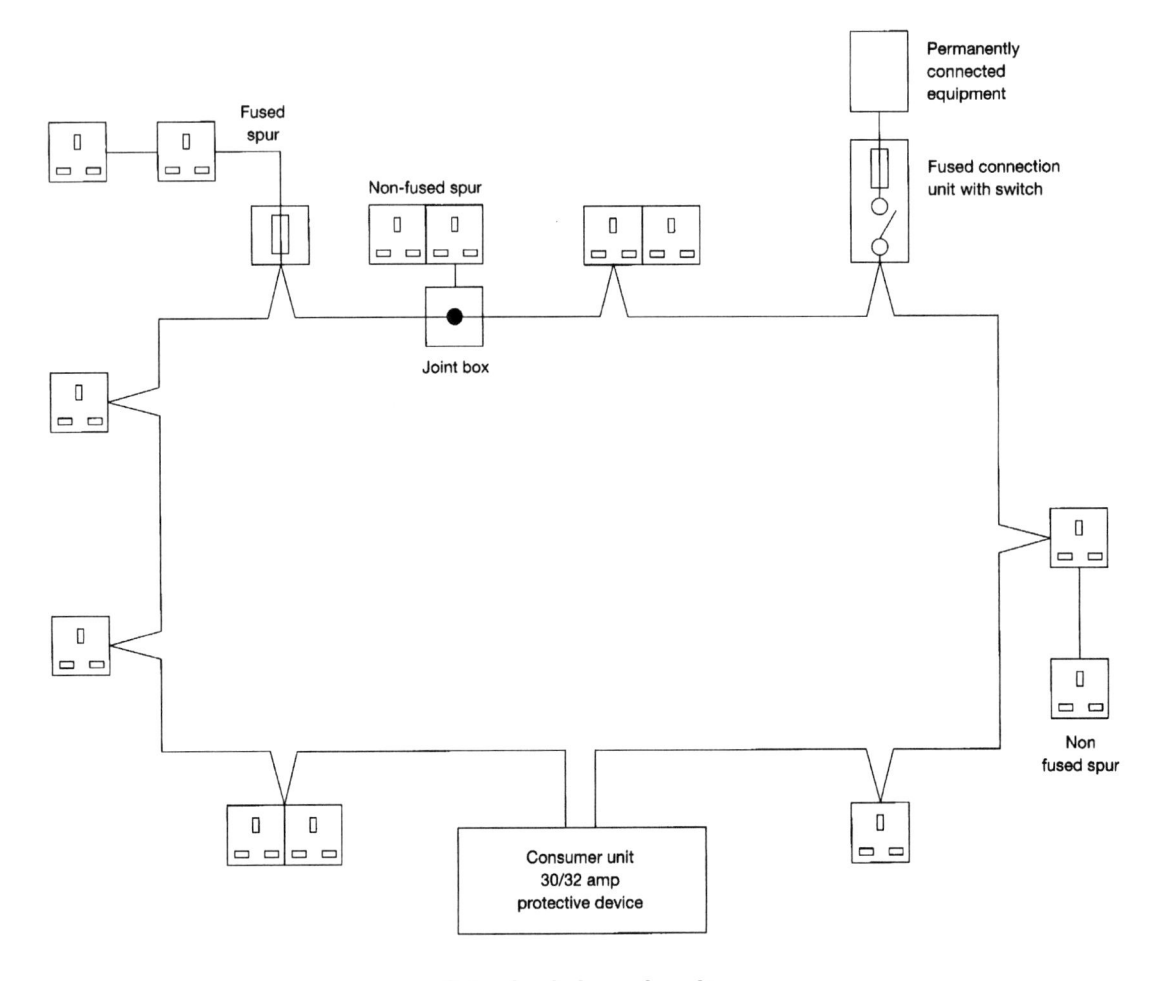

A typical ring circuit

Spurs

The total number of fused spurs is unlimited, but the number of non-fused spurs must not exceed the total number of socket-outlets and any stationary equipment connected directly to the circuit.

Non-fused spurs

A non-fused spur may supply only one single or one twin socket-outlet or one item of permanently connected equipment.

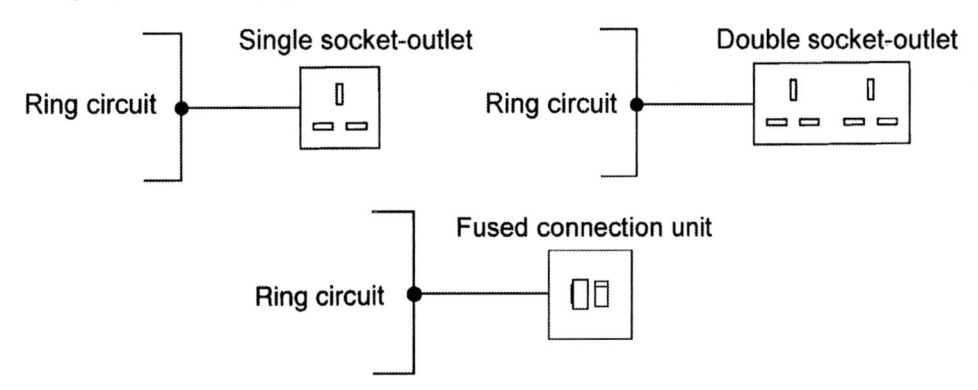

Note: The cable size of non-fused spurs must not be less than that of the ring circuit.

Method of connecting spurs to circuit

(a) At the terminal of accessories on the ring

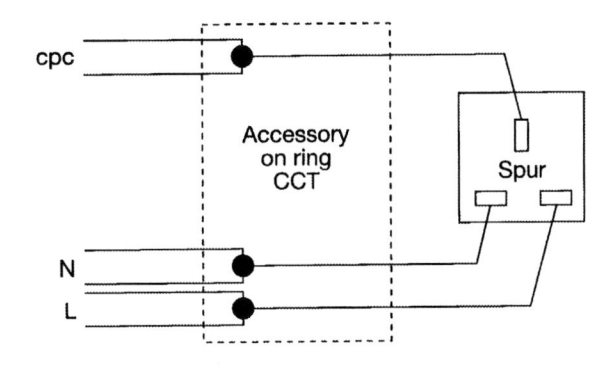

(b) At a joint box

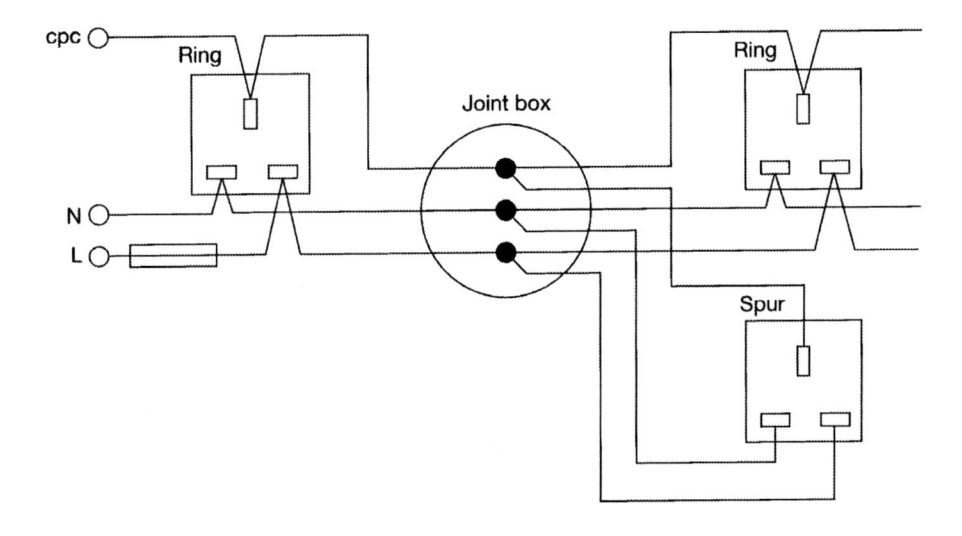

Fused spurs

A fused spur is connected to a circuit through a fused connection unit. The fuse incorporated must be related to the current-carrying capacity of the cable used for the spur, but must not exceed 13 A.

When outlets are wired from a fused spur, the minimum size of conductor is:

- 1.5 mm² for PVC or thermosetting insulated cables with copper conductors
- 1 mm² for mineral insulated cable copper conductors.

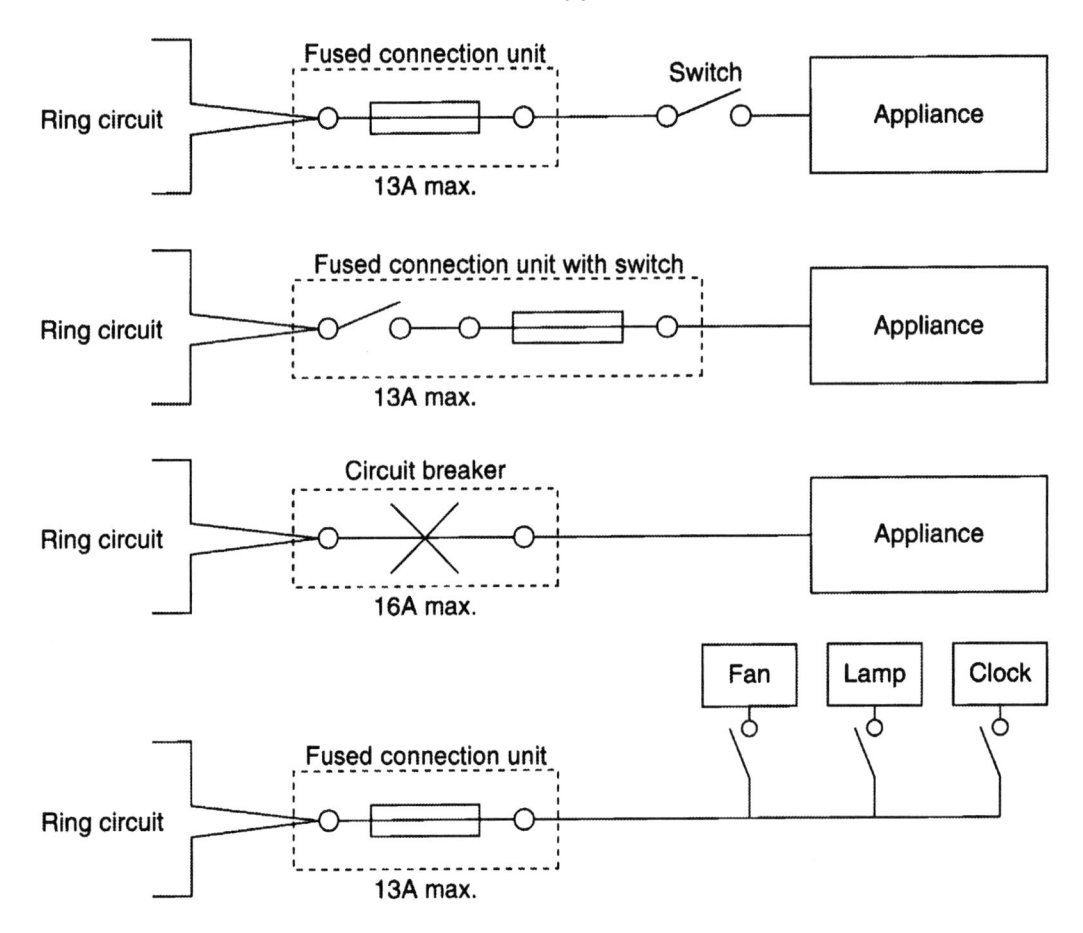

Note: The cable size for the fused spur is dependent on the magnitude of the connected load.

Radial circuits

An unlimited number of socket-outlets may be supplied, but the floor area that may be served by the outlets is limited to either 50 m² or 75 m² depending on the size and type of the cable used and size of overcurrent protection afforded.

Using 4 mm² cables with thermosetting (PVC) insulation and copper conductors.

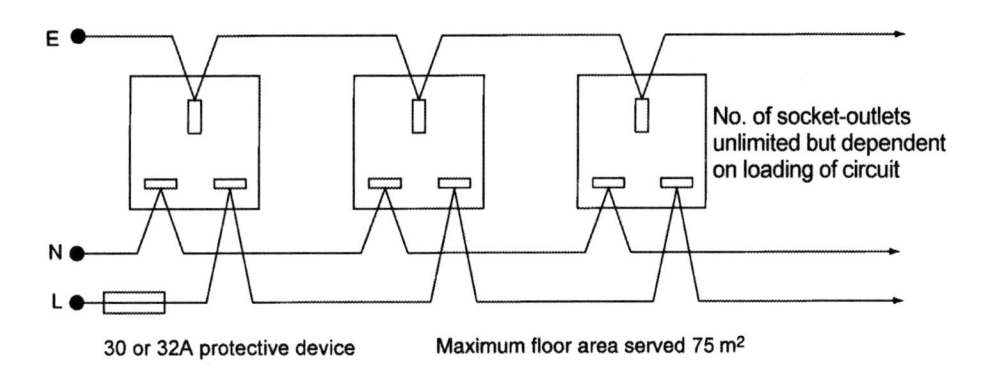

30 or 32A protective device   Maximum floor area served 75 m²

No. of socket-outlets unlimited but dependent on loading of circuit

Using 2.5 mm² cables with thermosetting (PVC) insulation and copper conductors.

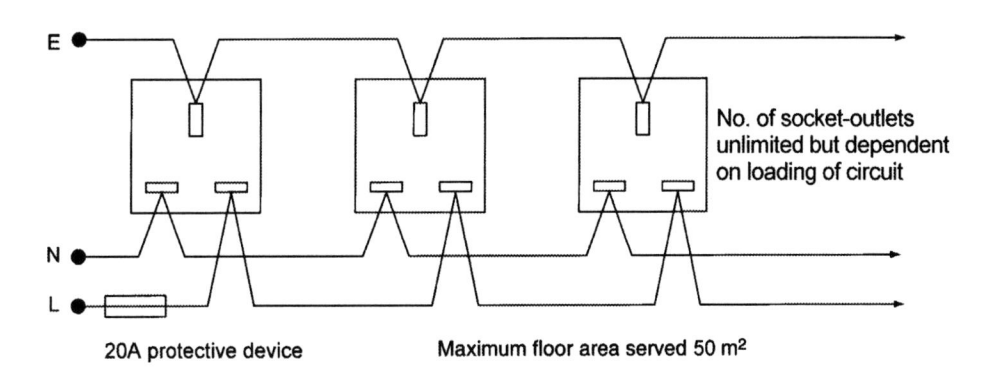

20A protective device   Maximum floor area served 50 m²

No. of socket-outlets unlimited but dependent on loading of circuit

Typical radial circuits using BS 1363 socket-outlets

Note: For a fuller explanation of the requirements for radial circuits, reference should be made to Appendix 8 of the IEE On-site Guide.

Circuits for immersion heaters and space heating

Where immersion heaters are installed in storage tanks with a capacity in excess of 15 litres, or a fixed comprehensive space heating installation is to be installed, for example, electric fires or storage radiators, a separate circuit must be provided for each heater.

Cooker circuits

A circuit supplying a cooking appliance must include a control switch or cooker control unit, which may include a socket-outlet. The rating of the circuit must be determined by an assessment of the current demand in accordance with Table 1A of the IEE On-site Guide.

Final circuits using 16 A BS EN 60309-2 socket-outlets

Plugs and socket-outlets to BS EN 60309-2 are available in 16 A, 32 A, 63 A and 125 A ratings and are typically used in industrial circuits and construction site installations.

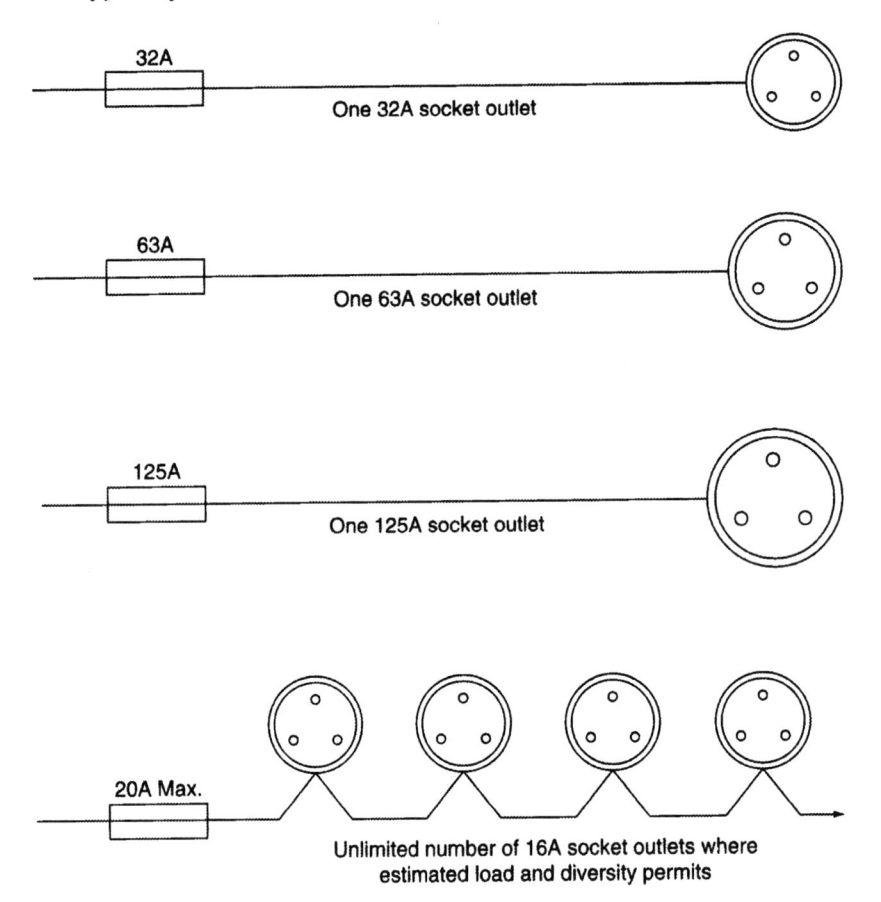

BS EN 60309-2 socket-outlet circuits

The 32 A, 63 A and 125 A rating sockets must be wired on radial circuits each supplying one socket-outlet only. The 16 A version can be wired in unlimited numbers on radial circuits where the estimated load and diversity permits.

The maximum rating of an overcurrent device for a BS EN 60309-2 16 A socket-outlet radial circuit is 20 A.

The size of conductor is determined by applying the relevant correction factors in Appendix 4 of BS 7671.

BS EN 60309-2 socket-outlets

The range of accessories consists of:

- plugs
- sockets
- cable coupler
- appliance inlets.

Accessories are available for single-phase and three-phase supplies with a voltage between phases not exceeding 750 V at a rated current of up to 125 A.

Discrimination between different voltages

This is achieved in two ways:

- by the positioning of the earth contact in relation to a keyway
- by colour codes.

Both of these are as illustrated below.

Discrimination between accessories of different voltages by position of the earth contact in relation to the keyway:

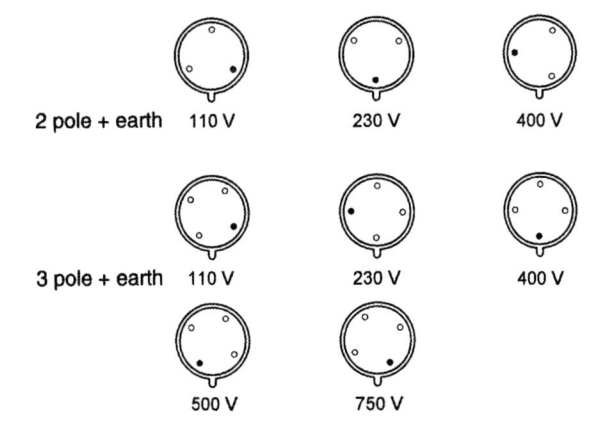

2 pole + earth   110 V   230 V   400 V

3 pole + earth   110 V   230 V   400 V

500 V   750 V

Types of BS EN 60309-2 plugs and sockets

BS 7373 Electricity on Construction Sites recommends that plugs, socket-outlets and cable couplers are identified by colour according to operating voltage as shown below:

Operating voltage	Colour
25 V	Violet
50 V	White
110 V to 130 V	Yellow
220 V to 240 V	Blue
380 V to 415 V	Red
500 V to 750 V	Black

PROTECTION FOR SAFETY **BS 7671**
Part 4

Protection against electric shock *(Chapter 41)*

Introduction *(410)*

Chapter 41 of the Regulations require that measures be taken to protect persons or livestock by ensuring that hazardous, live parts are not accessible and that accessible, conductive parts are not hazardous when live, under normal or single-fault conditions.

The above are now defined as:

- **basic protection** – protection under normal conditions
  (previously described in 16th Edition as protection against direct contact)

- **fault protection** – protection under fault conditions
  (previously described in 16th Edition as protection against indirect contact).

Protection may be afforded by using combined measures for both types of protection.

General requirements *(410.3)*

A protective measure shall consist of:

- a combination of appropriate basic protection and independent provision for fault protection

- an enhanced protective provision which provides both basic and fault protection.

The following protective measures are permitted:

- automatic disconnection of supply
- double or reinforced insulation
- electrical separation for the supply to one item of equipment
- extra-low voltage (SELV and PELV).

Different protective measures applied to an installation or equipment shall have no influence on each other, so that the failure of one measure shall not impair the other measure or measures.

Fault protection may be omitted for the following:

- inaccessible lengths of metal conduit, 150 mm maximum
- metal supports of overhead line insulators, placed out of arm's reach and attached to a building
- inaccessible steel reinforcement used in concrete poles for overhead lines
- unearthed street furniture, which has been supplied from an overhead line and is inaccessible during normal use
- small conductive parts (of approximately 50 mm x 50 mm dimensions), where connection with a protective conductor and significant contact could only be made with difficulty or would be unreliable, e.g. bolts, rivets, name plates and cable clips.

Automatic disconnection of supply *(411)*

Automatic disconnection of supply is provided by:

- basic protection
- fault protection.

Protective earthing *(411.3.1.1)*

Exposed-conductive-parts and simultaneously accessible exposed-conductive-parts should be connected to a protective conductor of the supply system earthing arrangement.

A circuit protective conductor shall be installed and terminated at each accessory. Lampholders with no exposed-conductive-parts, suspended from a ceiling rose or fitting, do not require earthing.

Protective equipotential bonding *(411.3.1.2)*

Equipotential bonding conductors must connect any extraneous-conductive-parts to the main earthing terminal of the installation. Extraneous-conductive-parts may include the following items:

- water pipes
- gas pipes
- other pipes and ducting
- central heating and air-conditioning systems
- exposed metallic parts of the building structure
- lightning protection systems.

When an installation serves more than one building, these requirements apply to each building.

To comply with BS 7671, metal sheaths of telecommunication cables require equipotential bonding but this must only be carried out with the consent of the owner or operator of the cable.

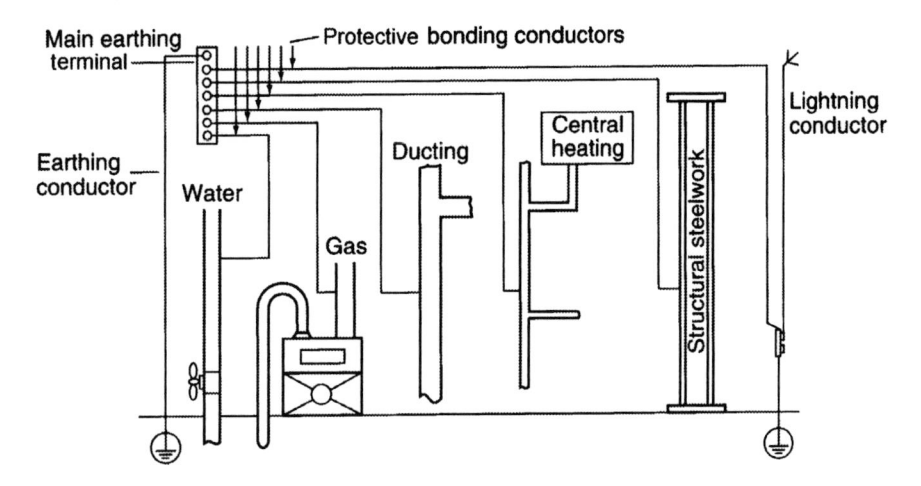

For lightning protection systems, the system designer must determine the position of the bonding connection(s). This is due to the complex requirements of BS 6651: Code of Practice for protection of structures against lightning.

Typical earthing and bonding conductor arrangements for domestic installations

Note: In the case of TN-S and TT supplies the earthing conductor may be sized using the adiabatic equation (see Regulations 543.1.1 to 543.1.3). This can result in a smaller size of earthing conductor, resulting in smaller protective bonding conductors. However, the smallest size of protective bonding conductor permitted is 6 mm² (see Regulation 544.1.1).

(a)

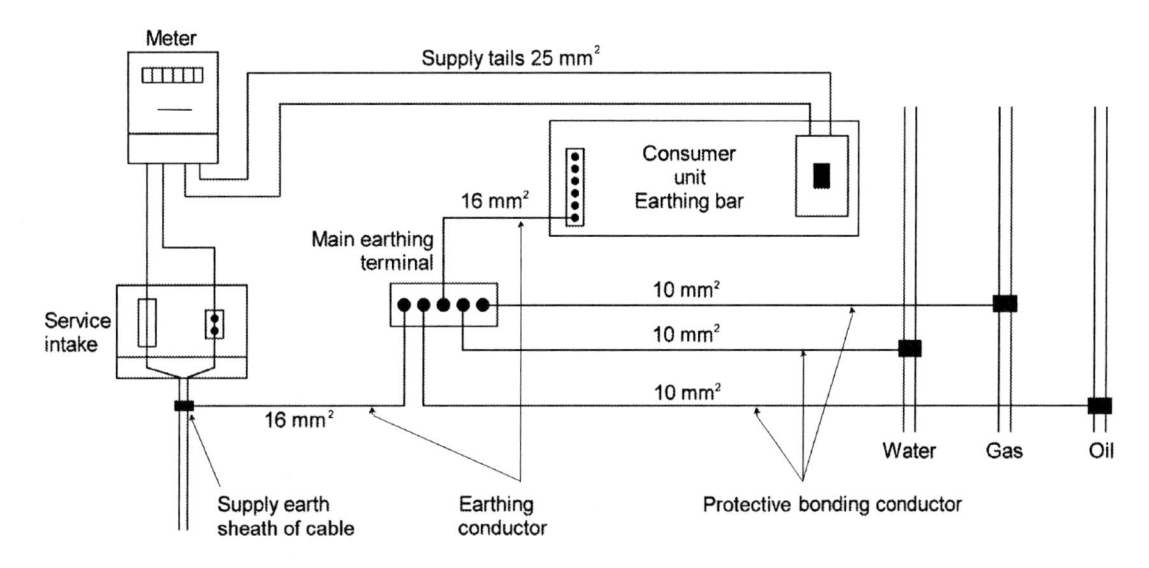

TN-S earthed to armour or metal sheath of the distributor's cable

(b)

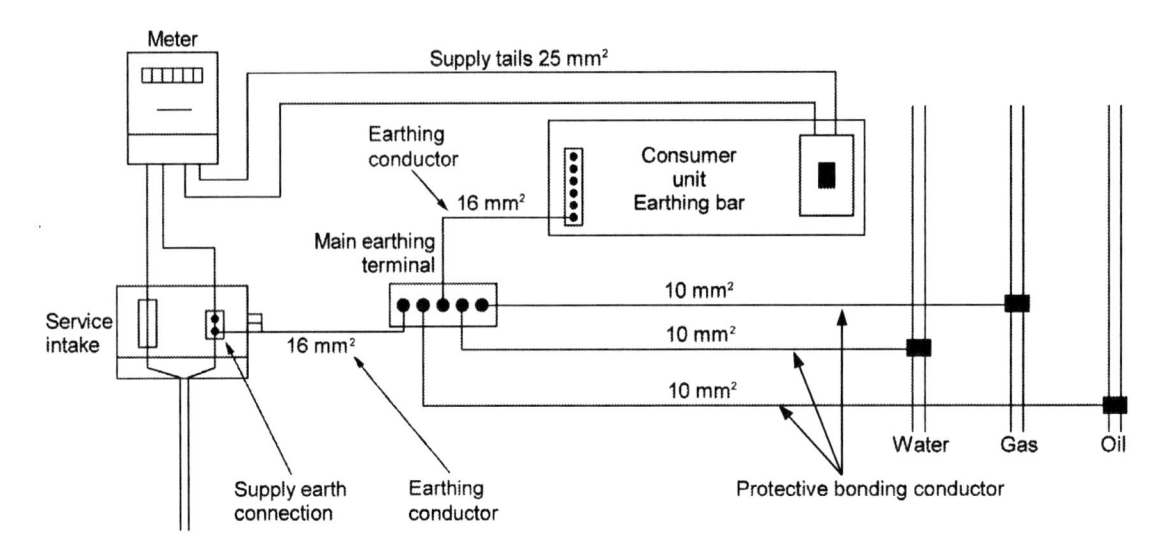

TN-C-S earthed using combined neutral and earth conductor of the distributor's cable

(c)

TT earthed via an earth electrode

Protective bonding conductors can be installed separately or as an unbroken loop.

Note: The local distributor's network conditions may require larger conductors.

Automatic disconnection in case of a fault *(411.3.2)*

A protective device shall automatically interrupt the supply of a circuit or equipment when a fault of negligible impedance occurs between a line conductor and an exposed-conductive-part or a protective conductor in a circuit or equipment (411.3.2.1).

The maximum disconnection times for final circuits not exceeding 32 A are given in Table 41.1 of BS 7671 (411.3.2.2).

Extract of Table 41.1 maximum disconnection times

Supply system	U_o nominal a.c. line voltage to earth			
TN	50 to 120 V a.c. 0.8 s	120 to 230 V a.c. 0.4 s	230 to 400 V a.c. 0.2 s	> 400 V a.c. 0.1 s
TT	0.3 s	0.2 s	0.07 s	0.04 s

Note: When disconnection is achieved in a TT system by an overcurrent protective device and equipotential bonding of all extraneous-conductive-parts has been carried out, the times for TN systems may be used.

For TN systems disconnection times not exceeding 5 s are permitted for distribution circuits and not exceeding 1 s for distribution circuits in TT systems (411.3.2.3 and 411.3.2.4).

If automatic disconnection cannot be achieved in the time required by the above Regulations, supplementary equipotential bonding shall be provided (411.3.2.6).

Additional protection *(411.3.2)*

In a.c. systems additional protection must be provided by a 30 mA residual current device for:

- socket-outlets not exceeding 20 A rating that are intended to be used by ordinary persons and are for general use, and
- mobile equipment not exceeding 32 A rating for use outdoors.

Both the above requirements apply to domestic premises.

Note: Exceptions exist when the socket-outlets are for use only under the supervision of a skilled or instructed person, such as those situations found in some commercial or industrial locations.

Another exception is when the socket-outlet is provided for the connection of a particular item of equipment only and is suitably labelled.

The requirements for RCDs which are used to provide additional protection are given in Regulation 415.1 of BS 7671:2008 (see page 4/15). It should be noted, however, that cables concealed in walls or partitions, including lighting circuits, must either meet the requirements of Regulation 522.6.6 for mechanical protection or they must be provided with additional protection by means of a 30 mA RCD, as required by Regulation 522.6.7 (see pages 5/17 and 5/18).

TN system *(411.4)*

The earth fault loop impedance (Z_s) is made up of the impedance of the consumer's line and protective conductors R_1 and R_2 respectively, and the impedance external to the installation Z_e (i.e. impedance of the supply).

$$Z_s = Z_e + (R_1 + R_2)$$

Where:

Z_s = system phase-earth loop impedance

Z_e = external impedance (impedance of the supply)

R_1 = resistance of circuit line conductor

R_2 = resistance of circuit protective conductor

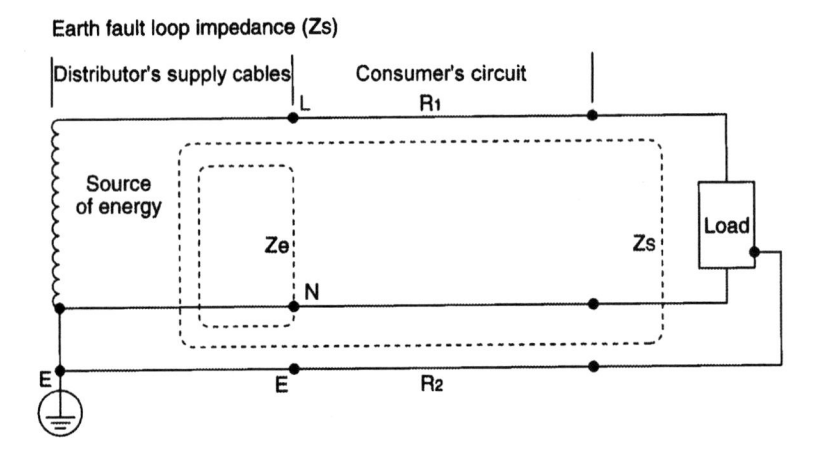

Note:

1) *Values of R_1 + R_2 for different sizes of cable are given in Appendix 9 of the IEE On-site Guide (or Appendix 1 of IEE Guidance Note 3 Inspection and Testing).*

2) *Further references and details of the external impedance value are explained in Module 8 – Protection Against Overcurrent.*

Regulation 411.4.5 states that the requirements of Regulation 411.3.2 for automatic disconnection are considered to be satisfied when the characteristic of each protective device, and the earth fault loop impedance of each circuit protected, are such that automatic disconnection of the supply under fault conditions occurs within a specified time. The requirement is met when:

$$Z_s \times I_a \leq U_o$$

Where:

Z_s = earth fault loop impedance

I_a = current causing the automatic operation of the protective device within the time given in Table 41.1

U_o = nominal voltage to earth

BS 7671 gives maximum values of Z_s for each type and rating of overcurrent device (see Tables 41.2, 41.3 and 41.4). Where the measured value of Z_s is less than the maximum value of Z_s, then the disconnection time is assured. However, the note at the end of Tables 41.2, 41.3 and 41.4 states that these maximum Z_s values pertain to the normal operating temperature of the cables (i.e. 70° C). When testing is carried out at a different temperature (say, an ambient temperature of 20°C) then an adjustment needs to be made. It is normal to assume a temperature of 20°C when testing and to adjust the Z_s values by multiplying them by 0.8 (known as the 80% rule). (Refer to Appendix 14 of BS 7671.)

TT systems *(411.5)*

Exposed-conductive-parts in TT systems that are protected by a single protective device must be connected via the main earthing terminal (MET) to a common earth electrode. If two or more protective devices are in series, the exposed-conductive-parts may be connected to separate earth electrodes relating to each protective device.

One or more of the following types of protective device must be used (411.5.2):

- RCD (preferred)
- overcurrent protective device.

When an RCD is used to provide earth fault protection, the following must be achieved for each circuit (411.5.3):

$$R_A \times I_{\Delta n} \leq 50 \text{ V}$$

Where:

R_A = sum of the resistances of the earth electrode and the protective conductor that connects it to the exposed-conductive-part

$I_{\Delta n}$ = rated residual operating current of the RCD

Note: If R_A is not known it may be replaced by Z_s.

Table 41.5 of BS 7671 provides maximum values of earth fault loop impedance, to ensure RCD operation and satisfy the requirements of Regulation 411.5.3 for non-delayed RCDs in final circuits not exceeding 32 A. An extract of Table 41.5 is illustrated below.

Maximum earth loop impedance in ohms			
RCD rated residual operating current in mA	**Voltage U_o**		
	50 to 120 V	120 to 230 V	230 to 400 V
30	1,667	1,667	1,533
100	500	500	460

In practice, however, values above 200 ohms may be unstable. Where overcurrent protective devices are used the following condition shall be met:

$$Z_s \times I_a \leq U_o$$

Therefore

$$Z_s \leq \frac{U_o}{I_a}$$

IT system *(411.6)*

An IT system is used for industrial applications, for example mines and quarries. Live parts of the installation are insulated from earth or connected to earth through a high impedance, the connection being made at the neutral or midpoint of the system, as illustrated.

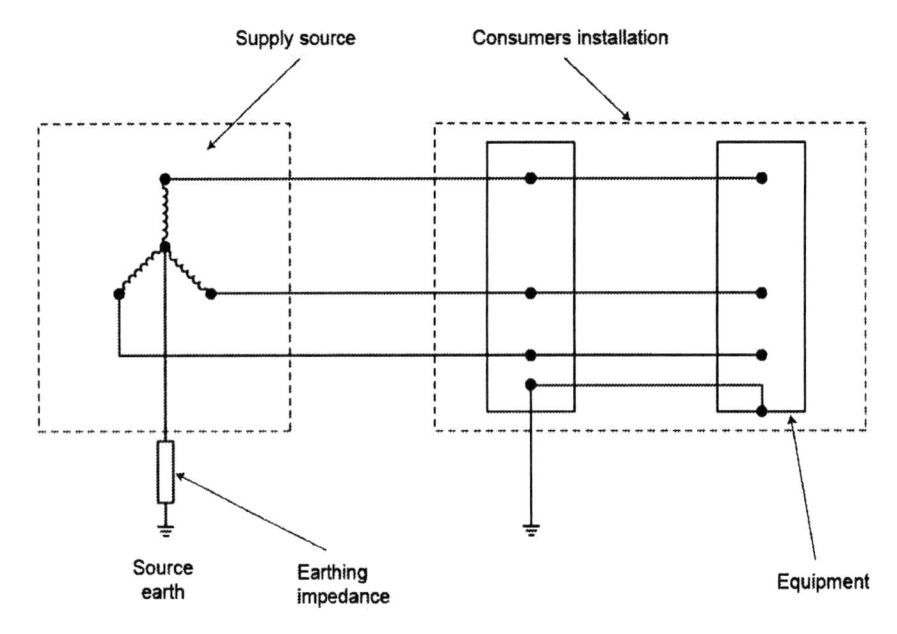

Connection can also be made at an artificial neutral point, which may be connected directly to earth if the resulting impedance to earth is high enough.

When no neutral or midpoint exists, a line conductor can be connected to earth through a high impedance.

BS 7671 makes strong recommendations that IT systems with distributed neutrals should be avoided.

When a single fault occurs in an IT system the resulting fault current to exposed-conductive-parts or to earth is low and the requirements of Regulation 411.3.2 for automatic disconnection are not needed, provided the following requirements from Regulation 411.6.2 are met.

Any exposed-conductive-parts shall be earthed and the following condition met (411.6.2):

$R_A \times I_d \leq 50$ V for a.c. systems

$R_A \times I_d \leq 120$ V for d.c. systems

Where:

R_A = The sum of the resistances of the earth electrode and the protective conductor connecting it to exposed-conductive-parts

I_d = The fault current of the first fault of negligible impedance between a line conductor and any exposed-conductive-part

Provisions must be taken to avoid any harm to persons in contact with simultaneously accessible exposed-conductive-parts if two faults exist simultaneously.

The following monitoring and protective devices can be used with IT systems:

- insulation monitoring devices (IMDs)
- residual current monitoring devices (RCMs)
- insulation fault location systems
- overcurrent protective devices
- residual current devices (RCDs).

In some installations, continuity of supply must be maintained. In these situations an insulation monitoring device shall be provided, capable of indicating when a first fault from a live part to an exposed-conductive-part or to earth occurs. The device used shall give an audible and/or visual signal for as long as the fault persists (411.6.3.1).

Except where a protective device is used to interrupt the supply in the event of a first fault, an RCM or insulation fault location system may be used. This will give an audible and/or visual signal for as long as the fault exists (411.6.3.2).

The conditions to satisfy the requirements for automatic disconnection of supply after a second fault occurring on a different live conductor are as follows: If the neutral conductor is not distributed in a.c. systems and the midpoint conductor is not distributed in d.c. systems:

$$Z_s \leq \frac{U}{2I_a}$$

When the neutral or midpoint conductor is distributed:

$$Z^1_s \leq \frac{U_o}{2I_a}$$

Where:

- U = nominal a.c. or d.c. voltage between line conductors

- U_o = nominal a.c. rms or d.c. voltage in volts between a line conductor and the neutral conductor/midpoint conductor

- Z_s = impedance of the fault loop comprising the line and protective conductors of the circuit

- Z^1_s = impedance of the fault loop comprising the neutral conductor and the protective conductor of the circuit

- I_a = current causing automatic operation of the protective device within the time specified in Table 41.1 of Regulation 411.3.2.2, or Regulation 411.3.2.3 for TN systems (411.6.4).

Functional extra-low voltage (FELV) *(411.7)*

When a nominal voltage not exceeding 50 V a.c. or 120 V d.c. is used for circuits or equipment, and the requirements of Section 414 covering SELV and PELV are not fulfilled or not necessary, the following requirements shall be applied to provide basic and fault protection.

Basic protection *(411.7.2)*

Basic protection shall be provided by basic insulation corresponding to the nominal voltage of the primary circuit of the source or by barriers and enclosures complying with Regulation 416.2.

Fault protection *(411.7.3 to 411.7.5)*

Exposed-conductive-parts of equipment of the FELV circuit shall be connected to the protective conductor of the primary circuit of the source, provided Regulations 411.3 to 411.6 have been met for automatic disconnection of the supply.

The source of FELV systems is usually a transformer with separation between the windings (often referred to as a double wound transformer), which complies with Regulation 414.3.

Socket-outlets and luminaire supporting couplers shall use a plug type which is not compatible with any of those used in the same premises.

Reduced low-voltage systems *(411.8)*

Nominal voltages of reduced low-voltage circuits shall not exceed 110 V a.c. rms between phases:

- for three-phase systems 63.5 V to an earthed neutral
- for single-phase 55 V to the earthed midpoint.

Reduced low voltage systems are often used on construction sites, using a double wound isolating transformer.

Basic protection *(411.8.2)*

Basic protection is provided by basic insulation in accordance with Regulation 416.1, which corresponds to the maximum nominal voltage of the system or by barriers or enclosures complying with Regulation 416.2.

Fault protection *(411.8.3)*

Fault protection shall be provided by an overcurrent protective device in each line conductor or by an RCD. Exposed-conductive-parts of the reduced voltage system shall be connected to earth.

The disconnection time for every point of utilisation, including socket-outlets, shall not exceed 5 s.

When a circuit breaker is used the maximum value of earth fault loop impedance Z_s shall be determined by the formula:

$$Z_s \times I_a \leq U_o$$

Alternatively, the Z_s values for 5 s disconnection time given in Table 41.6 of BS 7671 may be used.

Where a fuse is used the maximum value of earth fault loop impedance Z_s for a disconnection time of 5 s is given in Table 41.6 of BS 7671. An extract from this table, for circuit conductors operating at their normal operating temperature, is illustrated below.

Maximum Z_s values for 5 s disconnection time and U_o of 55 V (single-phase) and 63.5 V (three-phase)								
	Circuit breakers to BS EN 60898 and RCBOs to BS EN 61009 Type						General purpose (gG) fuses to BS 88-2.1 and BS 88-6	
	B		C		D			
U_o (Volts)	55	63.5	55	63.5	55	63.5	55	63.5
Rating amperes								
6	1.83	2.12	0.92	1.07	0.47	0.53	3.20	3.70
16	0.69	0.79	0.34	0.40	0.18	0.20	1.00	1.15

Plugs, socket-outlets, cable couplers, etc. of a reduced low-voltage system, shall have a protective conductor contact and not be compatible with any other plugs, socket-outlets or cable couplers used at any other voltage or frequency in the same installation. To achieve this, type BS EN 60309-2 plugs, socket-outlets and cable couplers are generally used (411.8.5).

Protection by double or reinforced insulation *(412)*

This protective measure is provided by supplementary insulation or reinforced insulation between live parts and accessible parts.

The electrical equipment used shall be of the following types and when tested this shall be to the relevant standards:

- Class II equipment
- equipment declared in the product standard as being equivalent to Class II.

With Class II equipment (double or reinforced insulation), the first line of defence for protection against electric shock is its basic insulation. The second line of defence takes the form of additional safety precautions such as supplementary insulation. The outer case of such equipment need not necessarily be insulated.

An exposed metal case of such a Class II appliance is not likely to become live in the event of a fault in the basic insulation and does not, therefore, come within the definition of an exposed-conductive-part.

Exposed metalwork of Class II equipment must be mounted such that it is not in electrical contact with any part of the installation which is connected to a protective conductor. However, Regulation 412.2.3.2 requires that where a circuit supplies items of Class II equipment, a circuit protective conductor must be run to and terminated at such points. This is to allow for the possibility that Class II equipment may be replaced at a later date by Class I equipment, except where Regulation 412.1.3 applies.

Note: *Class II equipment is identified by the symbol* ▢ .

The following symbol should be fixed in a visible position on both the interior and exterior of any enclosure to identify electrical equipment which has had supplementary or reinforced insulation applied in the process of erecting the electrical installation:

Enclosures *(412.2.2)*

When using insulating enclosures, they shall provide protection to IPXXB or IP2X. The enclosure must not contain any screws or fixing means which might need to be removed during installation or maintenance, as they could be removed and replaced by metallic screws or fixings, which could impair the enclosure insulation.

When a lid or door in an insulated enclosure can be opened without using a key or tool, any conductive parts shall be behind an insulating barrier to IPXXB or IP2X, which can only be removed using a key or tool (412.2.2.3).

Any conductive parts enclosed in an insulating enclosure shall **not** be connected to a protective conductor. No exposed-conductive-parts or an intermediate part shall be connected to any protective conductor unless the equipment specification allows it (412.2.2.4).

Protective measure: Electrical separation *(413)*

Electrical separation is achieved by using:

- for basic protection, insulation of live parts, barriers and enclosures in accordance with Section 416
- for fault protection, simple separation of the separated circuit from other circuits and from earth.

This arrangement is limited to one item of equipment, supplied from one unearthed source, unless Regulation 413.1.3 applies.

Basic protection *(413.2)*

All equipment shall comply with the requirements of Section 416 or Section 412.

Fault protection *(413.3)*

The separated circuit shall not exceed 500 V. Live parts of the separated circuit shall not be connected to any other circuit or earth.

Any flexible cables or cords used shall be visible where they are liable to mechanical damage (413.3.4).

Separate wiring systems are recommended where separated circuits are used. Where separated circuits and other circuits are installed in the same wiring system, the following wiring methods can be used as long as the rated voltage is not less than the highest voltage present and each circuit has protection against overcurrent (413.3.5):

- mulit-core cable without a metallic covering
- insulated conductors in an insulating conduit, ducting or trunking.

No exposed-conductive-parts of separated circuits shall be connected to protective conductors or exposed-conductive-parts of other circuits or to earth (413.3.6).

Protective measure: SELV or PELV *(414)*

In order to achieve protection by SELV or PELV the following requirements must be satisfied:

- nominal voltage of the circuit shall not exceed the upper limit of voltage Band 1, 50 V a.c. or 120 V d.c.
- the supply source must be one of the following:
  - a safety isolating transformer complying with BS EN 61558-2-6 with no connections between the output winding and the body of the transformer or the protective earthing circuit, if any
  - a motor generator with windings providing electrical separation equivalent to that of a safety isolating transformer

- an electrochemical source (e.g. battery)
- an engine-driven generator
- certain electronic devices complying with appropriate standards which ensure that in the event of an internal fault the output voltage cannot exceed the maximum value for extra-low voltage 50 V a.c. or 120 V d.c..

SELV and PELV supplies

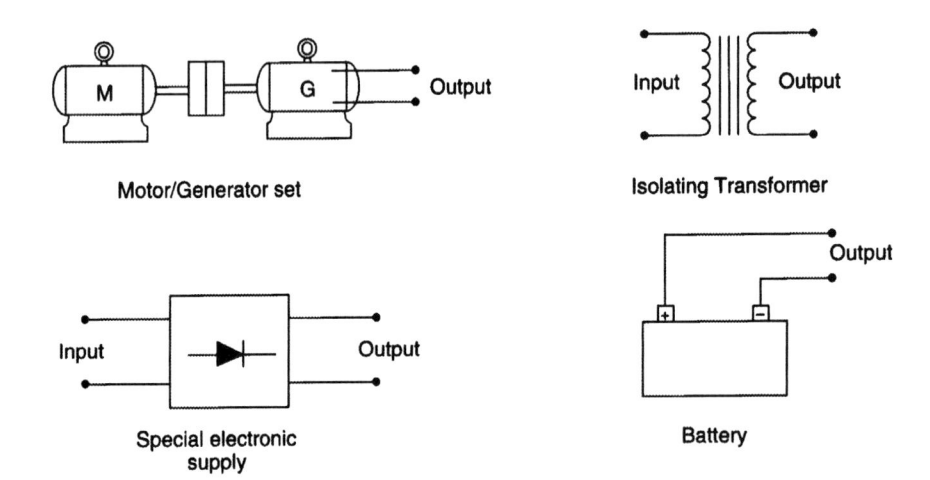

Motor/Generator set

Isolating Transformer

Special electronic supply

Battery

Any system supplied from a system at a higher voltage which does not provide the necessary electrical separation (such as autotransformers, potentiometers and semiconductor devices, etc.) is not classed as a SELV or PELV system.

SELV and PELV electrical installations must be installed so that circuits connected to the system are electrically separate from any other circuit and have no interconnection with any other electrical system, including earth. Where physical separation of circuits is impracticable, one of the following methods of installation can be used for circuits:

- conductors insulated to the highest voltage present in the installation
- circuit cable must be enclosed in an insulating sheath or insulating enclosure additional to their basic insulation
- conductors of circuits operating at different voltages must be separated from those at SELV or PELV by earthed metal screens (such as found in some trunking systems) or by an earthed metallic sheath (such as MICC or SWA cables)
- a multicore cable may be used to supply SELV and PELV circuits as well as circuits operating at different voltages, provided the SELV or PELV conductors are insulated for the highest voltage present.

Basic insulation of conductors needs to be sufficient for the voltage of only the circuit of which it is a part.

For all electrical equipment in a SELV or PELV system, i.e. switches, relays, etc., the electrical separation necessary between the live parts of the SELV or PELV system and any other systems shall be maintained.

Under no circumstances must an exposed-conductive-part of a SELV system be connected to any of the following:

- earth
- exposed-conductive-parts of other systems
- protective conductors of other systems
- an extraneous-conductive-part, except where electrical equipment is inherently required to be connected to that part, in which case measures must be incorporated to prevent parts exceeding the maximum value for extra-low voltage.

Note: PELV circuits and exposed-conductive-parts of equipment supplied by a PELV circuit, can be earthed.

On installations where the nominal voltage exceeds 25 V a.c. or 60 V d.c., protection against electric shock must be provided by barriers or enclosures affording protection to IP2X, IPXXB or by insulation complying with the relevant standard for such equipment.

Note: IP2X provides that any aperture should not be so large as to allow a British Standard 'finger' through to touch live parts; and never more than 12 mm wide. Adequate clearance must be maintained between live parts inside the enclosure.

If a SELV circuit does not exceed 25 V a.c. or 60 V d.c., the measures described above are unnecessary and no measures need to be taken to provide basic protection. If there is a risk of electric shock because the body resistance is lower than normal (for example, where there are wet conditions or in a confined location, such as when working in a large metal pipe or boiler which could act as a conductor) then the voltage limit will need to be further reduced.

Plugs and socket-outlets for SELV or PELV systems must **not** be capable of use with other voltage systems in the same premises. Plugs and socket-outlets in SELV systems must not have a protective conductor contact.

Additional protection *(415)*

Residual current devices (RCDs) *(415.1)*

Additional protection for installation circuits can be achieved using RCDs with a rated residual operating current ($I_{\Delta n}$) not exceeding 30 mA and an operating time not exceeding 40 ms at a residual current of 5 $I_{\Delta n}$.

For further information on RCDs refer to Technical Data Sheet 4A.

Supplementary equipotential bonding *(415.2)*

Additional protection can also be provided by means of supplementary equipotential bonding. Supplementary equipotential bonding includes all simultaneously accessible exposed-conductive-parts of fixed equipment and extraneous-conductive-parts. The equipotential bonding shall be connected to the protective conductors of socket-outlets and any other equipment.

If any doubt exists regarding the effectiveness of supplementary bonding it shall be confirmed that the resistance (R) between simultaneously accessible exposed-conductive-parts and extraneous-conductive-parts satisfies the following:

$$R \leq \frac{50 \text{ V}}{I_a} \quad \text{for a.c. systems}$$

Where:

I_a = operating current of the protective device

For RCDs, $I_{\Delta n}$ is used

For overcurrent devices Ia is the current causing automatic disconnection in 5 s

Provisions for basic protection *(416)*

For basic protection one or more of the following measures must be used:

- basic insulation of live parts *(416.1)*
- barriers or enclosures *(416.2)*
- protected by obstacles *(417.2)*
- placing out of reach *(417.3)*.

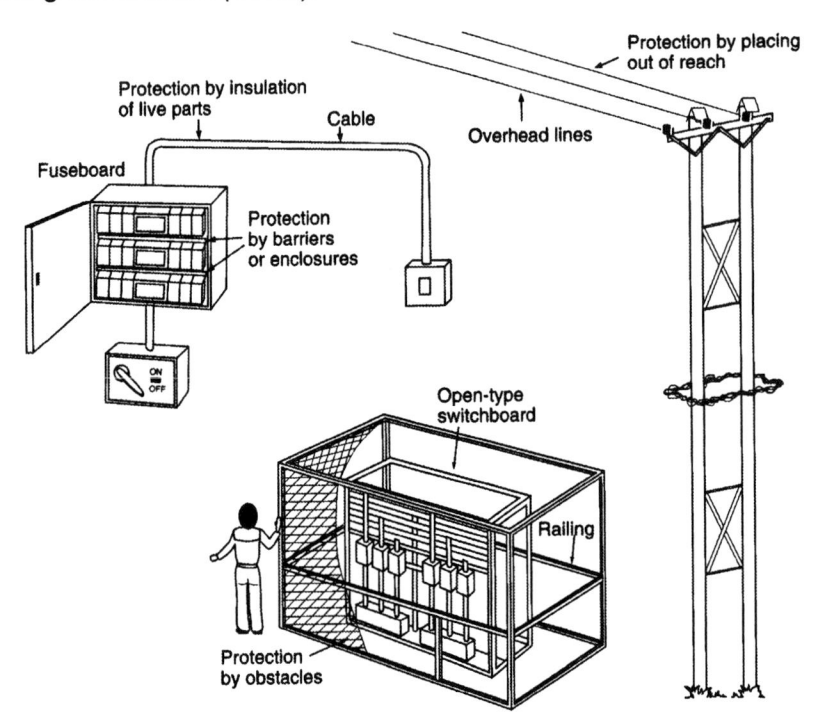

Protection by insulation of live parts *(416.1)*

Live parts must be completely covered with insulation that can only be removed by destruction and be capable of withstanding all stresses (electrical, mechanical, thermal and chemical) in service.

The insulation of factory-built equipment must comply with the relevant standards for that electrical equipment. For other equipment, insulation must durably withstand the stresses to which it may be subjected in service. Paint, varnish, lacquer, etc. are not generally considered to offer adequate protection.

Insulated cables give protection by insulation of live parts. If insulation is applied during the erection of the installation (such as a switch panel where shrink-on type of insulation is applied to bare copper conductors) the quality of the insulation must be confirmed by tests equivalent to those which ensure the quality of the insulation of type-tested equipment, e.g. flash-tested.

Protection by barriers or enclosures *(416.2)*

For this form of protection, live parts must be situated inside enclosures or behind barriers. The barriers and enclosures must be firmly secured in place, have sufficient stability and durability for the known conditions of normal service and provide protection to at least IP2X or IPXXB.

For the top surfaces of barriers and enclosures that are readily accessible, the minimum degree of protection is IP4X or IPXXD.

When it is necessary to remove a barrier or open an enclosure this shall only be possible using a key or tool (416.2.4).

If an item of equipment, such as a capacitor, is installed in an enclosure or behind a barrier, and may retain its charge after the supply is switched off, a warning label should be fitted.

Protection by obstacles *(417.2)*

Protection by obstacles is intended to prevent unintentional bodily approach to and unintentional contact with live parts when energised equipment is operating in normal service.

The obstacles must be secured so as to prevent unintentional removal but they may be removable without the use of a key or a tool.

This measure can be used in areas accessible only to skilled persons or to instructed persons under direct supervision of a skilled person.

Protection by placing out of reach *(417.3)*

Overhead lines for distribution between buildings and structures must be installed in accordance with the Electricity Safety, Quality and Continuity Regulations 2002.

Bare live parts of any installation (other than overhead lines) must not be within arm's reach or within 2.5 m of any:

- exposed-conductive-parts
- extraneous-conductive-parts
- bare live parts of other circuits.

The out-of-reach limits required by this measure must be increased where bulky or long conducting objects are normally handled, e.g. aluminium ladders or scaffold poles.

This measure can be used in areas accessible only to skilled persons or to instructed persons under direct supervision of a skilled person.

For definition of 'arm's reach' refer to Regulation 417.3.3.

Protective measures for application only where the installation is controlled or under the supervision of skilled or instructed persons *(418)*

Protection by non-conducting location *(418.1)*

This is a special arrangement and has only very limited application. It can only be used in special locations which are under effective supervision and where specified by a suitably qualified engineer.

Compliance is achieved if the area has an insulating floor and insulating walls, and one or more of the following applies.

- Physical separation between any exposed-conductive-parts and between extraneous-conductive-parts to be not less than 2.5 m or, for parts out of the zone of arm's reach, to be not less than 1.25 m.

- Obstacles between exposed-conductive-parts and extraneous-conductive-parts to be the same as the distances above, not being connected to earth or to exposed-conductive-parts. They must be constructed as far as possible of insulating material.

- The insulation of extraneous-conductive-parts to be of adequate electrical and mechanical strength.

These situations require strict supervision as they can become dangerous due to the possible introduction of earthed metal into the location, in the form of a portable appliance fed by leads connected outside the location. The safety of this system lies in the fact that a potential reached by metalwork in such a situation is unimportant because it is never possible to touch two pieces of metalwork at different potentials at the same time.

Protection by earth-free local equipotential bonding *(418.2)*

Earth-free local equipotential bonding can only be used in an area under effective supervision and where specified by a suitably qualified electrical engineer. It relies on the fact that all simultaneously accessible exposed and extraneous-conductive-parts are connected together by bonding conductors, but not to earth. Inside the area there can be no danger, even if the potential is high, because all internal exposed metalwork is at the same potential. A warning notice complying with Regulation 514.13.2 must be prominently displayed adjacent to every point of access to the location.

Protection by electrical separation to more than one item of equipment *(418.3)*

All equipment shall comply with at least one provision for basic protection from Section 416.

Protection by electrical separation for the supply to more than one item of equipment shall be achieved by complying with the requirements of Section 413 and Regulations 418.3.3 to 418.3.8.

Note: Regulation 413.1.2 is not applicable.

Any exposed-conductive-parts of separated circuits shall be connected by insulated equipotential bonding conductors, which are not connected to earth and any protective conductors or exposed-conductive-parts of other circuits or any extraneous-conductive-part.

The recommended circuit values for the product of the nominal voltage and length in metres should not exceed 100,000 V/m. In addition, the length of the wiring system must not exceed 500 m (418.3.8).

Protection against thermal effects *(Chapter 42)*

Scope *(420.1)*

This chapter applies to electrical installations and equipment and the measures necessary to protect people, property and livestock against:

- harmful effects of heat or thermal radiation
- ignition, combustion or degradation of materials
- flames and smoke when a fire hazard could be propagated from an electrical installation to other nearby fire compartments of a building
- safety services being cut off due to failure of the electricity supply.

Protection against a fire caused by electrical equipment *(421.2)*

Fixed electrical equipment shall be selected and installed so that its normal operating temperature is not likely to cause a fire. Consideration shall be given to the harmful effects on adjacent fixed materials and any other materials which may be in proximity to the equipment.

Where arcing or high temperature particles could be emitted from fixed equipment, one or more of the following methods should be used (421.3):

- enclosed in arc-resistant material
- screened by arc-resistant material
- mounted at a suitable distance from any material which could be harmfully affected, allowing the safe extinction of any emissions
- comply with the standard for that particular item of equipment.

Note: *Non-ignitable, low thermal conductivity and adequate thickness material shall be used as the arc-resistant material.*

Special consideration must be given to enclosures for oil-cooled transformers and switchgear when the quantity of liquid in a single location is in excess of 25 litres. The equipment should be housed with a drainage pit surround or in a chamber constructed of adequate fire-resistant materials, with means such as sills to prevent the liquid spreading and with vents to the external atmosphere (421.5).

Every termination of a live conductor or connection shall be contained within an enclosure complying with Regulation 526.5.

Precautions when a particular fire risk exists *(422)*

All electrical equipment shall be selected and installed so that a fire is not caused by its normal temperature rise and any other rise caused by a fault occurring.

Conditions for evacuation in an emergency *(422.2)*

Authorities that are responsible for building construction, public gatherings and fire prevention may specify a BD condition for this situation. BS 7671 provides the following BD classifications:

- BD2 – low density occupation, difficult conditions of evacuation
- BD3 – high density occupation, easy conditions of evacuation
- BD4 – high density occupation, difficult conditions of evacuation.

(Appendix 5 of BS 7671:2008 gives further information.)

Where a fire alarm situation for BD2, BD3 or BD4 classifications of a building exist, wiring systems shall not be installed in an escape route unless the cable or wiring system is suitable. A suitable cable sheath or enclosure is authorised by national Regulations for building elements, e.g. Building Regulations for England and Wales or the Scottish Building Regulations. In the absence of any such Regulation, building elements must have a fire rating of two hours.

When a wiring system is installed in an escape route it shall not be within arm's reach, unless provided with mechanical protection, and the run shall be as short as possible (422.2.1).

For classifications BD2, BD3 or BD4 any switchgear or control gear shall only be accessible to authorised persons and, if installed in an escape route, it shall be enclosed in a cabinet or enclosure of non-combustible material (422.2.2).

Electrical equipment which contains any flammable liquids shall not be installed in escape routes where conditions BD3 and BD4 exist. Individual capacitors used in discharge luminaires and motor starters are allowed (422.2.3).

Locations with risk of fire due to the nature of processes or stored materials *(422.3)*

These requirements are in addition to Section 421 and apply to:

- locations with risks of fire due to the nature of processed or stored materials, including the presence of dust, e.g. woodworking and textile factories, paper mills, etc., now classified as BE2 conditions
- installations in areas built from combustible materials. These locations may be regulated by national authorities.

Electrical equipment must be selected and erected so that its normal operating temperature (or any temperature rise due to fault) will not cause a fire. This can be achieved by the equipment's construction or the provision of additional protection.

This section does not apply to locations with explosion risks (see BS EN 60079-14).

In the absence of other recommendations from manufacturers, spotlights or projectors must be installed at a distance from combustible materials of at least:

- lamp rating up to 100 W   – 0.5 m
- lamp rating 100 W–300 W   – 0.8 m
- lamp rating 300 W–500 W   – 1 m

Any luminaires must be of a design that prevents flammable material ejecting from them, especially when they break down (422.3.1).

Enclosure temperatures of heaters, etc. must not exceed 90°C under normal conditions and 115°C under fault conditions (422.3.2).

Note: Luminaires marked $\triangledown$*D are designed to provide surface temperatures that are limited.*

Switchgear and control gear shall always be installed outside a flammable area, unless it is suitable for that location or is installed to provide protection to at least IP4X, or when dust is present IP5X (422.3.3).

Cables not completely embedded in non-combustible materials (e.g. plaster) or protected from fire by other means must meet the flame propagation requirements of BS EN 60332-1-2. In high risk areas, bunched cables or long vertical runs must meet the flame propagation requirements of the BS EN 50266 series (422.3.4).

When a wiring system passes through a location but does not supply any electrical equipment in that location it shall meet the following requirements (433.3.5):

- specified in Regulation 422.3.4
- that no joints or connections shall be made in that location, unless enclosed in an enclosure that does not affect the flame propagation characteristics of the wiring system
- to be adequately protected against overcurrent in accordance with Regulation 422.3.10
- that no bare conductors shall be used.

Motors that are not constantly supervised or are automatically or remotely controlled shall have manual reset overloads to protect them against excess temperature. Star-delta connected motors must have excess temperature protection in both the star and delta configurations (422.3.7).

All luminaires shall be appropriate for use in their respective locations, be provided with suitable enclosures providing protection to at least IP5X, have limited surface temperature to BS EN 60598-2-24 and be constructed to prevent lamp components falling from them (422.3.8).

Except for busbar trunking systems, MICC or powertrack systems, wiring systems are to be protected from insulation faults as follows:

- for TN and TT systems, protection by an RCD that has a residual operating current ($I_{\Delta n}$) not greater than 300 mA according to Regulation 531.2.4
- in IT systems, an audible and visual insulation monitoring device must be used and adequately supervised to allow manual disconnection. In the event of a second fault the overcurrent device must operate within the times given in Chapter 41 (422.3.9).

Circuits supplying equipment in the location, or a circuit traversing the location, shall be protected against overload and faults with a protective device located outside of the location on the supply side (422.3.10).

Safety extra-low voltage circuit live parts shall be contained in enclosures providing protection to at least IP2X or IPXXB or be provided with insulation capable of withstanding a test voltage of 500 V d.c. for one minute (422.3.11).

No PEN conductors are allowed (this does not apply to wiring systems passing through the area but the requirements of 422.3.5 must be met) (422.3.12).

A means of isolation shall be provided for every circuit from all live supply conductors using a linked switch or linked circuit breaker (422.3.13).

Flexible cables and cords are to be heavy duty 450/750 V or protected against mechanical damage (422.3.14).

Heating appliances must be securely fixed, barriers must be provided to prevent ignition of combustable materials and heat storage appliances shall prevent ignition of combustible dust or fibres by the heat storage core (422.3.15 and 16).

Combustible construction materials *(422.4)*

Where CA2 conditions exist, that is where buildings are mainly constructed of combustible materials, such as wood, precautions shall be taken to ensure electrical equipment cannot ignite any walls, floors or ceilings.

Electrical equipment (such as distribution boards) fixed to or in combustible material shall comply with the relevant standard for enclosure temperature rise. When equipment does not comply, it shall be enclosed with non-flammable material of suitable thickness. The material used must not adversely affect the dissipation of heat from the electrical equipment.

Any cables and cords installed shall comply with BS EN 60332-1-2. Conduit and trunking systems shall comply with BS EN 61386-1 and BS EN 50085-1 respectively.

Note: Luminaires marked $\triangledown{F}$ in accordance with BS EN 60598-1 are classed as suitable for mounting on a normally flammable surface.

Fire propagating structures *(422.5)*

The requirements of BS 7671 relate again to where CB2 conditions exist, relating to the propagation of fire that exists in buildings where the spread of fire will be facilitated (such as high rise buildings or buildings with forced ventilating systems). Precautions shall be taken to ensure the electrical installation does not provide for propagation of fire. In some installations this can be achieved by fire detectors closing fire-proof shutters in ducts, troughs and trunking.

Selection and erection of installations in locations of national, commercial, industrial or public significance *(422.6)*

Locations would include national monuments, museums, railway stations, airports, laboratories and computer centres and any rooms or buildings containing assets of significant value.

Installations in such locations shall comply with Regulation 422.1.

Protection against burns *(423)*

Any accessible part of an enclosure of fixed equipment liable to reach a temperature that would cause burns (i.e. in excess of the values in Table 42.1) must be guarded so as to prevent accidental contact. An exception is made for any equipment manufactured to a British Standard which has a specific limiting temperature.

Equipment within arm's reach must comply with Table 42.1 of BS 7671.

Typical examples of maximum permissible temperatures:

Metallic hand-held equipment	55°C
Non-metallic hand-held equipment	65°C
Metallic equipment which can be touched	70°C
Non-metallic equipment which can be touched	80°C
Metallic equipment that does not need to be touched	80°C
Non-metallic equipment that does not need to be touched	90°C

Protection against overcurrent *(Chapter 43)*

General requirement *(430.3)*

The device used must be capable of breaking the circuit before any overload current could cause a rise in temperature or mechanical effect which might damage insulation, terminations, joints or the surroundings of the conductors.

These requirements mean that the circuit protective device is selected to ensure that overheating does not occur to the degree of insulation breakdown.

Scope *(431)*

Chapter 43 of BS 7671 gives the requirements for live conductors to be protected against overcurrent, fault current and makes a distinction between overload current (Section 433) and fault current (Section 434).

An overload current is an overcurrent occurring in a circuit which is electrically sound, such as the starting current of an electric motor.

Short-circuit current is an overcurrent resulting from a fault of negligible impedance between live conductors differing in potential under normal operating conditions.

Protection according to the nature of the circuit and distribution system *(431)*

Detection of overcurrent shall be provided for all line conductors and shall cause disconnection of the conductor in which the overcurrent has been detected. In circumstances where the disconnection of one phase could cause damage (for example, a three-phase motor), precautions shall be taken to prevent this.

When circuits are supplied by three-phase only and the neutral has not been distributed on TN and TT systems, overcurrent detection can be omitted from one of the line conductors as long as a protection system is provided on the supply side to detect unbalanced loads, and to disconnect all three line conductors, and the neutral conductor is not distributed from an artificial neutral point of the circuit on the load side (431.1.2).

Protection of the neutral conductor *(431.2)*

For TN and TT supply systems the neutral conductors shall be protected against fault currents, unless they have the same cross-sectional area as the line conductors. When a reduced cross-sectional area neutral conductor is used, overcurrent protection shall be provided, appropriate to its cross-sectional area. The resultant disconnection of the associated line conductors need not necessarily disconnect the neutral conductor (431.2.1).

For an IT system neutral conductors shall not normally be distributed unless overcurrent detection for every circuit neutral is provided, which disconnects all live conductors of the circuit and the neutral conductor. This measure is not necessary when short-circuit protection of the neutral is provided by a device on the supply side (e.g. at the origin of the installation) or the circuit is protected by an RCD with a rated residual current ($I_{\Delta n}$) not exceeding 0.2 times the current-carrying capacity of the neutral conductor, which disconnects all the live and neutral conductors of the circuit (431.2.2).

Disconnection and reconnection of the neutral conductor *(431.3)*

The neutral conductor shall not disconnect before any line conductors and shall always reconnect at the same time or before any line conductors.

Nature of protective devices *(432)*

Protection against both overload and fault current *(432.1)*

To be able to give this protection there is a requirement (except for that allowed by Regulation 434.5.1) that a protective device must be capable of breaking, and for a circuit breaker making, any overcurrent up to and including the maximum prospective fault current (PFC) at the point where the device is installed.

Protection against overload current only *(432.2)*

Devices providing this type of protection are generally inverse-time-lag devices whose rated short-circuit breaking capacity may be below the value of the maximum prospective fault current where the protective device is installed.

Protection against fault current only *(432.3)*

Devices providing this type of protection shall be capable of breaking and, in the case of circuit breakers making, the fault current up to and including the prospective fault current. Suitable devices may be a circuit breaker with a short-circuit release or a fuse.

Protection against overload current *(433)*

Co-ordination between conductors and overload protective devices *(433.1)*

Circuits shall be designed so that small overloads of long durations are unlikely to occur.

Quantities to be considered are:

- I_n = nominal current or current setting of device
- I_b = design current of the circuit
- I_z = current-carrying capacity of any of the circuit conductors
- I_2 = the current which ensures effective operation of the device.

The co-ordination of requirements are that:

- $I_b \leq I_n$ i.e. design current must not exceed current rating of protective device
- $I_n \leq I_z$ i.e. current setting of device must not exceed the lowest conductor rating
- $I_2 \leq 1.45 \times I_z$ i.e. the current causing effective operation of the protective device must not exceed 1.45 times the lowest of the current-carrying capacities (I_z) of any of the conductors in the circuit.

These three conditions are summarised by two basic rules:

$I_b \leq I_n \leq I_z$ (a)

$I_2 \leq 1.45 \times I_z$ (b)

when the device is one of the following:

- an HBC fuse to BS 88 Part 2.2
- an HBC fuse to BS 88 Part 6
- a cartridge fuse to BS 1361
- a circuit breaker to BS EN 60898
- a circuit breaker to BS EN 60947-2
- an RCBO to BS EN 61009-1.

If the conditions in expression (a) are satisfied then the conditions in (b) will also be satisfied.

When the device is a semi-enclosed fuse to BS 3036, in order to satisfy expression (b), $I_n \leq 0.725 \times I_z$. The reason for this is due to the possibility of such a fuse having a fusing factor as high as two, then

$$\text{fusing current} = \text{fusing factor} \times \text{fuse rating}$$
$$= 2 \times \text{fuse rating}$$

then $I_2 = 2 \times I_n$

and to satisfy (b) $\quad 2 \times I_n \leq 1.45 \times I_z$

therefore $I_n \leq \dfrac{1.45 \times I_z}{2} \leq 0.725 \times I_z$

Position of devices for overload protection *(433.2)*

In the application of overload protective devices particular attention must be given to the position of a device in a circuit.

A device for overload protection must be placed at a point where reduction occurs in the current-carrying capacity of the conductors. Examples of how this reduction may occur are:

- reduced conductor csa (e.g. a fused spur on a ring circuit)
- installation method changed (e.g. overhead to underground)
- type of cable changed (e.g. PVC cables in conduit changed to MICC)
- ambient temperature changed (e.g. from a boiler house to normal room conditions).

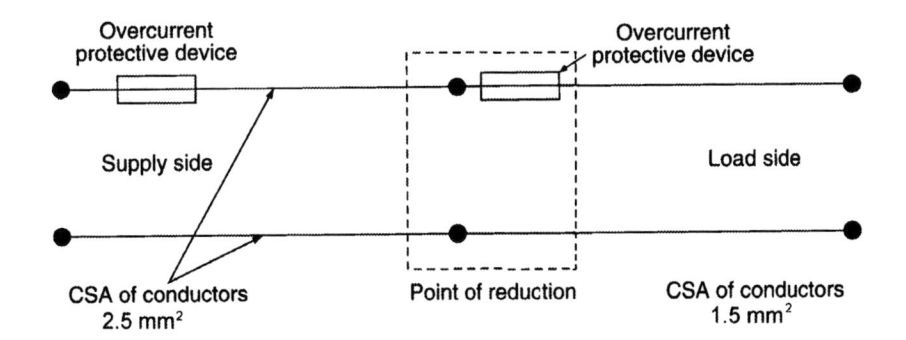

If there are no outlets or spurs along a section, the protective device may be placed along the cable run, provided that:

- it is protected against fault current (see Section 434)

or, between the reduction in current-carrying capacity and the protective device, each conductor shall:

- not exceed 3 m in length
- be erected in a manner to minimise the risk of fault current
- be erected in a manner to minimise the risk of fire or danger to persons.

Omission of devices for protection against overload *(433.3)*

Regulations in this section shall not be applied to installations where the location is a fire or explosion risk.

Overload protective devices need not be provided:

- when a conductor on the load side of a position, where the current-carrying capacity is reduced, and protection of that conductor against an overload is installed on the supply side

- for a conductor where the characteristics of the load or supply are not likely to carry any overload current and where the conductor is protected against a fault current in accordance with Section 434

- at the origin of an installation when the supply distributor provides an overload device and has confirmed that it provides protection between the origin and main distribution point of the installation.

Position or omission of devices for protection against overload in IT system *(433.3.2)*

The provisions in Regulation 433.2.2 and 433.3 are not applicable to IT systems unless each circuit not protected against an overload is protected by one of the following:

- protective measures in Regulation 413.2

- RCDs installed in each circuit that operate immediately when a second fault occurs.

Systems which are permanently supervised or have an insulation monitoring device need to disconnect the relevant circuit on first fault or give a signal indicating the presence of a fault.

In IT systems which do not have a neutral conductor the overload device can be omitted in one of the line conductors if RCDs are installed in each circuit (433.3.2.2).

Omission of devices for protection against overload for safety reasons *(433.3.3)*

The omission of an overload device for circuits supplying current using equipment is allowed if unexpected disconnection could cause danger or damage. Examples are:

- exciter circuit of a rotating machine

- supply circuit of a magnet used for lifting

- secondary circuit of a current transformer

- circuit to a fire extinguishing device

- circuit supplying safety devices such as fire alarms and gas alarms.

Overload protection of conductors in parallel *(433.4)*

Where a single protective device is used to protect conductors in parallel, there must be **no** branch circuits or devices for isolating or switching in any of the parallel conductors.

Note: This does not apply to ring final circuits, where spurs are allowed.

Equal current sharing by conductors in parallel *(433.4.1)*

Parallel conductors that are equally loaded and protected by a single protective device will have a value of I_z (Regulation 433.1.1) that is, the sum of the current-carrying capacities of the parallel conductors concerned.

It is considered that the currents are equal in the parallel conductors if the requirements of Regulation 523.8(i) are met. This Regulation includes requirements, such as the need for conductors to be of the same construction, material, size, length and have no branch circuits along the length, etc.

Unequal current sharing by conductors in parallel *(433.4.2)*

Where the currents in parallel conductors are unequal, both the design current and overload current protection must be considered separately for each conductor.

Note: When the difference in currents is more than 10% of the design current for each conductor then they are considered to be unequal. Further guidance is provided in Appendix 10 of BS 7671.

Protection against fault current *(434)*

The prospective fault current shall be determined at every relevant point in an installation between conductors belonging to the same circuit. This can be done by calculation, measurement or enquiry from the supply provider. An example of a calculation is as follows.

Calculation

Calculating the PFC requires knowledge of the impedance values of the supply conductors and supply transformer.

The prospective fault current is the value of overcurrent at a given point in the circuit resulting from a fault of negligible impedance between live conductors or between a live conductor and an exposed-conductive-part (see diagram on next page).

It can be calculated as follows:

$$PFC = \frac{V}{Z_t + Z_1 + Z_2}$$

Where: V = source voltage

Z_t = impedance of supply transformer

$Z_1 Z_2$ = conductor impedances ($Z_2 = Z_n$ for single-phase and neutral circuit)

Note: The term 'negligible impedance' means there is, in effect, a 'fault' between the line conductors, or between line and neutral conductors in the installation (see diagram below).

The determination and application of the fault current is dealt with later.

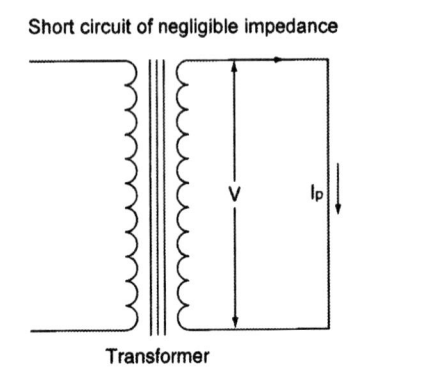

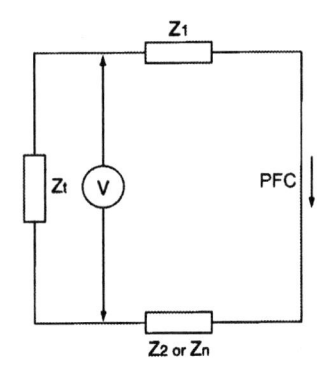

Enquiry

Enquiry involves the supply distributor quoting the maximum value of PFC and this must be recorded since it will form part of the installation design data. The value will also be required on the Electrical Installation Certificate.

Measurement

To measure the PFC:

- for single-phase supplies, the highest value of either the line-to-earth or line-to-neutral fault current must be recorded

- for three-phase installations, the maximum PFC current would normally be the line-to-line value.

*Note: **Unless a test instrument suitable for use on 400 V is available, you will need to calculate the value of PFC.** Doubling the value of the line-to-neutral fault current would give an approximate value of the line-to-line fault current.*

You must not attempt to measure line-to-line fault current with a 230 V test instrument, as damage could be caused to the instrument.

Position of devices for fault protection *(434.2)*

As for overload protection, there are certain conditions to be satisfied when deciding the position of fault current protective devices.

The devices may be positioned on the load side of a point where there is a reduction in the value of the current-carrying capacity, subject to a restriction of 3 m maximum length and erected to reduce to a minimum the risk of a fault or of a fire or danger to persons (434.2.1).

An example of the use of this relaxation is when connecting items of switchgear, e.g. switch fuse to busbars.

There is no limit stated to the size of the conductor that can be used. The risks mentioned may be reduced by enclosing the conductors in trunking or conduit.

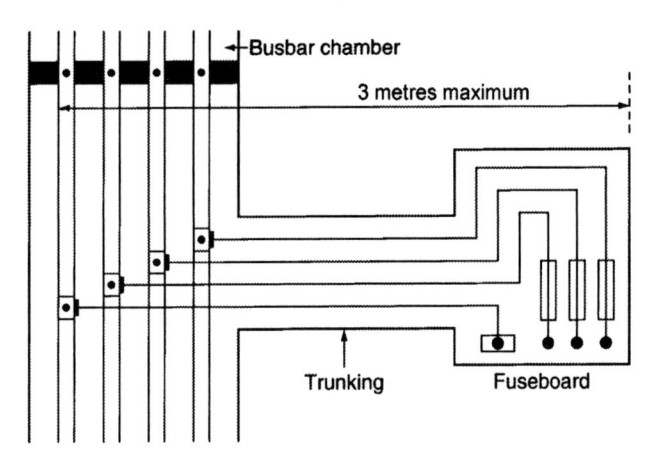

Fault protection of conductors in parallel *(434.4)*

A single protective device to protect conductors wired in parallel against fault current is allowed, provided the characteristics of the device will ensure an effective operation under the most onerous fault condition in any one of the parallel conductors.

Consideration must be given to the sharing of the fault currents between the conductors. Faults can be fed from both ends of parallel conductors.

If the effective operation of a single device cannot be guaranteed, then one or more of the following must be employed:

- the risk of fault current is reduced to minimum by the nature of the wiring e.g. protection against mechanical damage
- a fault current protective device is installed at the supply end of each parallel conductor (where two conductors are wired in parallel)
- fault current protection is provided at the supply and load ends of each parallel conductor when there are more than two conductors in parallel.

Omission of devices for protection against fault current *(434.3)*

Devices providing protection against fault current need not be provided:

- for conductors connecting a generator, transformer, rectifier or an accumulator battery to an associated control panel, if the protective device is placed in the panel
- for circuits where disconnection could cause danger (see those listed in Regulation 433.3.3)
- for certain measuring circuits

- at the origin of installations, where the distributor installs one or more devices for protection against fault current and agrees protection is afforded to that part of the installation, between the origin and the main distribution point of the installation, where fault protection has been provided.

Short-circuit protection of conductors in parallel *(434.4)*

A single device may be used to protect conductors connected in parallel against fault current, if the operating characteristics of the device used provide effective operation when a fault occurs at the most onerous position in one of the parallel conductors.

If the effective operation of a single protective device cannot be guaranteed, one or more of the following measures shall be used:

- wiring shall be installed in a way that reduces the risk of fault current. One example would be to provide protection against mechanical damage. In addition conductors shall be installed to reduce to a minimum risk of fire or danger to persons

- when two conductors are installed in parallel, a fault current protective device shall be installed at the supply end of each parallel conductor

- when more than two conductors are installed in parallel, a fault current protective device shall be provided at both the supply and load ends of each parallel conductor.

Note: See Appendix 10 for more information.

Characteristics of a fault current protective device *(434.5)*

The short-circuit breaking capacity of devices shall not be less than the maximum prospective fault current at the point of installation, except where the following apply:

- when another protective device with the necessary breaking capacity is installed on the supply side and the characteristics of the devices are co-ordinated so that any energy let-through by both devices does not exceed what can be withstood without damage by the device(s) on the load side

- a fault occurring at any point in a circuit shall be interrupted within a time that does not cause conductor or cable permitted limiting temperatures to be exceeded

- for faults of less than 0.1 s, for current limiting devices, k^2S^2 shall be greater than the let-through energy (I^2t) for BS EN 60898-1, BS EN 6098-2 or BS EN 61009-2.1 protective devices or as quoted by the manufacturer.

The time taken (t) by any given fault current to increase the temperature of the live conductors from the highest permissible temperature in normal operation to the limiting temperature can be obtained from the following formula:

$$t = \frac{k^2 S^2}{I^2}$$

Where:

- t = duration of short circuit in seconds
- S = conductor cross-sectional area (csa) in mm²
- I = the value of fault current in amperes, expressed for a.c. as the rms value, due account being taken of the current limiting effect of the current impedance
- k = factor for a particular type of cable (Table 43.1 of BS 7671 gives values of k for common materials). The k value is dependent upon material of conductor, the conductor size, its initial and final temperature and insulation. Because it takes account of the resistivity, temperature coefficient and heat capacity of the conductor.

In order to determine the time (t) and to show that it comes within the limit of 5 s when a fault occurs, it must be calculated. If greater accuracy is required, refer to BS 7454.

Example

Cable size and type – 25 mm² PVC insulated and sheathed with copper conductors operating at 70°C.

Value of k = 115 (from Table 43.1)

Take the effective fault current as 4 kA = 4,000 A

Time permitted for the fault current to exist before damage to the cable insulation could occur is:

$$t = \frac{k^2 S^2}{I^2}$$

$$= \frac{115^2 \times 25^2}{4,000^2}$$

$$= 0.52 \text{ S}$$

Taking a 50 A Type B circuit breaker as the protective device, check that the circuit breaker will clear the fault current within 0.52 S. Refer to the time/current characteristic (Fig. 3.4, Appendix 3 of the Regulations) and read off the time value when the current is 4,000 A.

Time obtained = 0.1 s. Therefore the circuit breaker will prove satisfactory.

When installing busbar trunking complying with BS EN 60439-2, or a powertrack system complying with BS EN 61534, one of the following requirements shall apply:

- the rated short-time withstand current (I_{cw}) and the rated peak withstand current of these systems shall not be lower than the rms value of the prospective fault current and prospective fault peak current value. The maximum time for which the (I_{cw}) is defined for a busbar system shall be higher than the maximum operating time of the protective device.

- the rated conditional short-circuit current of both systems associated with a particular protective device shall not be lower than the prospective fault current.

Co-ordination of protection *(435)*

There is a requirement to co-ordinate overload and fault current protection by assessing the characteristics of the protective devices used, so that the energy let-through by the fault current protective device does not exceed that which can be withstood without damage by the overload current protective device.

This Regulation is also applicable to motor control circuits where the overload and fault current protection devices may be housed separately.

For further guidance, BS EN 60947-4.1 must be consulted and the advice of the manufacturer of the control gear must be sought.

For details of operational characteristics of devices, refer to the Technical Data Sheet 4D and 4E.

Limitation of overcurrent by characteristics of supply *(436)*

The conductors of an installation are regarded as being protected against overcurrent and fault current where they are fed from a source of supply incapable of supplying a current greater than the current-carrying capacity of the conductors. Examples include certain types of:

- bell transformers
- welding transformers
- thermoelectric generating sets.

Protection against voltage disturbances and electromagnetic disturbances *(Chapter 44)*

Scope

This chapter applies to the requirements for the safety of electrical installations when voltage and electromagnetic disturbances are generated.

Protection of low voltage installations

The following situations are covered:

- faults between a high voltage system and earth in transformer substations
- supply neutral loss
- short circuits between line and neutral conductors
- accidental earthing of a line conductor (IT system).

Rules for designers and installers of substations

The following information is required for the high voltage system:

- quality of system earthing
- maximum earth fault current
- earthing resistance.

Regulation 442.1.2 symbols

This Regulation provides symbols and their definitions to explain the four situations listed at the top of this page.

Magnitude and duration of power frequency fault voltage

The magnitude and duration of fault voltage, which appears in low voltage installations between exposed-conductive-parts and earth, shall not exceed the values in Figure 44.2 of BS 7671, which is taken from IEC document 61936-1.

Magnitude and duration of power frequency stress voltages

The magnitude and duration of power frequency stress of low voltage equipment in the low voltage installation, due to an earth fault in the high voltage system, shall not exceed the requirements of Table 44.2 of BS 7671.

Requirements for calculation of limits

The requirements of the above are fulfilled when installations receive a low voltage supply from a system for distribution of electricity to the public.

In order to fulfil the above requirements co-ordination is required between the high voltage system operator and the low system installer.

The responsibility for compliance is usually with the installer/owner/operator of the substation.

Measures suitable to fulfil the requirements are:

- earthing arrangements which are separate for the high and low voltage supplies
- change of low voltage system earthing
- reduction of earth resistance.

Protection against overvoltages of atmospheric origin or due to switching

This section covers protection of electrical installations against transient overvoltages:

- transmitted by the supply distribution system
- switching overvoltages generated by equipment within the installation.

This information is mainly for designers of substation installations. Electricians will not normally be involved in such activities unless under the supervision of an experience electrical engineer.

TECHNICAL DATA SHEET 4A

Residual current device (RCD)

This term refers to a complete range of devices including:

  RCCB – Residual current circuit breaker

  RCBO – Residual current breaker with overcurrent device

  SRCD – Socket-outlet with residual current device

  PRCD – Portable residual current device.

An RCD is designed to give protection against shock risk and against fire. The basic circuit for a single-phase device is as illustrated.

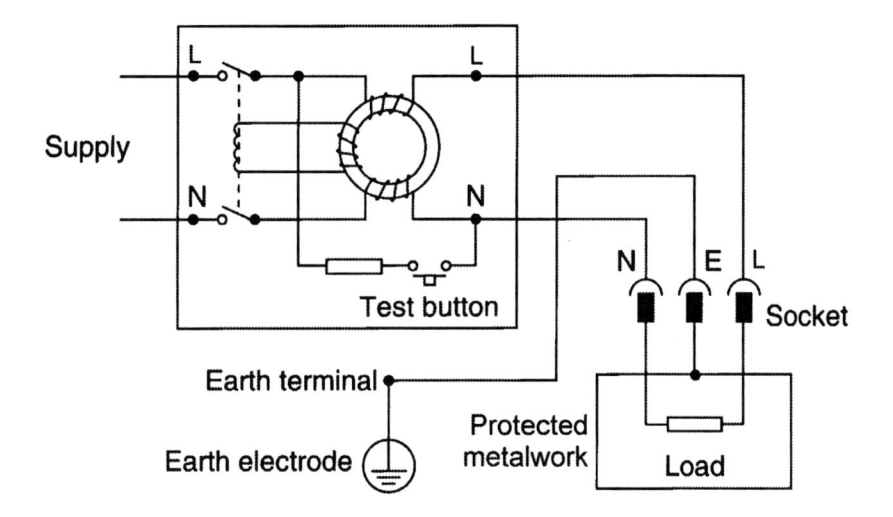

Basis of operation (single-phase)

Two conductors pass through, or are wound on, a common transformer core. A third coil, the detector coil, is wound onto the transformer core and is connected directly to a sensitive relay (trip coil).

When the phase and neutral currents are balanced, as they should be in a healthy circuit, they produce equal and opposing fluxes in the transformer core and therefore result in no current in the detector coil.

If more current flows in the line conductor than the neutral, an out of balance flux will be produced, thus inducing a current in the detector coil which will supply the trip coil, causing the relay to operate the tripping mechanism and open the main contacts.

Normally, the reason for more current flowing in the line than the neutral conductor is because some current has returned to earth via an earth fault.

Residual current devices

Where an RCD may be operated by other than a skilled or instructed person, it should be designed or installed so that adjustment of the setting or calibration of its rated trip current or time delay is not possible without the use of a key or tool. Visible indication of its setting or calibration is required.

Test button

The operation of an RCD by depressing the test button proves that the mechanical parts of the device are working correctly. The users of an installation are advised to carry out this test quarterly.

Selectivity

RCDs are completely selective in operation of the circuit they protect and are unaffected by parallel earth paths.

Application

The device must be capable of disconnecting all line conductors of the circuit. The live conductors of the circuit must be contained within the magnetic field of the transformer of residual current devices. Any protective conductor must be outside this magnetic field, eliminating any possibility of an induced EMF in the protective conductor.

If the operation of the device relies upon an auxiliary supply that is external to the device, it must be of a type that will automatically operate if the auxiliary supply fails. Alternatively, the device may be provided with a supply which will automatically become available upon failure of the auxiliary supply.

RCDs must be installed outside the magnetic fields of other equipment (unless the manufacturer gives instructions to the contrary).

When a device is fitted as additional protection, the device must be capable of withstanding, without damage, any thermal and mechanical stress which may occur under fault current conditions on the load side of the device.

If the disconnection times are to be achieved by the use of an RCD in an installation supplied from a TN or TT system, the product of the rated residual operating current ($I_{\Delta n}$) and the earth fault loop impedance (Ω) must not exceed 50 V (25 V for the construction site or agricultural installation).

When installing RCDs circuits must be arranged to avoid danger and minimise inconvenience caused by operation of the device.

Note: The use of an RCD is excluded for automatic disconnection when the system is TN-C. In such a case there is no difference between line and neutral currents because there is no separate path for neutral and earth leakage currents.

TECHNICAL DATA SHEET 4B

Consumer units

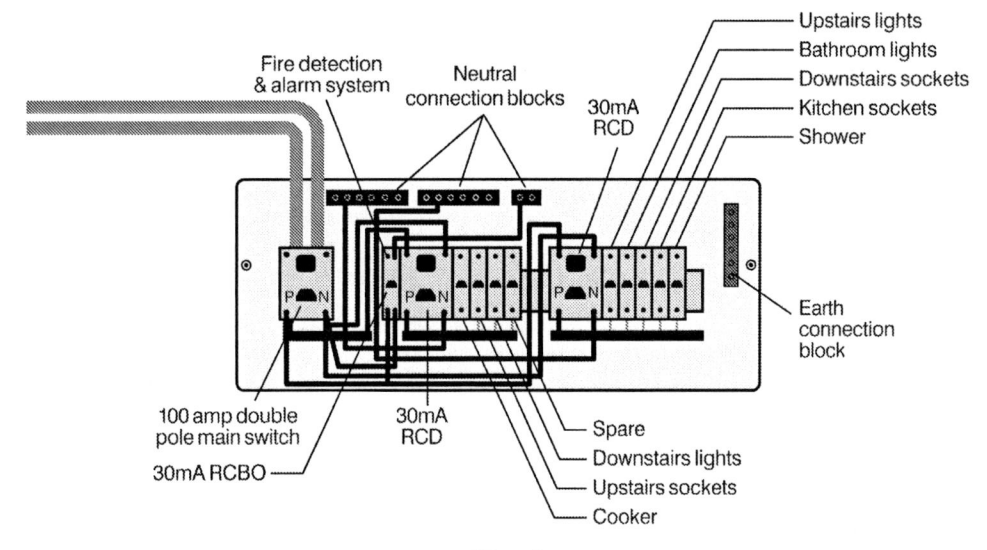

Fig. 1

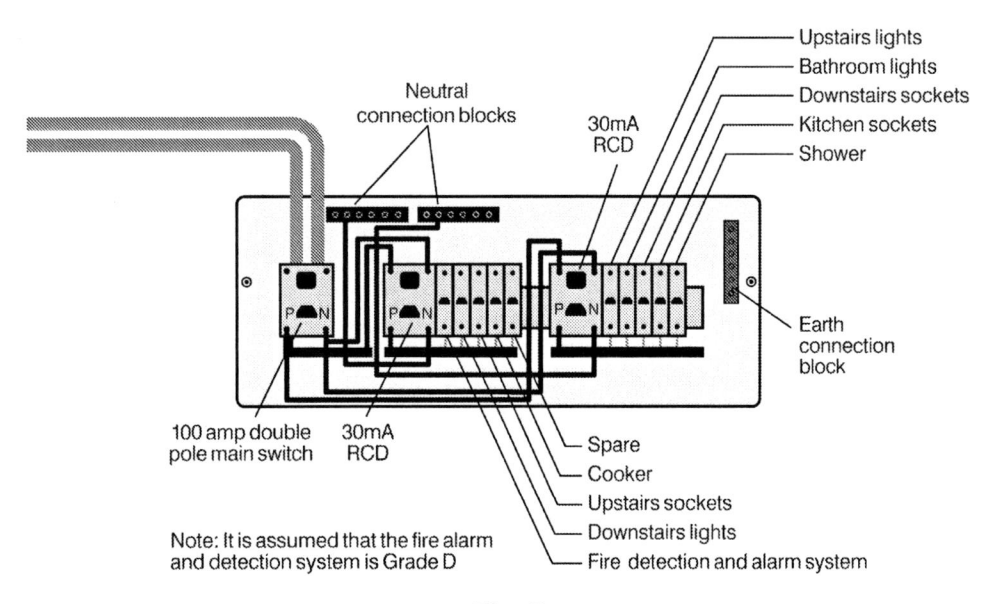

Labels: Fire detection & alarm system; Neutral connection blocks; 30mA RCD; Upstairs lights; Bathroom lights; Downstairs sockets; Kitchen sockets; Shower; Earth connection block; 100 amp double pole main switch; 30mA RCBO; 30mA RCD; Spare; Downstairs lights; Upstairs sockets; Cooker

Fig. 2

Labels: Neutral connection blocks; 30mA RCD; Upstairs lights; Bathroom lights; Downstairs sockets; Kitchen sockets; Shower; Earth connection block; 100 amp double pole main switch; 30mA RCD; Spare; Cooker; Upstairs sockets; Downstairs lights; Fire detection and alarm system

Note: It is assumed that the fire alarm and detection system is Grade D

Modified consumer unit arrangements to comply with BS 7671: 2008

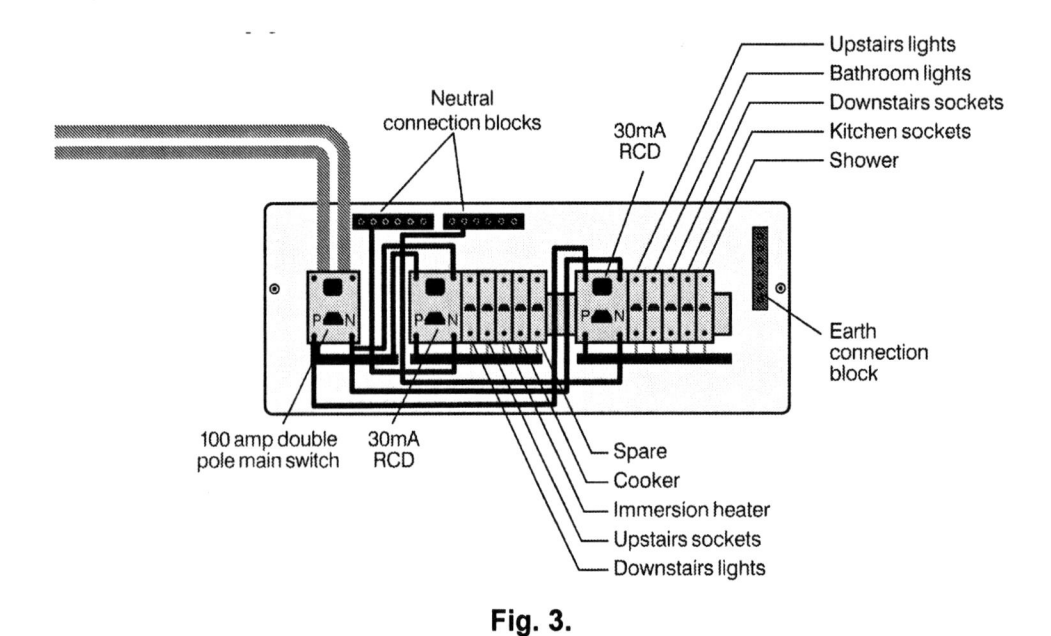

Fig. 3.

Modified consumer unit arrangement for a rewire to comply with BS 7671:2008

TECHNICAL DATA SHEET 4C

Fuses

Type of fuses

- Semi-enclosed, often referred to as re-wireable (BS 3036).
- Cartridge (BS 1361) and (BS 1362).
- High breaking capacity, referred to as HBC (BS 88 Part 2).
- High breaking capacity, referred to as HBC (BS 88 Part 6).

Semi-enclosed or rewireable fuses

Advantages

- Cheap.
- No mechanical moving parts.
- Simple to observe whether element has melted.

Disadvantages

- Danger on insertion with fault on installation.
- Can be repaired with incorrect size fuse wire.
- Element cannot be replaced quickly.
- Deteriorate with age.
- Lack of discrimination.
- Can cause damage in conditions of severe short circuit.
- BS 7671 prefers cartridge fuses (see Regulation 533-01-04).

The diameter of copper wires to be used as fuse elements in these fuses is given in Table 53.1 of the Regulations.

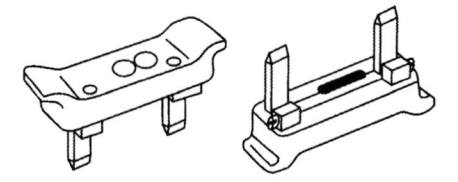

Cartridge fuses (BS 1361)

The body of the fuse can be either ceramic (low grade) or glass with metal end caps to which the fuse element is connected. The fuse is sometimes filled with silica sand.

Advantages

- Small physical size.
- No mechanical moving parts.
- Accurate current rating.
- Not liable to deterioration.

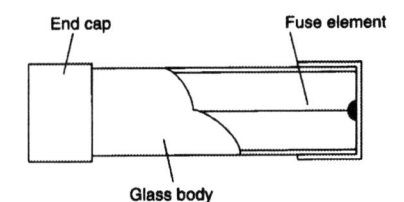

Disadvantages

- Not suitable where extremely high fault current may develop (typically 16.5 kA max).
- Can be shorted out by the use of silver foil or wire.

HBC fuses (BS 88)

The barrel of the high breaking capacity (HBC) fuse is made from high grade ceramic to withstand the mechanical forces of heavy current interruption.

Plated end caps afford good electrical contact.

An accurately machined element, usually made of silver, is shaped to give precise characteristic.

The barrel is filled with quartz silica sand to ensure an efficient arc extinction in all conditions of operation.

Some fuses are fitted with an indicator bead to show when it has blown.

Advantages

- Discriminates between overload currents of short duration (e.g. motor starting) and high fault currents.
- Suitable where high levels of PFC could occur (typically 40–80 kA duty rating).
- Consistent in operation.
- Reliable.

Disadvantage

- Could be replaced with a fuse of incorrect current rating.

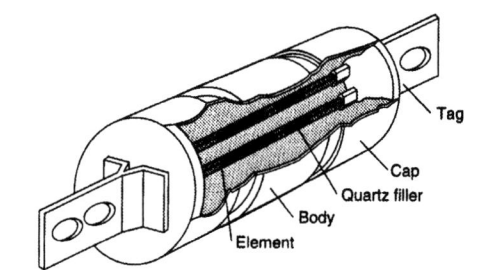

TECHNICAL DATA SHEET 4D

Circuit breakers

Circuit breakers to BS EN 60898 and miniature circuit breakers to BS 3871

Types

- Thermal and magnetic.
- Magnetic hydraulic.

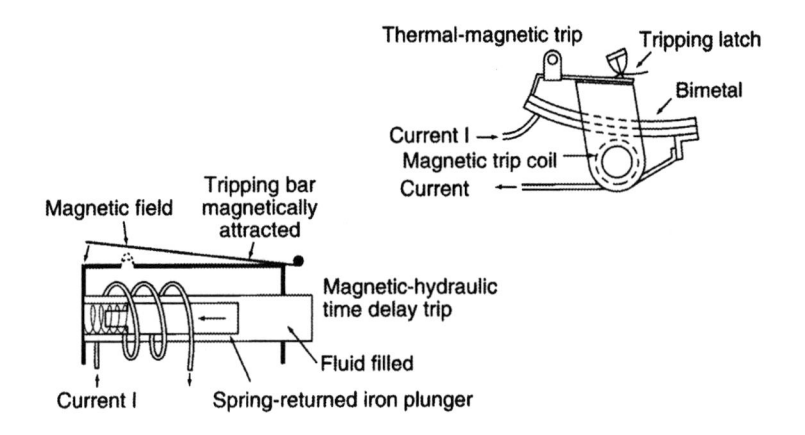

Advantages

- Tripping characteristic set during manufacture; cannot be altered.
- They will trip for a sustained overload but not for transient overloads.
- Faulty circuit is easily identified.
- Supply quickly restored.
- Tamper-proof.
- Multiple units available.

Disadvantages

- Have mechanically moving parts.
- Characteristics affected by ambient temperature.

Classification of circuit breakers

Circuit breakers are now classified according to BS EN 60898. This has replaced BS 3871 Part 1 1965: the classification of MCBs. The technical data for MCBs has, however, been retained alongside that for circuit breakers since, for the foreseeable future, MCBs will continue to be widely found in existing installations.

BS EN 60898 and BS 3871 instantaneous tripping currents

Type	Amperes			
BS 3871				
1	>	$2.7\,I_n$	≤	$4\,I_n$
2	>	$4\,I_n$	≤	$7\,I_n$
3	>	$7\,I_n$	≤	$10\,I_n$
4	>	$10\,I_n$	≤	$50\,I_n$
BS EN 60898				
B	>	$3\,I_n$	≤	$5\,I_n$
C	>	$5\,I_n$	≤	$10\,I_n$
D	>	$10\,I_n$	≤	$20\,I_n$

Typical applications for circuit breakers

Type	Application
1 B	General domestic and commercial installations with little or no surges
2 3 C	General commercial and industrial installations where fluorescent lighting and small motors produce switching surges
4 D	For use where transformers, industrial welding equipment, x-ray machines, etc. where high inrush currents could happen

Duty ratings

Miniature circuit breakers (BS 3871–Part 1)		Circuit breakers (BS EN 60898)	
Category of duty	Prospective current of the test circuit (A)	I_{cn} kA	I_{cs} kA
M 1	1,000	1.5	1.5
M 1.5	1,500	3	3
M 3	3,000	6	6
M 4.5	4,500	10	7.5
M 6	6,000	15	7.5
M 9	9,000	20	10
		25	12.5

Duty ratings of circuit breakers (BS EN 60898)

Two rated short-circuit capacities may be quoted: I_{cn} and I_{cs}.

I_{cn} The rated ultimate short-circuit capacity. The maximum fault current the device can safely interrupt, although the device may be damaged and no longer usable.

I_{cs} The rated service short-circuit breaking capacity. The maximum level of fault current operation without loss of performance of the circuit breaker.

The circuit breaker will be marked with the I_{cn} value (rated short-circuit capacity).

The value will appear inside a rectangle without a unit symbol, e.g. $\boxed{10{,}000}$

BS EN 60898 circuit breakers have rated short-circuit values of 1.5 kA–25 kA, in practice the smallest rating will be 3 kA. The larger 25 kA values are new and take into account modern current limiting technology. The old BS 3871 and the new BS EN 60898 standard conditions for test and the test duty sequences are entirely different. It is therefore difficult to assess if a 10 kA rating to BS EN 60898 is better or worse than a M9 rating from BS 3871. The user must be aware of these variations when comparing devices manufactured to the different standards.

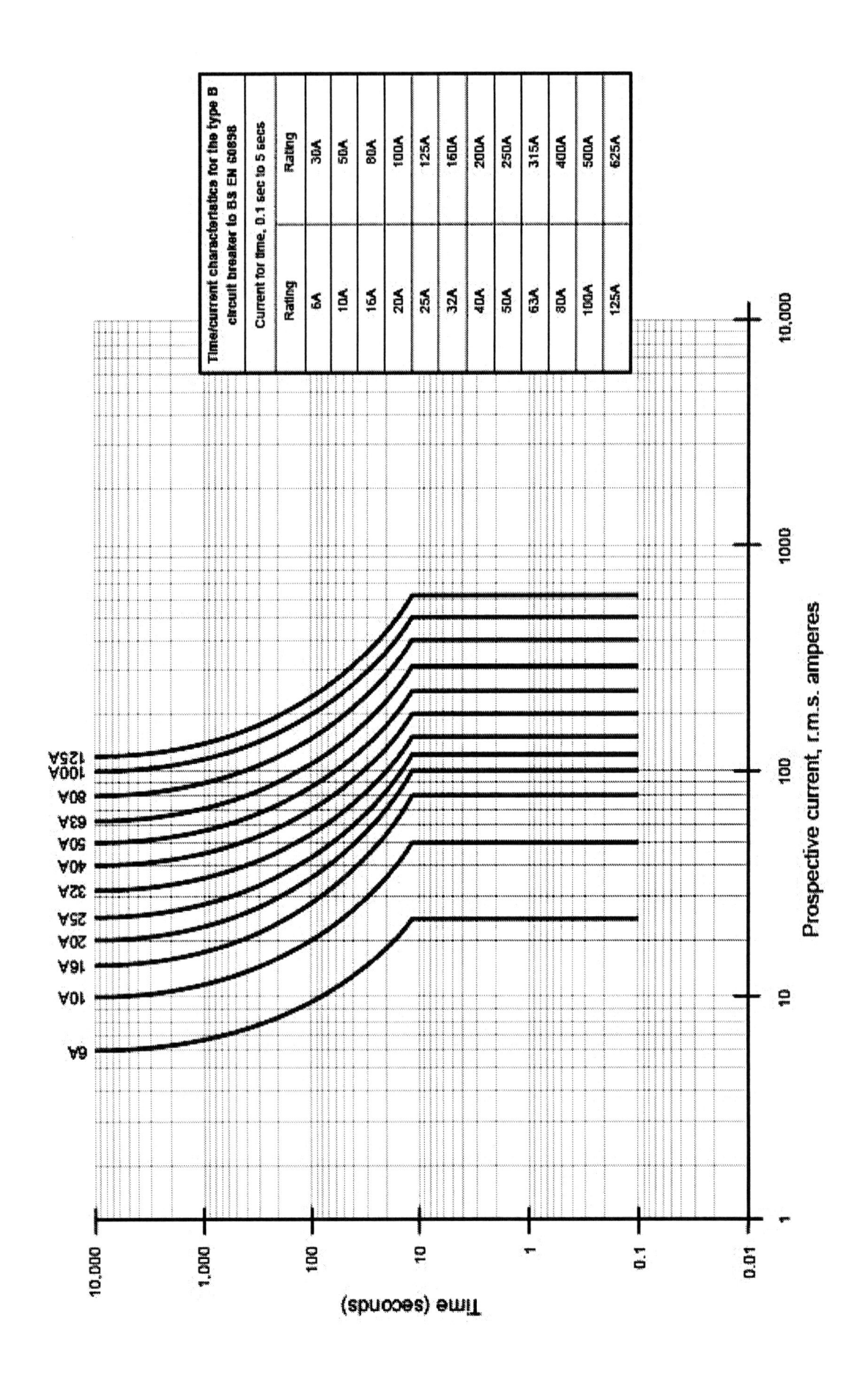

Time/current characteristics for the type B circuit breaker to BS EN 60898

Current for time, 0.1 sec to 5 secs

Rating	Rating
6A	30A
10A	50A
16A	80A
20A	100A
25A	125A
32A	160A
40A	200A
50A	250A
63A	315A
80A	400A
100A	500A
125A	625A

E1: BS 7671
(February 2008)

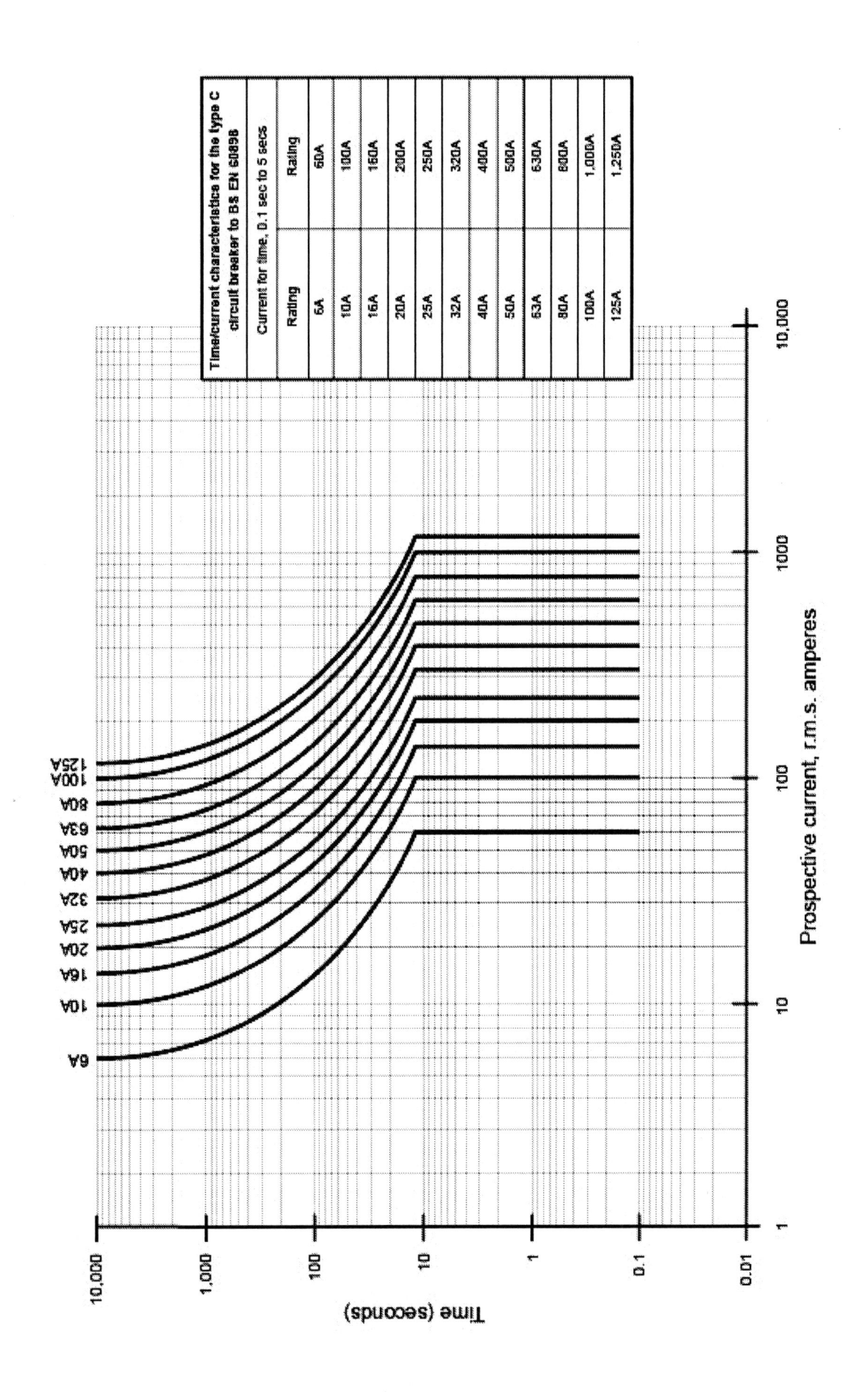

Time/current characteristics for the type C circuit breaker to BS EN 60898

Current for time, 0.1 sec to 5 secs

Rating	Rating
6A	60A
10A	100A
16A	160A
20A	200A
25A	250A
32A	320A
40A	400A
50A	500A
63A	630A
80A	800A
100A	1,000A
125A	1,250A

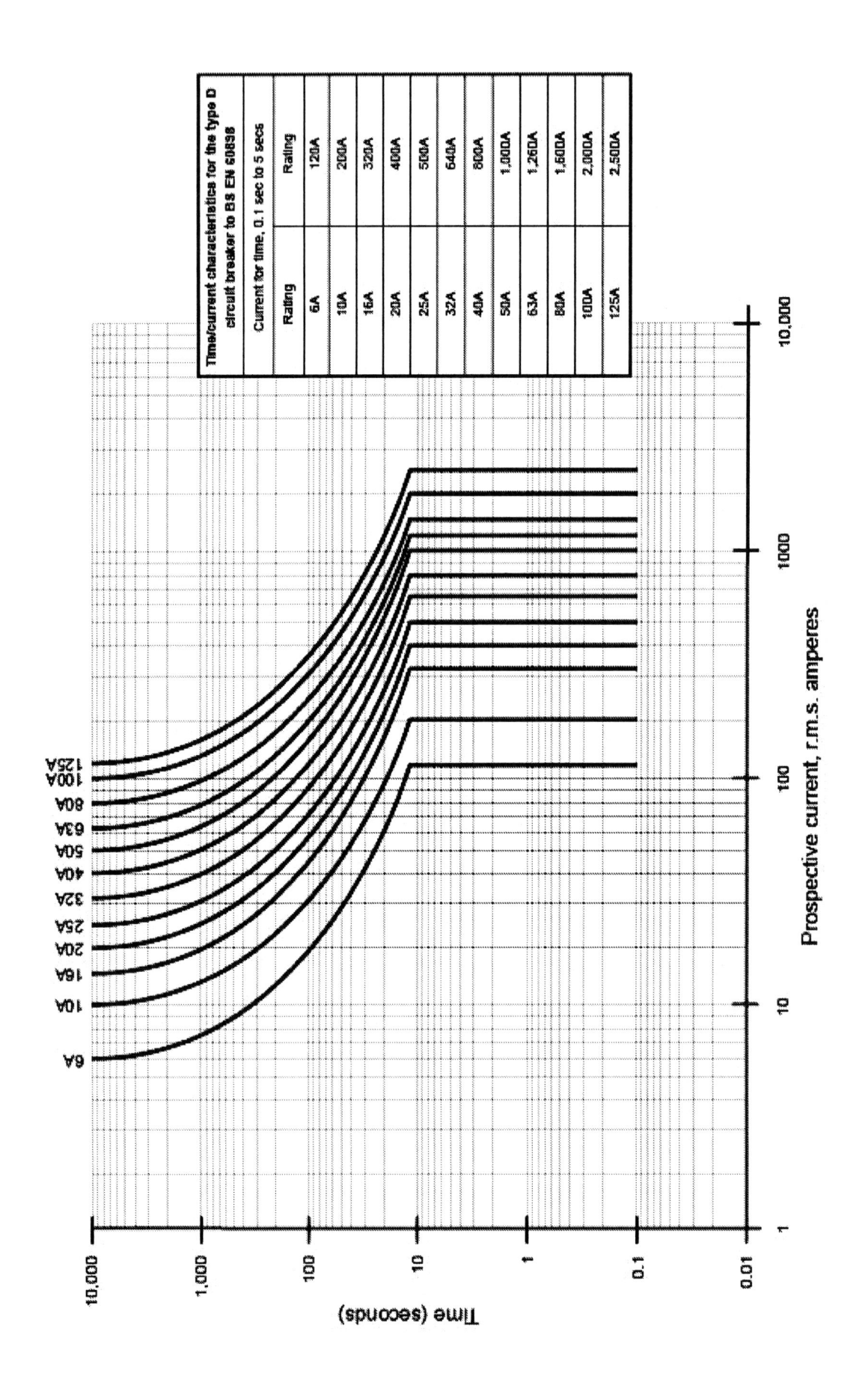

Time/current characteristics for the type D circuit breaker to BS EN 60898

Current for time, 0.1 sec to 5 secs

Rating	Rating
6A	120A
10A	200A
16A	320A
20A	400A
25A	500A
32A	640A
40A	800A
50A	1,000A
63A	1,260A
80A	1,600A
100A	2,000A
125A	2,500A

TECHNICAL DATA SHEET 4E

Operating characteristics of overcurrent protective devices

The effects of a fault current are:

- risk of causing a shock
- heat energy (proportional to I^2t)
- magnetic forces (proportional to I^2).

Operation of an HBC fuse under fault current conditions

An HBC fuse will normally operate under fault current conditions sooner than a rewireable fuse or circuit breaker, and the HBC fuse has the advantage of being totally enclosed and self-extinguishing.

The operating characteristics of an HBC fuse are its melting time and arcing time. The molten metal of the fuse element arcs and rapidly disperses into the silica sand, whereas a rewireable fuse or circuit breaker takes longer to operate, resulting in the production of greater heat energy and electromagnetic stress.

Operation of circuit breakers under overload conditions

A circuit breaker can detect a sustained overload of about 25%. An overload of 60–70% is needed to 'blow' an HBC fuse and 100% overload is needed to 'blow' a semi-enclosed (rewireable) fuse.

Discrimination

When designing a distribution system it is necessary to consider effective discrimination. Protective devices in an installation must be graded so that when a fault occurs the device nearest to the fault comes into operation. Other devices must remain intact.

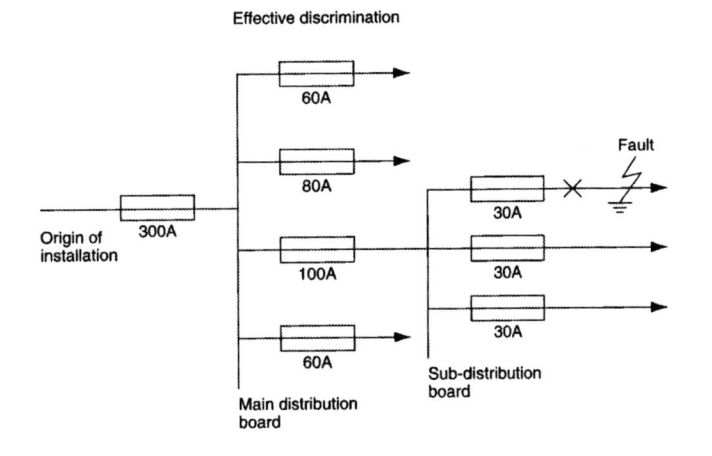

Effective discrimination

[Diagram: Origin of installation (300A) feeds Main distribution board with 60A, 80A, 100A, 60A fuses. The 80A feeds a Sub-distribution board with 30A, 30A, 30A fuses. A fault is shown on the first 30A circuit.]

E1: BS 7671
(February 2008)

Operating characteristics of protective devices are given in manufacturer's literature to which reference is necessary to achieve effective discrimination in an installation.

The following illustration shows an I^2t characteristic for one manufacturer's HBC link. When total I^2t of the minor fuse in an installation does not exceed the pre-arcing I^2t of the major fuse, discrimination has been achieved.

I^2t characteristics – HBC fuse

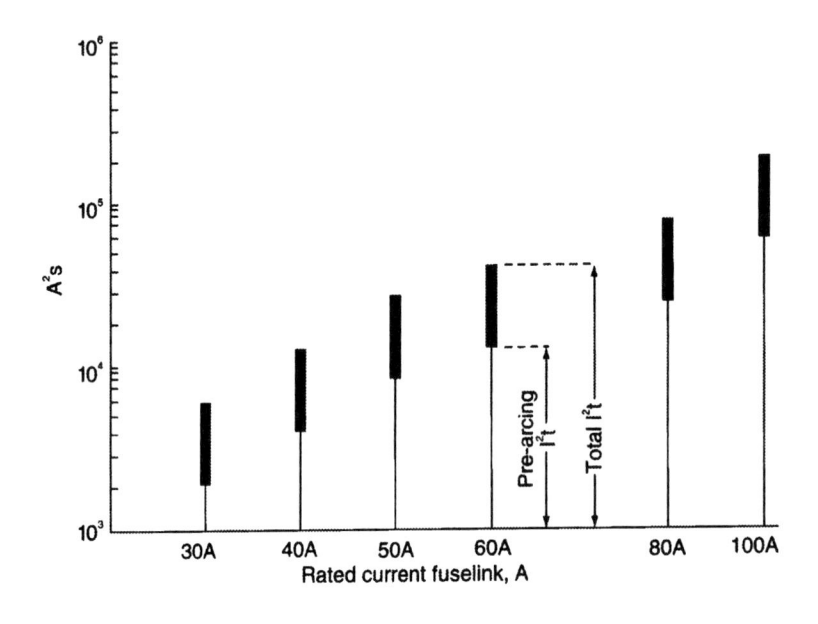

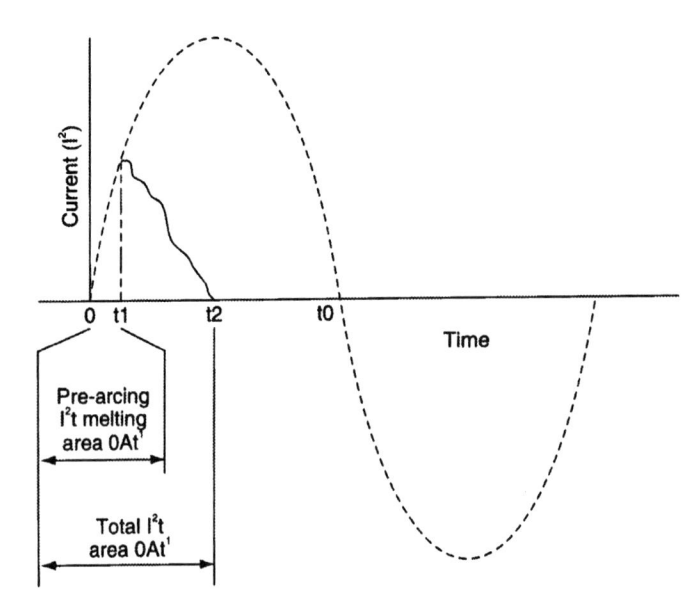

SELECTION AND ERECTION OF EQUIPMENT BS 7671 Part 5

Common rules *(Chapter 51)*

General *(510.1)*

All items of equipment used in an electrical installation must be selected and erected so as to comply with the requirements of the Regulations.

Compliance with standards *(511)*

Every item of equipment used in an electrical installation must comply with the relevant requirements of the applicable British Standard or Harmonised European Standard.

Note: A list of publications by the British Standards Institution to which reference is made in BS 7671 is given in Appendix 1 of BS 7671.

Where a particular item of equipment complying with a foreign national standard based on an IEC standard has been specified for a use, the designer or other person responsible for specifying the installation must verify that the equipment will not result in a lesser degree of safety than that afforded by compliance with the British Standard.

Similarly, where the equipment to be used is not covered by a British Standard or Harmonised Standard, or is used outside the scope of its standard, the designer or other person responsible for specifying the installation must confirm that the equipment provides the same degree of safety as that given by compliance with the Regulations.

Operational conditions and external influences *(512)*

All equipment must be suitable for the:

- nominal voltage U_o (rms value for a.c.)
- design current allowing for capacitive and inductive effects
- current likely to flow in abnormal conditions (the duration of which is dependant on the characteristic of the protective devices concerned)
- frequency
- power characteristics
- compatibility
- impulse withstand voltage.

External influences *(512.2)*

From Appendix 5 of BS 7671, external influences are described by the use of two capital letters and a number.

The first letter indicates the general category, as follows:

    A = Environment

    B = Utilisation (who is using the building, materials stored or processed, etc.)

    C = Building construction.

The second letter refers to the type of influence, for example (AD) water.

The number describes the class, for example (AD4) splashes.

Some examples of external influences are:

- AA   Ambient temperature.
- AB   Atmosphere humidity.
- a.c. to AR   Other environmental conditions.
- B to C   Utilization and construction of buildings.

Electrical equipment shall be selected and erected in accordance with the requirements of Appendix 5. If for some reason the equipment's construction does not have the characteristics required for the external influence of its location, it may be provided with additional protection so that it can be used. The additional protection shall not adversely affect the operation of the equipment (512.2.2).

Accessibility *(513)*

All equipment must be installed so that it can be easily operated, inspected and maintained and provide ease of access to any connections. This Regulation does not apply to joints in cables, where such joints are permitted to be inaccessible in accordance with Section 526.

Identification and notices *(514)*

Except where there is no possibility of confusion, all switchgear and control gear in an installation must be labelled or identified by some other suitable method to indicate its purpose.

Where the operation of switchgear or control gear cannot be seen by the operator, and where this might lead to danger, a suitable indicator must be installed where the operator can see it.

As far as is reasonably practicable, all wiring shall be arranged or marked so it can be readily identified for inspection, testing, repairs or alterations.

Except where there is no possibility of confusion, clear marking must be provided at the interface between conductors identified in accordance with the harmonised cable core colours of BS 7671 and conductors identified to earlier versions of the Regulations. A notice in accordance with 514-04 must be displayed.

Where electrical conduits need to be easily distinguished from the pipelines of other services (such as gas, oil, water, steam, etc.), they must be painted orange.

Identification of conductors (514.3)

Every cable core must be identified at its terminations and preferably throughout its length. Methods of identification may include coloured insulation applied to conductors during manufacture or the application of binding or sleeves (to BS 3858). The colours used must be those specified in Table 51 of BS 7671.

Note: Table 51 of BS 7671 covers the identification of both fixed wiring and flexible cords and cables.

Cores of cables must be identified by:

- colour
- lettering and/or numbering.

This does **not** apply to:

- concentric conductors of cables
- metal sheath or armour used as a cpc
- bare conductors (where identification is not practicable)
- extraneous or exposed-conductive-parts used as protective conductors.

Switchboard busbars or conductors when identified must comply with Table 51, as applicable.

Identification of conductors by colour (514.4)

Neutral or mid-point conductors

For identification by colour, a neutral or mid-point conductor must be **blue**.

Protective conductors

The two-colour combination of **green** and **yellow** is exclusively used to identify a protective conductor. This colour combination **must not** be used for any other purpose.

A bare conductor or busbar used as a protective conductor must have equal **green** and **yellow** stripes 15–100 mm wide. If tape is used it must be bi-colour.

Green and **yellow** single-core cables must only be used as protective conductors and must not be overmarked at their terminations. This does not apply to insulated PEN conductors.

PEN conductors

A PEN conductor that is insulated can be marked either of the following:

- **green** and **yellow** throughout its length with **blue** at the terminations
- **blue** throughout its length with **green** and **yellow** at its terminations.

Other conductors

- All other conductors should be as Table 51 of BS 7671.
- The single colour **green** must not be used.

Harmonised cable core colours *(Appendix 7)*

The requirements of BS 7671 have been harmonised with the requirements of the European Electrical Standards Body, CENELEC, regarding the identification of conductors and the identification of cores in cables and flexible cords. (See BS 7671, Table 51 of BS 7671 and Tables 7A to 7E in Appendix 7 of BS 7671.)

These standards specify the cable core marking, including the cable core colours, for the CENELEC countries.

For single-phase installations:

The **red** phase and **black** neutral

are replaced by

brown phase and **blue** neutral

The protective conductor remains **green** and **yellow**.

An example of identification of conductors and core colours to Table 51 of BS 7671 is shown below.

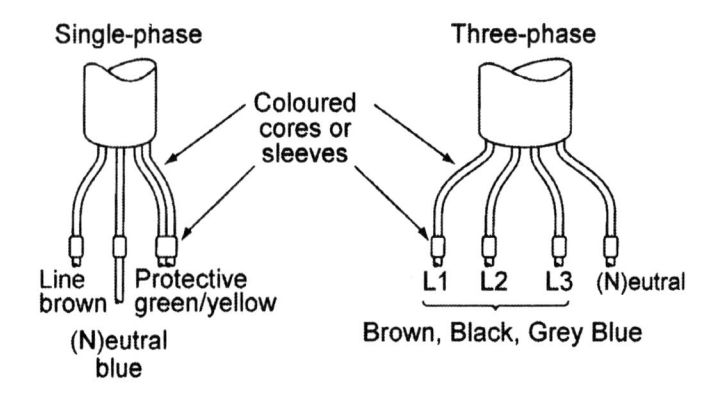

Single-phase

Coloured cores or sleeves

Line brown | Protective green/yellow

(N)eutral blue

Three-phase

L1 L2 L3 (N)eutral

Brown, Black, Grey Blue

The colour of the conductor for any functional earthing conductor is CREAM. A typical example would be for a telecommunications system.

Therefore, fixed wiring for single-phase installations now adopt the same colours that three-core flexible cables and cords have used for many years.

Note: Wherever an interface (connection) occurs between old and new cable core colours, a warning notice as 514.14.1 should be displayed.

Alterations or additions to existing installations

Single-phase installations

Alterations or additions to a single-phase installation do not need marking at the interface where old wiring is connected to new, providing that the:

- old cables are coloured **red** for phase and **black** for neutral
- new cables are coloured **brown** for phase and **blue** for neutral.

Two-phase or three-phase installations

At a wiring interface between old core colours and new core colours, clear identification is required as follows:

Old and new conductors

Neutral conductors = **N**

Line conductors = **L1, L2, L3**

BS 7671 Table 7A gives examples of conductor marking at an interface for additions and alterations to an a.c. installation identified with the old cable colours (see below).

(From Table 7A of BS 7671)

Function	Old conductor colour	Old/new marking	New conductor colour
Phase 1	Red	L1	Brown
Phase 2	Yellow	L2	Black
Phase 3	Blue	L3	Grey
Neutral	Black	N	Blue
Protective conductor	Green and Yellow		Green and Yellow

For a three-phase installation, as an alternative to the **brown**, **black** and **grey** identification of the line conductors shown above, three **brown** or three **black** or three **grey** conductors may be used. However, they must be marked L1, L2 and L3 or oversleeved **brown**, **black** and **grey** at their terminations.

Switch wires

New installations or modification to existing installations

Where a PVC-insulated and sheathed (twin and earth) cable is used as a switch wire and the cores, both being used as line conductors, are coloured **brown** and **blue**, the:

- **blue** conductor must be oversleeved **brown** or marked L at its terminations
- bare cpc must be oversleeved **green** and **yellow** as normal.

Intermediate and two-way switch wires

New installations or modification to existing installations

Where a three-core and earth cable with core colours **brown**, **black** and **grey** is used, and all three conductors are used as line conductors, the:

- **black** and **grey** conductors must be oversleeved **brown** or marked L at their terminations
- bare cpc being oversleeved **green** and **yellow**, as normal.

Line conductors

New installations or modification to existing installations

The colours and markings of line conductors should be as BS 7671, Table 51.

For control circuits, extra-low voltage and other applications the:

- line conductor could be coloured **brown, black, red, orange, yellow violet, grey, white, pink** or **turquoise** or marked L
- neutral or mid-wire must be coloured **blue** or marked N or M.

Note: The mid-wire (marked M) is the earthed conductor of a two-wire earthed d.c. circuit or the mid-wire of a three-wire d.c. circuit. Only the mid-wire of three-wire circuits may be earthed.

An earthed protective extra-low voltage (PELV) conductor must be **blue**.

Alterations or additions to a d.c. installation

Clear identification is required where an installation wired in the new core colours is connected to wiring in the old core colours. The cores must be marked as follows:

Old and new conductors

Neutral and mid-point conductors = M

Line conductors = **brown** or **grey** or L, L+ or L-

Bare conductors

Bare conductors must be identified (where necessary) by use of tape, sleeve, disk or paint of the appropriate colour specified in Table 51 of BS 7671.

Identification of conductors by letters and/or numbers *(514.5)*

This system applies to the identification of individual conductors and of conductors in a group. The identification must be clear and legible, and in strong contrast to the colour of the insulation. To avoid confusion, the numbers 6 or 9 must be underlined.

Protective conductors must not be numbered other than for circuit identification.

The alphanumeric system is as Table 51 of BS 7671.

Identification by numbers is permitted with 0 (zero) being the neutral or mid-point conductor.

Identification of a protective device *(514.8)*

All protective devices in an installation must be arranged and identified so that their respective circuits may be easily recognised.

Diagrams *(514.9)*

Diagrams and charts must be provided for every electrical installation indicating:

- the type of circuits
- the number of points installed
- the number and size of conductors
- the type of wiring system
- the location and types of protective devices and isolation and switching devices
- details of the characteristics of the protective devices for automatic disconnection, the earthing arrangements for the installation and the impedances of the circuit concerned
- circuits or equipment vulnerable to a typical test.

Note: For simple installations the above information may be given in a schedule. If symbols are used they must conform to BS EN 60617.

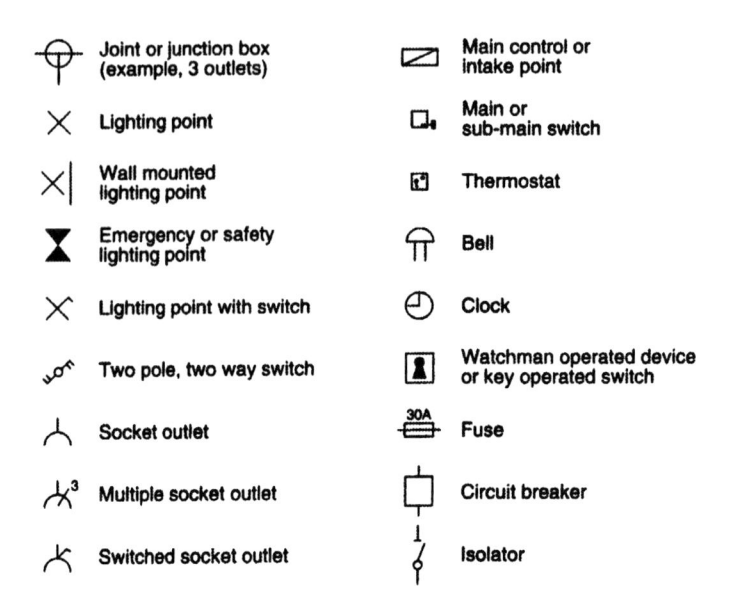

Symbol	Description	Symbol	Description
⊕	Joint or junction box (example, 3 outlets)	▱	Main control or intake point
×	Lighting point	▫	Main or sub-main switch
×|	Wall mounted lighting point	[t°]	Thermostat
⊠	Emergency or safety lighting point	⌂	Bell
×	Lighting point with switch	⊕	Clock
ϟ	Two pole, two way switch	▨	Watchman operated device or key operated switch
⊥	Socket outlet	30A fuse	Fuse
⊥³	Multiple socket outlet	▯	Circuit breaker
⊥	Switched socket outlet	⎓	Isolator

Some commonly used symbols to BS EN 60617

The purpose of providing diagrams, charts and tables for an installation is so that it can be inspected and tested in accordance with Chapter 61 of BS 7671 and to provide any new occupier/owner of the premises with the fullest possible information concerning the electrical installation.

For simple installations, this information could be contained in a schedule. It is essential that diagrams, charts and tables are kept up to date and a durable copy placed within or adjacent to each distribution board.

A typical chart and diagram for a simple installation, such as a domestic premises, are shown here.

Schedule for modified consumer unit arrangement

CCT No.	Type of circuit	Points served	Line conductor mm²	Protective conductor mm²	Protective devices	Type of wiring
1	Fire alarm	5	1.5	1	6 A RCBO	PVC/PVC
2	Cooker	1	10	4	40 A MCB	PVC/PVC
3	U/S sockets	8	2.5	1	32 A MCB	PVC/PVC
4	D/S lights	7	1.5	1	6 A MCB	PVC/PVC
5	Spare					
6	U/S lights	6	1.5	1	6 A MCB	PVC/PVC
7	Bathroom light	1	1.5	1	6 A MCB	PVC/PVC
8	D/S sockets	9	2.5	1	32 A MCB	PVC/PVC
9	Kitchen sockets	8	2.5	1	32 A MCB	PVC/PVC
10	Shower	1	10	4	40 A MCB	PVC/PVC

See Technical Data Sheet 4B for the accompanying consumer unit configuration (Figure 1).

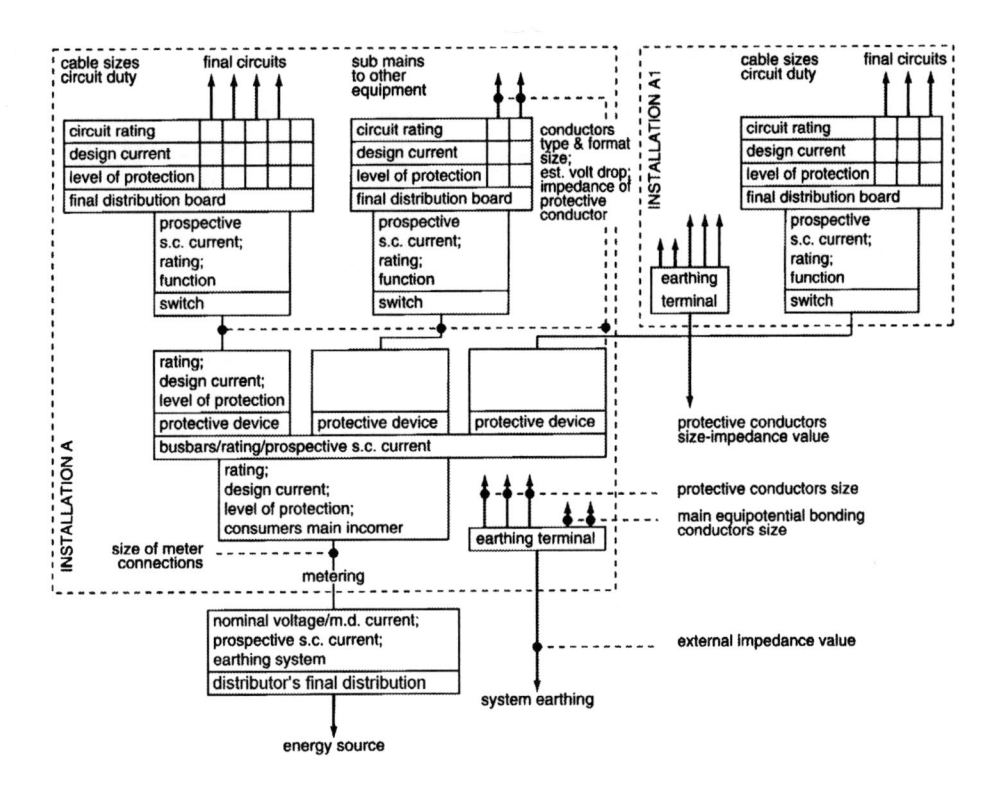

Warning notice – voltage *(514.10)*

A warning notice stating the maximum voltage present must be fixed to every item of equipment (or enclosure) which contains circuits operating at voltages in excess of 230 V and where the presence of such a voltage would not normally be expected. The notice must be fixed in a position where it can be seen before access is gained to a live part.

Where separate enclosures or items of equipment are wired on different phases of a three-phase supply and can be touched simultaneously, a notice must be placed in a position such that anyone gaining access to live parts is warned of the maximum voltage which exists between such live parts.

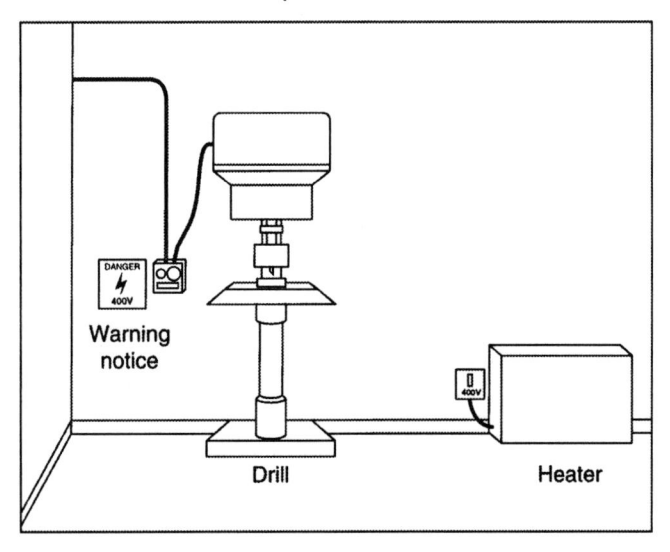

Warning notice – isolation *(514.11)*

A notice made of durable material in accordance with Regulation 537.2.1.3 shall be fixed in each location where live parts of an installation cannot be isolated by a single device. The location of each isolator shall be indicated unless there is no possibility of confusion.

Inspection and testing *(514.12)*

Upon completion of an electrical installation, the electrical contractor must fix a label with details of the date of the last inspection and the recommended date of the next inspection. This label must be fixed in a prominent position, at or near the origin of the installation upon completion of the work.

The notice must be inscribed with characters (not smaller than 11 point), as illustrated below.

IMPORTANT

This installation should be periodically inspected and tested, and a report on its condition obtained, as prescribed in the IEE Wiring Regulations BS 7671 Requirements for Electrical Installations.

Date of last inspection: ....

Recommended date of next inspection: ....

Residual current device – notices *(514.12.2)*

When an installation incorporates a residual current device, a notice must be fixed in a prominent position, at or near the origin of the installation. It must be printed in indelible characters, not less than 11 point in size and must read as follows:

This installation, or part of it, is protected by a device which automatically switches off the supply if an earth fault develops. Test quarterly by pressing the button marked 'T' or 'Test'. The device should switch off the supply and should then be switched on to restore the supply. If the device does not switch off the supply when the button is pressed, seek expert advice.

Earthing and bonding *(514.13)*

A warning notice (as illustrated on the next page) must be fitted in a visible position near to the point of connection of an earthing conductor to an earth electrode, a bonding conductor to an extraneous-conductive-part and the main earth terminal when it is separate from main switchgear.

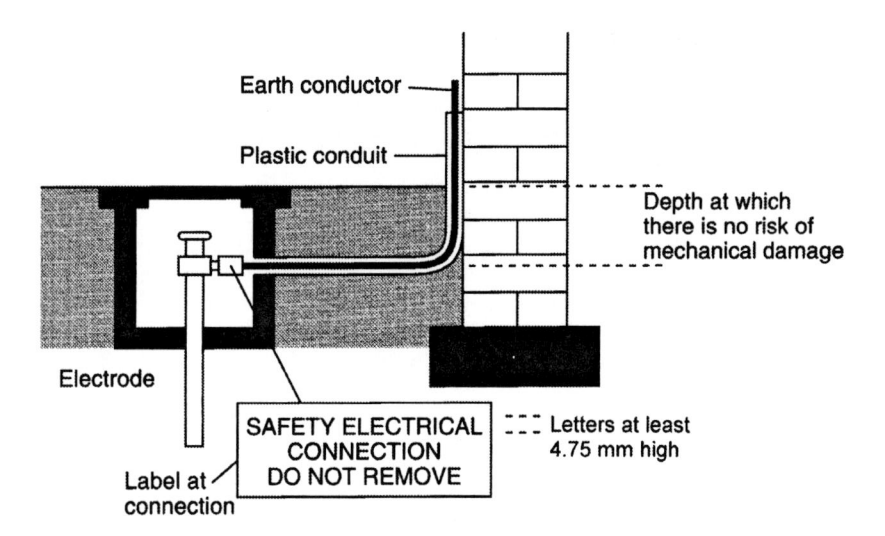

Earth conductor
Plastic conduit
Depth at which there is no risk of mechanical damage
Electrode
Label at connection
SAFETY ELECTRICAL CONNECTION DO NOT REMOVE
Letters at least 4.75 mm high

Where protection against indirect contact is achieved by earth-free local equipotential bonding covered by Regulation 418.2.5 and electrical separation covered by Regulation 418.3, a notice that is durably marked in letters not less than 4.75 mm high must be attached and must read as follows (514.13.2):

> The equipotential protective bonding conductors associated with the electrical installation in this location MUST NOT BE CONNECTED TO EARTH.
>
> Equipment having exposed-conductive-parts connected to earth must not be brought into this location.

Warning notice – non-standard colours *(514.14)*

If alterations or additions are made to the wiring of an installation so that some wiring complies with the current version of BS 7671 (the harmonised cable core colours), but the other wiring complies with an earlier version of BS 7671, a warning notice must be fixed at or near the distribution board concerned with the wording:

> **CAUTION**
>
> This installation has wiring colours to two versions of BS 7671.
>
> Great care must be taken before undertaking extension, alteration or repair that all conductors are correctly identified.

Warning notice – dual supply

When a generating set is used in an installation as an additional supply source in parallel a warning notice as illustrated shall be fixed at the locations indicated below:

- origin of the installation
- meter position if remotely located
- consumer unit or distribution board
- points of isolation of both supply sources.

> **WARNING – DUAL SUPPLY**
>
> Isolate both mains and on-site generation before carrying out work.
>
> Isolate the mains supply at: ....
>
> Isolate the generator at: ....

Mutual detrimental influence *(515)*

All electrical equipment must be selected and erected so as to avoid any harmful influences between the electrical installation and any non-electrical services. When equipment carrying currents of different types (a.c. or d.c.) or at different voltages is grouped in a common assembly, e.g. switchboard, control desk or cubicle, all equipment using any one type of current or any one voltage must be effectively segregated from equipment of any other type, to avoid mutual detrimental influence, e.g. the segregation of Band II circuit cables from Band I telephone cables by use of multi-compartment trunking.

Electromagnetic – compatibility *(515.3)*

Equipment must be suitable with regard to its immunity levels for any electromagnetic influences present. The equipment itself **must not** cause electromagnetic interference. Where this is a possibility, precautions must be taken to minimise the effects of these emissions. Always refer to the equipment standard or BS EN 50081.

Selection and erection of wiring systems *(Chapter 52)*

Types of wiring system *(521)*

The methods of installation of wiring systems is covered in BS 7671. Reference should be made to the following tables in Appendix 4:

- types of conductor or cable   Table 4A1
- situation concerned   Table 4A2.

Busbar trunking and powertrack systems *(521.4)*

Busbar trunking and powertrack systems shall comply with the following standards and shall be installed in accordance with the manufacturer's instructions:

- busbar trunking   BS EN 60439-2
- powertrack   BS EN 61534 series.

A.C. circuits – electromagnetic effects *(521.5)*

All conductors or cables shall have adequate strength and be installed to withstand any electromechanical forces caused by any current they have to carry.

Single-core cables with a steel wire armour or tape shall not be used on a.c. circuits. When conductors of a.c. circuits are installed in a ferromagnetic enclosure, all line, neutral and protective conductors shall be contained in the same enclosure. No conductor shall be individually surrounded by ferromagnetic material (521.5.2).

Several circuits are allowed to be wired using the same conduit, ducting, trunking or multicore cable so long as the requirements of Section 528 are complied with (521.6 and 521.7).

Line and neutral conductors of each final circuit shall be electrically separated from any other circuits to prevent the energisation of circuits which are intended to have been fully isolated (521.8.2).

When several circuits are terminated using a single junction box this should comply with BS EN 60670 or BS EN 60947-7 (521.8.3).

Use of flexible cables or cords *(521.9)*

Stationary equipment which is temporarily moved in order to clean it or connect it to the electrical supply point may be connected using non-flexible cable, e.g. a cooker. Equipment which is subject to vibration must be connected by means of a flexible cable or cord.

Installation of cables *(521.10)*

Where non-sheathed cables are being installed they shall be enclosed in conduit, trunking or ducting, with protection to IP4X or IPXXD and any trunking lids only being removable using a tool. The exception to this is for protective conductors, which comply with Section 543.

Ambient temperature (AA) *(522.1)*

The type of conductor, cable, flexible cord and joint used in the wiring of circuits must be suitable for the highest operating temperature likely to occur in normal service. Account should be taken of the minimum temperature likely to occur so as to avoid the risk of mechanical damage to those cables susceptible to low temperatures. For example, PVC insulated cables crack if installed in refrigeration plants where the temperature is consistently below freezing point.

External heat sources *(522.2)*

In order to avoid the effects of heat from external sources (such as solar gain), any of the following actions must be taken:

- shielding
- placing at a distance from the heat source
- selecting a suitable wiring system
- reinforcing or substitution of the insulating material.

Where cables or flexible cords are in contact with equipment or accessories which transfer heat, such as immersion heaters and luminaires, termination to this equipment must be made using heat-resisting flexible cable or cord, or a suitable supplementary insulated sleeve or insulation must be applied to the conductors (522.2.2).

Enclosures must be selected so that they are suitable for the extremes of ambient temperature that will be encountered in normal service. A typical example might be PVC conduit which distorts in hot weather if expansion couplings have not been correctly fitted.

Presence of water (AD) or high humidity (AB) *(522.3)*

Wiring systems must not be exposed to rain, water or condensation, but where this cannot be avoided, the wiring system must be selected so that no damage is caused. PVC conduit and trunking, galvanised conduit etc., are normally used in these circumstances.

Care must be taken in damp and wet conditions to ensure that the composition of the wiring system and its fixings and accessories are chosen so that electrolytic action is prevented, e.g. an aluminium conductor must not be placed in contact with brass or copper. Copper clad aluminium conductors must not be used if these are likely to be exposed to damp or wet conditions.

Where conduit systems are not designed to be sealed, the system must be provided with drainage outlets at any point where moisture is likely to collect (522.3.2).

Where a wiring system could be subject to waves, etc. (AD6) mechanical damage protection as 522.6 to 522.8 must be provided.

Dust, solid foreign bodies (AE) *(522.4)*

Wiring systems must be installed to minimise the ingress of solid foreign bodies. Where dust or similar may accumulate, precautions must be taken to prevent any adverse effects on the heat dissipation of wiring or equipment.

Corrosive and polluting substances (AF) *(522.5)*

Where a corrosive or polluting environment cannot be avoided the wiring system used must be of a type (or be suitably protected) so as to withstand exposure to the corrosive or polluting substances. Non-metallic materials must not be placed in contact with oil/creosote or similar hydrocarbon substances likely to cause chemical deterioration.

There must be no contact with materials likely to cause electrolytic action, deterioration or hazardous degradation.

Materials likely to cause corrosion of wiring systems are:

- materials containing magnesium chloride (used in the construction of floors and dados)
- plaster undercoats containing corrosive salt
- lime, cement and plaster
- oak and other acidic woods
- dissimilar metals liable to set up electrolytic action, e.g. copper and aluminium.

Where conductors require termination involving soldering (as in the sweating on of cable lugs) the soldering flux used must not remain acidic or corrosive after the completion of the soldering process.

Polystyrene

Thermoplastic (PVC) cables must be separated from expanded polystyrene materials (e.g. polystyrene granule insulation) to prevent the plasticiser migrating to the polystyrene. This makes the cable sheathing less flexible and the polystyrene becomes soft and sticky.

Timber treatment

PVC cables must not come into contact with wood preservatives during their application until any solvents have evaporated. Also, creosote must never be allowed near PVC cables as it will cause decomposition of the cable (e.g. swelling and loss of flexibility).

Impact (AG) *(522.6)*

Wiring systems must be designed to minimise mechanical damage, i.e. impact, abrasion, penetration, compression or tension during installation, use and maintenance.

The practice of covering cables with plaster is widespread and cases do occur of nails and other objects penetrating cables and causing damage. This gives rise to the risk of electric shock. It is necessary to reduce this risk as much as possible.

In an installation where a medium (usual industrial conditions) or high severity (severe industrial conditions) impact could occur then protection can be provided by one or more of the following:

- mechanical characteristics of the wiring system
- location
- additional local or general mechanical protection.

Underground and buried cables

Cables buried directly in the ground must be armoured or metal sheathed, or both, or be of the PVC insulated concentric type. Such cables must be marked by cable covers or marking tape and installed at a depth sufficient to avoid damage by foreseeable disturbance of the ground. It is recommended that cables be buried at a depth of at least 500 mm.

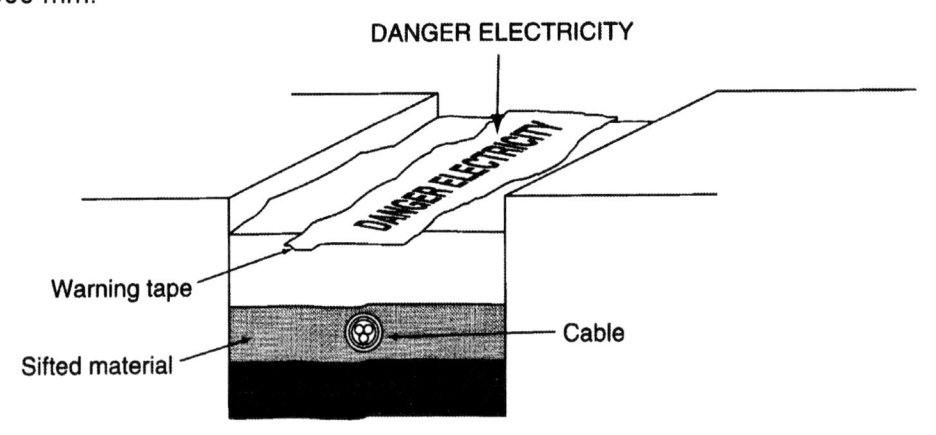

Cables installed under floors

Cables installed under floors and over ceilings must be routed so that they will be undamaged through contact with the floor or ceiling, or by the method of fixing. This involves careful routing and clipping of cables.

Cables run in the space under floors and over ceilings must be installed at least 50 mm from the top or bottom, as appropriate, of the joist or batten to prevent penetration by the nails or screws used in fixing flooring and ceiling materials. Alternatively, the cable shall incorporate an earthed metallic covering, e.g. MICC cable, or cables must be installed in an earthed steel conduit, earthed trunking or ducting that is securely supported, or provided with equivalent mechanical protection that will prevent penetration by nails or screws, etc. (522.6.5).

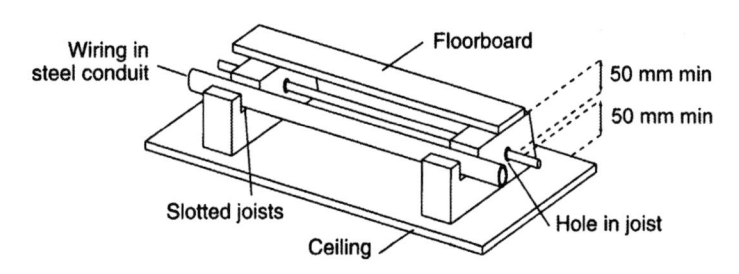

Support and protection for cables run under floorboards

Cables concealed in walls or partitions

Regulation 522.6.6 requires that where cables are installed in walls or partitions at depths of less than 50 mm from the surfaces of the walls or partitions, the risk of damage is minimised. The permitted methods are as follows:

- Cables with an earthed metallic covering that is suitable as a protective conductor complying with BS 5467, BS 6346, BS 6724, BS 7846, BS EN 60702-1 or BS 8436.

- Cables installed in earthed conduit, trunking or ducting that is suitable as a protective conductor, or provided with adequate mechanical protection to prevent damage to the cable by screws, nails, etc.

- Cables installed in a 150 mm zone from the top of a wall or partition, or within 150 mm of an angle created by adjoining walls or partitions.

- Cables may be run in safe zones, i.e. run horizontally or vertically to accessories installed on walls or partitions. Note: If the location of an accessory can be determined from the reverse side and the wall or partition is of 100 mm thickness or less, the zone extends to the reverse side of the wall or partition. This means that a concealed cable is permitted to be less than 50 mm from the surface of the wall or partition on the reverse side. Before drilling a wall or partition, the other side must be checked to determine the location of concealed cables.

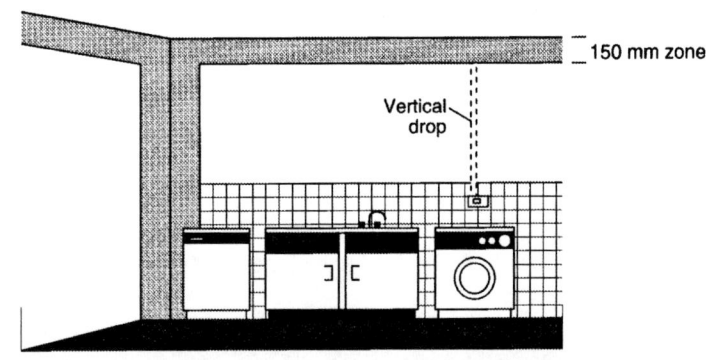

Cables should be run in permitted Zones or horizontally or vertically direct to accessory

Capping (metal or plastic) is used to protect the cable during the plastering operation, but gives very limited protection against nails and other objects driven into the plaster. Its use does **not** give compliance with Regulation 522.6.6.

Cables not installed to comply with methods for impact protection given in Regulation 522.6.6 shall be both:

- run in safe zones, and
- provided with additional protection by means of a 30 mA RCD.

Vibration (AH) *(522.7)*

When wiring systems are fixed to or supported by a building/structure, where the severity of vibration is AH2 medium or AH3 high, they must be suitable.

Other Mechanical Stresses (AJ) *(522.8)*

Conductors and cables must be installed so that they are protected against any risk of mechanical damage.

Where cables pass through holes in metalwork, such as metal accessory boxes and luminaires, bushes or grommets must be fitted to prevent abrasion of the cables on any sharp edge.

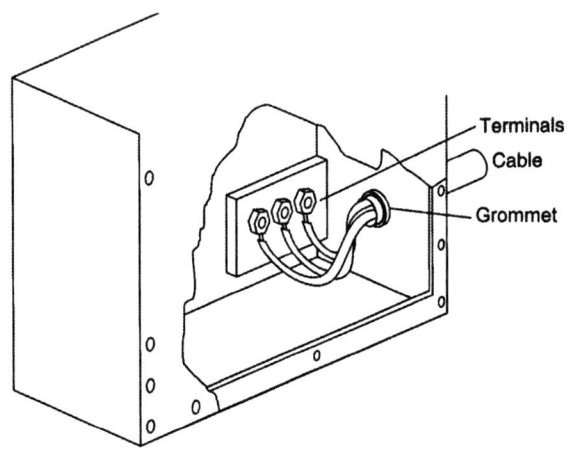

Conductors and cables must not be subject to damage from the incorrect radius of a bend (see Table 4E of IEE On-Site Guide), subject to inadequate support, including damage from its own weight, or be subject to any excessive mechanical strain. Tables for cable support distances are given in Tables 4A to 4D of the IEE On-site Guide.

Conduit and cable ducting systems, buried in the structure of a building other than pre-wired conduit systems, shall be completely erected before cables are drawn in.

Wiring systems where cables are drawn in and out shall have adequate means of access to allow this to take place, typical systems being conduit, trunking and ducting.

When cables and conductors are being fixed during installation, care shall be taken to avoid damage to them. For cables, busbars and any other electrical conductors that pass across an expansion joint, care shall be taken in their selection and installation to prevent damage to electrical equipment.

No wiring system shall penetrate any load bearing part of a building structure during installation, unless the integrity of the load bearing element of the building structure can be assured after such penetration.

Flora and/or mould growth (AK) *(522.9)*

Where conditions due to vegetation, plants or mould growth present a hazard (AK2), the wiring system must be suitably selected or protected by appropriate installation, location, use of closed wiring systems (e.g. conduit), and the use of easily cleanable wiring systems.

Fauna (AL) *(522.10)*

Wiring systems must be suitable or specially protected to avoid hazards (AL2) from insects, birds or small animals in harmful quantities or displaying an aggressive nature.

Solar radiation and ultraviolet radiation (AN) *(522.11)*

Where cable and wiring systems are installed in locations that will be subject to significant amounts of solar radiation (AN2) or ultraviolet radiation, a suitable wiring system must be used or adequate shielding must be provided.

Equipment subjected to ionising radiation may need the application of special precautions.

Seismic effects (AP) *(522.12)*

Wiring systems shall be selected and installed according to the seismic hazards of the installation location. Attention shall be paid to the method of fixing wiring systems to the building structure and connections between the fixed wiring and essential equipment.

Wind (AS) *(522.13)*

Refer to Vibration (AH).

Nature of processed or stored materials (BE) *(522.14)*

Follow the detail regarding wiring systems to minimize the spread of fire, Section 527.

Building Design (CB) *(522.15)*

If structural movement (CB3) occurs or is expected, cable supports and protection systems must be capable of permitting movement to prevent the conductors suffering excessive mechanical stress.

For flexible or unstable structures (CB4), a flexible wiring system must be installed.

Current-carrying capacities of cables *(523)*

When conductors carry current, including any harmonic current, for sustained periods the temperatures given in Table 52.1 shall not be exceeded (extract illustrated below).

Maximum operating temperatures for types of cable insulation	
Type of insulation	**Temperature limit**
Thermoplastic PVC	70°C

Groups containing more than one circuit *(523.5)*

Rating factors for groups of sheathed or non-sheathed cables with the same maximum operating temperatures can be found in Tables 4C1 to 4C5, Appendix 4 of BS 7671. When groups of cables containing sheathed and non-sheathed types, with different maximum operating temperatures are installed together, the current-carrying capacity of all cables shall be based on the one with the lowest maximum operating temperature, together with the appropriate group reduction factor.

When a sheathed and non-sheathed cable is not expected to carry a current greater than 30% of its grouped current-carrying capacity, it may be ignored when obtaining the reduction factor for the remaining group.

Number of loaded conductors *(523.6)*

When polyphase circuits carry balanced currents the neutral conductor does not need to be considered.

Where the neutral conductor carries the imbalanced current in a multicore cable, its size shall be selected based on the highest line current.

If the harmonic current produced in three-phase circuits is greater than 15%, the neutral conductor shall not be smaller than the line conductors; see Appendix 4 of BS 7671 for reduction factors.

Cables in thermal insulation (523.7)

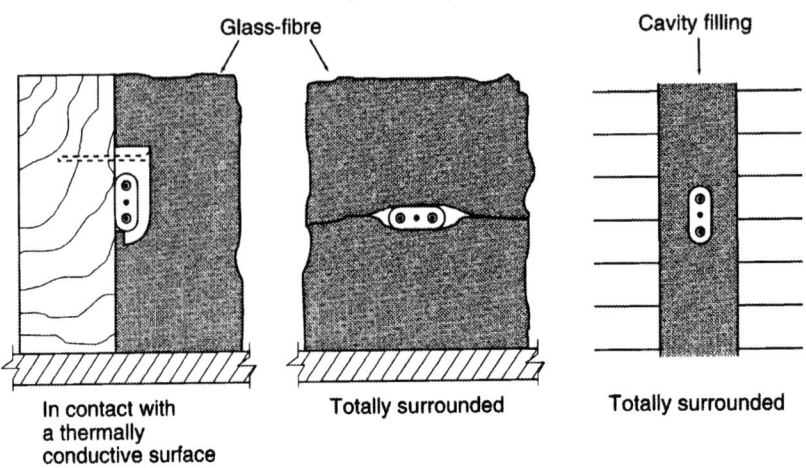

Examples of cables run in insulation

If a cable is to be run in a space in which thermal insulation is likely to be applied, the cable should be fixed, wherever practical, in a position where it will not be covered by the thermal insulation. Where this is not practicable the cable cross-sectional area must be appropriately increased.

For cable in a thermally insulated wall or above a thermally insulated ceiling, where the cable is in contact with the thermally conductive surface on one side, the current-carrying capacities are given in Appendix 4 of BS 7671.

Where a single cable is likely to be totally surrounded by insulating material over a length of more than 0.5 m then, in the absence of more precise information, the current-carrying capacity is to be taken as 0.5 times the current-carrying capacity for that cable if reference method C is used (clipped direct).

If a cable is surrounded by thermal insulation for less than 0.5 m, the current-carrying capacity is reduced according to the cable size, length of run in insulation and the thermal properties of the insulation. Table 52.2 of BS 7671 gives derating factors for conductors up to 10 mm² in insulation having a thermal conductivity greater than 0.04 $Wm^{-1}K^{-1}$. An illustration of the content of Table 52.2 is shown below.

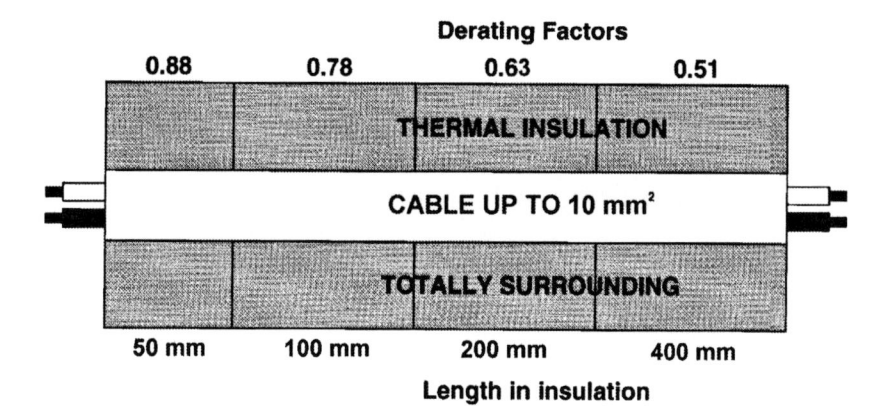

Derating Factors

0.88	0.78	0.63	0.51

THERMAL INSULATION

CABLE UP TO 10 mm²

TOTALLY SURROUNDING

50 mm	100 mm	200 mm	400 mm

Length in insulation

Conductors in parallel *(523.8)*

When two or more live conductors or PEN conductors are connected in parallel the load current must be shared. This can be achieved by ensuring the conductors are of the same material, cross-sectional area and length, without any branch circuits. In addition, the parallel cables must be either:

- multicore cables, twisted single-core cables or non-sheathed cables
- non-twisted single-core cables or non-sheathed cables in flat or trefoil formation. Where cable sizes of greater than 50 mm² for copper and 70 mm² for aluminium are used, attention must be paid to the cable configuration.

This Regulation does not exclude the use of ring final circuits.

Variation of installation conditions *(523.9)*

Where cables are installed along a route where heat dissipation differs, the current-carrying capacity of the cables shall be determined using the most adverse conditions.

Armoured single-core cables *(523.10)*

Where single-core metallic sheath or armoured cables are used, both ends of the cable run must be bonded together (solid bonding). However, this bonding might permit induced currents to circulate through the cable sheath and result in an increase in the temperature of the cable.

Alternatively, the sheaths or armour of cables having conductors with a cross-sectional area exceeding 50 mm² and a non-conducting outer sheath may be bonded together at one point (single point bonding) with suitable insulation at the un-bonded ends.

Cross-sectional areas of conductors of cables *(524)*

The cross-sectional area of conductors in either an a.c. or d.c. circuit shall have a cross-sectional area not less than those specified in Table 52.3 of BS 7671 (extract illustrated below).

Wiring system	Circuit	Material	Cross-sectional area
Fixed installation wires in sheathed or non-sheathed cable	Power and lighting	Copper	1 mm²

Voltage drop in consumers installations *(525)*

A further consideration when selecting cables is that of volt drop. This section of the Regulations requires that under normal service conditions the voltage at the terminals of any fixed current-using equipment shall be greater than the lower limit required by the product standard of the equipment. Where the equipment is not the subject of a product standard the voltage at the terminals must be such as not to impair the safe functioning of the equipment.

These requirements are satisfied if the volt drop between the origin of the installation (usually the supply terminals) and a socket-outlet or the terminals of fixed equipment does not exceed the voltage drop values in Table 12A in Appendix 12 of BS 7671. For a low voltage installation, supplied from a public low voltage distribution system, the following values are stated:

- lighting 3%
- other uses 5%.

Voltage drops can be determined from the information given in the tables in Appendix 4.

Electrical connections *(526)*

Every connection of a conductor must be electrically and mechanically sound.

When deciding on the method of connection, the following should be considered:

- conductor material/insulation
- conductor size, number and shape of wires
- the number of conductors being connected together
- the conductor insulation must not be affected by the temperature of the terminations in normal use
- soldered connections must be suitable for mechanical stresses and the temperature rise of fault current conditions
- protection against vibration or thermal cycling.

Every connection and joint must be accessible for inspection, testing and maintenance, with the following exceptions:

- a compound filled/encapsulated joint
- connections between the cold tail and a heating element (e.g. floor or ceiling heating systems, pipe trace heating)
- joints made by welding, soldering, brazing or compression tool
- a joint forming part of the equipment complying with the appropriate product standard.

Every termination or joint in a live or PEN conductor must be within one (or more) of the following:

- accessory or enclosure complying with the appropriate product standard
- suitable enclosure of material complying with the appropriate product standard
- enclosure formed or completed with non-combustible building material.

For all terminations made in enclosures, the enclosure must provide sufficient mechanical and external influence protection.

Connection of multiwire, fine wire conductors *(526.8)*

To prevent the separating or spreading of individual wires, conductor ends shall be treated or suitable terminals shall be used. Soldering or tinning of conductors shall not be carried out if screw terminals are used or where connection and junction points are subject to relative movement at the termination points.

Selection and erection of wiring systems to minimise the spread of fire (527)

Selecting and suitably installing appropriate materials can minimise the risk of fire spreading.

The wiring system must not affect building structure performance or fire safety by the installation methods utilised.

Cables to BS EN 60332-1-2 may be installed without special precautions. Where the risk of fire is high, cables must comply with the requirements of the BS EN 50266 series.

Cables not complying with BS EN 60332-1-2 must be only short lengths for connection of appliances to the fixed wiring system but they must not pass from one fire segregated compartment to another.

Conduit and trunking complying with the flame propagation requirements of their standards may be installed without special precautions.

Where a wiring system is required to pass through or penetrate material forming part of the construction of a building (e.g. cable trunking or busbar trunking systems), holes must be sealed and the wiring system must also be internally sealed in order to maintain the required fire resistance of the material.

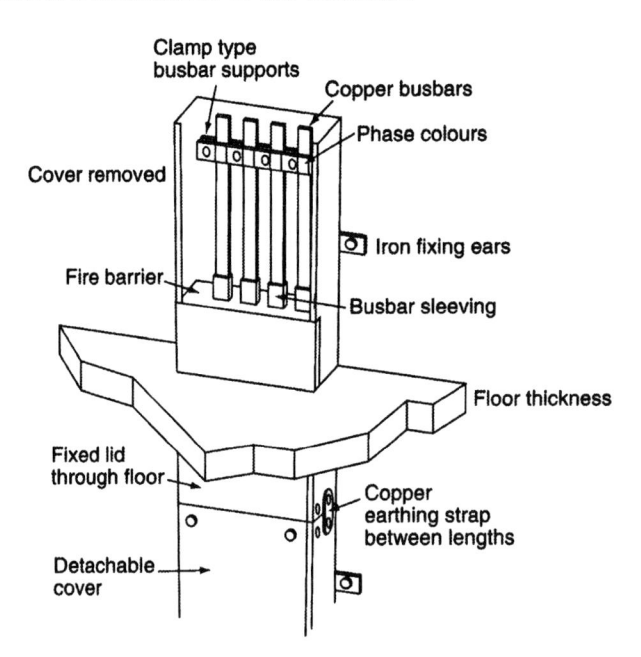

Wiring systems with non-flame propagating properties and an internal cross-section not exceeding 710 mm² need not be sealed internally, provided that the system meets the test for IP33 (BS EN 60529).

Any sealing arrangements shall:

- be resistant to products of combustion to the same degree as the building construction element penetrated
- provide protection from water penetration to the same degree as the building element in which it is installed
- be compatible with the wiring system concerned
- permit thermal movement of the wiring system without detriment to the sealing quality
- be removable without damage when additions to the wiring system are necessary
- be capable of resisting external influences to the same standards as the wiring system with which it is being used.
- have adequate mechanical stability if the wiring system supports are damaged by fire.

During the installation of wiring systems, temporary sealing arrangements must be made. In addition, any existing sealing which is disturbed or removed in the course of alterations to an installation must be reinstated as soon as possible.

It is essential that sealing arrangements are visually inspected during installation to verify that they conform to the manufacturer's instructions. Details of those parts of a building which are sealed and the methods used must be recorded.

Proximity of wiring systems to other services *(528)*

Voltage bands

The voltage levels of Band I and Band II voltages are contained in Part 2 (Definitions) of BS 7671.

Band I voltage

Installations where shock protection is provided under certain conditions by the value of voltage, and installations where, for operational reasons, the voltage value is limited (e.g. signalling bell, control, alarm and telecommunications installations) are Band I voltages. Extra-low voltage will normally be a Band I voltage.

Band II voltage

This is the voltage for supplies to industrial, commercial and domestic installations. Low voltage will normally be a Band II voltage.

The level of Band II voltage does not exceed 1,000 V a.c. rms or 1,500 V d.c..

Other considerations

The requirement is that Band I and Band II circuits are not to be contained in the same wiring system as a circuit exceeding low voltage, and Band I circuits must not be in the same wiring system as a Band II circuit unless one of the following methods is utilised:

- all cables are insulated for the highest voltage present

- in a multicore cable or cord, any Band I cores must be insulated either singly or collectively for the highest Band II voltage present

- Band I circuits must be separated from the cores of Band II circuits by an earth metallic screen of equal current-carrying capacity to the largest core of the Band II circuit.

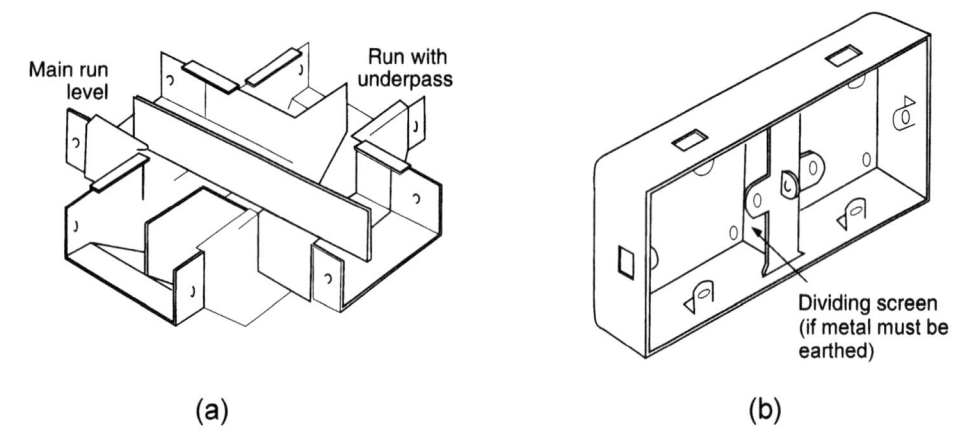

(a) (b)

- Cables are insulated for their voltage band and installed in separate compartments of a cable trunking or ducting system.

- Cables are installed on cable tray or ladder but physically separated by a partition.

- Cables are installed in separate trunking, conduit or ducting.

Where SELV and PELV circuits are used, the requirements of Regulation 414.4 must be complied with.

Proximity to communication cables *(528.2)*

Where underground telecommunication cables and underground power cables cross, a clearance of at least 100 mm shall be provided. Alternatively, a fire-retardant partition and some form of mechanical protection shall be provided.

Communication circuits must be segregated in compliance with the BS 6701 requirements and the BS EN 50174 series.

Proximity to non-electrical services *(528.3)*

Electrical circuits and exposed metalwork

A wiring system installed close to a non-electrical service must:

- be suitably protected from any likely hazards possible from the other service in normal use
- have fault protection provided in accordance with Section 411.

A wiring system must not be installed in the vicinity of a service which produces heat, smoke or fumes detrimental to the type of wiring used, unless suitably protected by shielding. It must also be suitably protected when installed near a service which could cause condensation, e.g. water, steam, gas, etc.

Wiring systems installed near non-electrical services must be located so that any work carried out on either service would not harm the other. Alternatively, a mechanical or thermal shield could be used.

No cable must be run in a lift shaft unless it forms part of the lift installation.

Selection and erection of wiring systems in relation to maintainability *(529)*

Any protective measures that must be removed to allow maintenance to be carried out shall be reinstated where practicable without reducing the original degree of protection.

Safe and adequate access shall be provided to all parts of the wiring system that require maintenance.

Protection, isolation, switching and monitoring *(Chapter 53)*

Requirements

A major requirement of this chapter is that every installation and all items of equipment must be provided with effective means to cut off every source of electrical energy.

The terms for isolation and switching in BS 7671 have specific meanings and will be explained.

Four main functions dealt with are:

- isolation
- switching off for mechanical maintenance
- emergency switching
- functional switching.

Table 53.2 of BS 7671 lists the devices suitable and provides the BS EN standard (extract of table illustrated below).

Device	Standard	Isolation	Emergency switching	Functional switching
Fuse	BS 88-2	Yes	No	No

Where more than one of these functions is to be performed by a common device, the arrangement and characteristics of the device must satisfy all of the relevant Regulations for the functions involved. BS 7671 accepts that one device may be used to fulfil two or more of these functions, which is often the case in practice.

Combined protective and neutral (PEN) conductors must not be isolated or switched. (Switching this type of conductor would result in danger due to the loss of earth continuity.)

Isolation *(537.2)*

Isolation is a function intended, for safety reasons, to make dead a whole or specific section of an electrical installation from every source of electrical energy.

Every circuit shall have a means of isolation provided to ensure it can be isolated from each live supply conductor.

Otherwise, in TN-S or TN-C-S systems, the neutral conductor need not be isolated or switched where the neutral is reliably at earth potential. For a supply complying with the Electricity Safety, Quality and Continuity Regulations 2002, the supply neutral (PEN or N) is considered to be connected to earth by a suitably low impedance.

Isolation is the cutting off of an electrical installation, a circuit or an item of equipment, from every source of electrical energy – in order to ensure the safety of those working on the installation – by making dead those parts which are live in normal service.

Every installation must be provided with a means of isolation, and attention needs to be given to the design, location and operation of any devices used.

An 'isolator' is a mechanical switching device, operated manually, which is used to open or close a circuit when there is no load on the circuit. In the open position, it complies with the requirements specified for isolation. An isolator is also called a disconnector.

Location and operation of isolation in installations

The means of isolation must be located at a point as near as practicable to the origin of the installation. In the case of equipment, the isolator must be adjacent to the equipment it controls unless the requirements stated below for remote location are satisfied.

A linked switch or linked circuit breaker must be provided as close as possible to the origin of every installation for switching the supply on load and as a means of isolation.

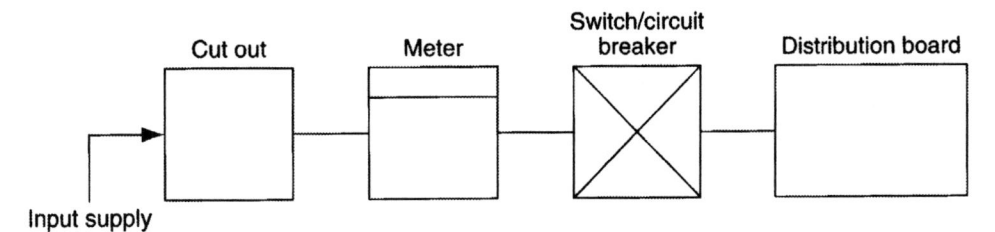

Switch/circuit breaker must be located at origin of the installation

When a main switch is to be operated by an ordinary person it shall interrupt both live conductors of a single-phase supply (phase and neutral). A typical example of this situation is the main switch of consumer unit in a domestic installation.

Where an isolator is used in conjunction with a circuit breaker as a means of isolating main switchgear for maintenance, the isolator must be interlocked with the circuit breaker so that it can be operated only when the circuit breaker is already open, or located and/or guarded where only a skilled person can operate it.

When isolating devices are to be placed at a location remote from the equipment to be isolated, the means of isolation must be capable of being secured in the **open** position. Where this is provided by lock or removable handle they must not be interchangeable with others used in the installation. The isolator must be properly identified.

Alternatively, a local (or secondary) means of isolation may be provided, such as that shown below at switch fuse B. In this case, the remote isolator, switch fuse A, does not need to be lockable.

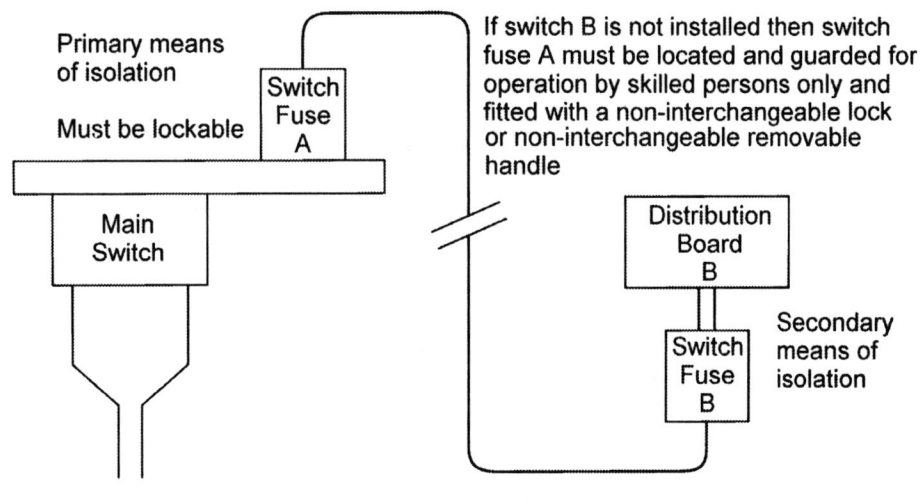

Remote location

There are additional requirements for discharge lighting installations operating at a voltage exceeding 1,000 V. One or more of the following means must be provided:

- An interlock on a self-contained luminaire arranged so that the supply is automatically disconnected before access to live parts can be made.
- An effective means of isolation of the circuit installed near to the luminaires, in addition to any switch normally used for controlling the circuit.
- Installation of a switch having a non-interchangeable lock or removable handle.

Devices for isolation *(537.2.2)*

Devices for isolation must be capable of being secured in the **off** or **open** position to prevent any equipment from being inadvertently or unintentionally energised.

The position of the contacts of an isolator (i.e. open or closed) must be visible or clearly and reliably indicated.

Devices for isolation must be installed so as not to be subject to shock or vibration which could cause unintentional closure of the device.

Semiconductor devices must not be used as isolators. A 'touch-switch' or a 'photoelectric switch' are not suitable devices.

The choice of equipment selected as isolators is likely to be governed by the particular conditions of the installation. Suitable isolation devices are listed in Table 53.2 of BS 7671.

Devices used for isolation must be identified by position or reliably labelled to clearly indicate the circuit or installation they isolate.

Switching off for mechanical maintenance *(537.3)*

The term 'mechanical maintenance' refers to the replacement, refurbishment and cleaning of lamps, and non-electrical parts of equipment, plant or machinery. The switch or device is intended to prevent danger and possible physical injury. Special attention must be given to the location and operation of the device.

The requirements are similar to those for isolation but generally the control switches will be local and must be capable of switching the full load current of the circuit or item of equipment and be continuously under the control of the person performing the maintenance.

The following measures may be used:

- located in a lockable space or enclosure
- padlocked
- warning notices.

Devices used for switching off for mechanical maintenance *(537.3.2)*

Devices must be selected and installed in such a manner as to prevent unintentional re-closure (e.g. by mechanical shock and vibration). Such a device must be capable of cutting off the full load current to the part of the installation it controls. It must have an external visible contact gap or a clearly and reliably indicated **off** or **open** position. The device must only indicate **off** or **open** when all poles of the supply have been isolated. The use of the symbol '0' for open and '1' for closed may be used.

Devices which are suitable for switching off for mechanical maintenance include:

- multipole switches
- circuit breakers
- control and protective switching devices
- control switches operating contactors
- plug and socket-outlets ≤ 16 A rating.

Whenever possible, the switch must be inserted in the main supply circuit although it is permissible to insert it in the control circuit if precautions are taken to provide a degree of safety equivalent to that of interruption of the main supply.

Devices for switching off for mechanical maintenance must be readily identifiable and convenient for their intended use.

Emergency switching *(537.4)*

Emergency switching is an operation that is intended to remove, as quickly as possible, danger which may have occurred unexpectedly. Devices for emergency switching must be installed in a readily accessible position where the hazard might occur and it must not be possible for the supply to be restored from another location.

Plugs and socket-outlets or similar devices must **not** be used for emergency switching.

Typically, emergency switching may be required in the event of a fire, accident, or explosion.

Any emergency switching shall act as directly as possible on the supply conductors. When electrically powered equipment is within BS EN 60204 the emergency switching requirement of that standard applies.

Devices for emergency switching *(537.4.2)*

A device (switch or circuit breaker) used for emergency switching must be capable of cutting off the full load current to that part of the installation affected. A means of emergency switching shall be provided in every place where an electrically driven machine may give rise to danger.

The device must be readily accessible and easily operated by the person in charge of the machine. Where more than one means of starting the machine is provided and danger might be caused by unexpected restarting, there must be a provision to prevent such restarting.

Plugs and socket-outlets or similar devices must not be used for emergency switching.

Due account must be taken where stalled motor conditions may occur. In some cases, it may be necessary to retain a supply to operate electric braking facilities.

The means of emergency switching must consist of a single switching device which cuts off the incoming supply, or a combination of several items of equipment operated by a single action and resulting in the cutting off of the supply (e.g. sequential control). Such an arrangement with two contactors is illustrated below.

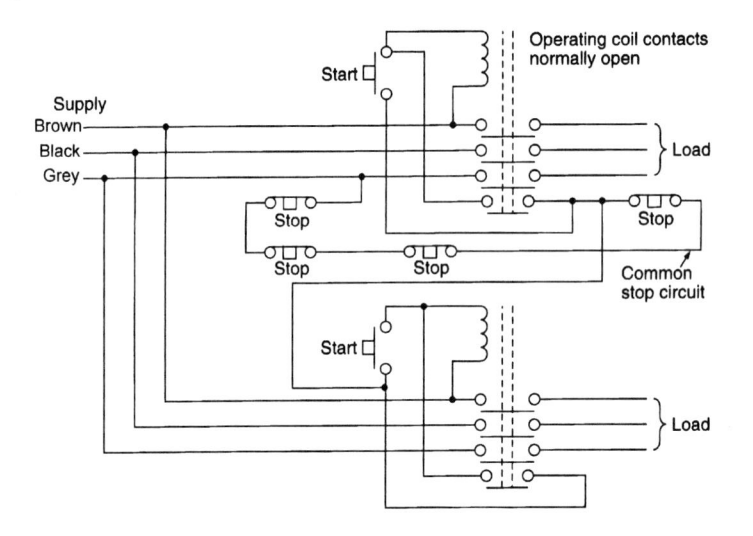

Emergency switching devices should, if possible, be manually operated, directly interrupting the main circuit. Where devices such as contactors or circuit breakers are operated by remote control, they should **open** on de-energisation of the coil.

The operating handle or push button must be clearly identifiable and preferably coloured **red**, e.g. emergency stop buttons installed in laboratories and training workshops. They must be readily accessible and located where the hazard may occur. (Where appropriate, additional devices must be provided at any remote operating positions.)

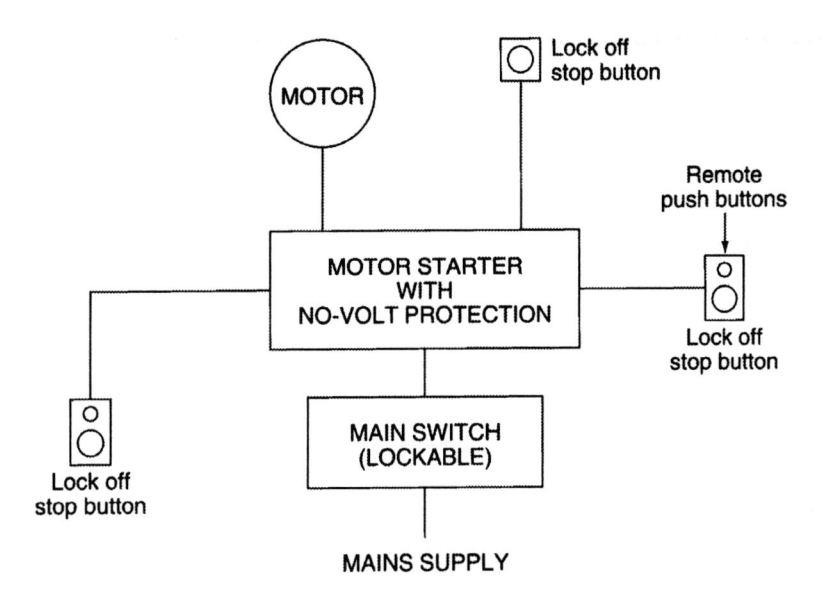

The switch must be of the latching type, capable of being held in the **stop** or **off** position. If the device is one which automatically resets itself, the operation and re-energisation of the device must be under the control of the same person. The release of any emergency operating device must not re-energise the equipment concerned.

Latching off – stop button

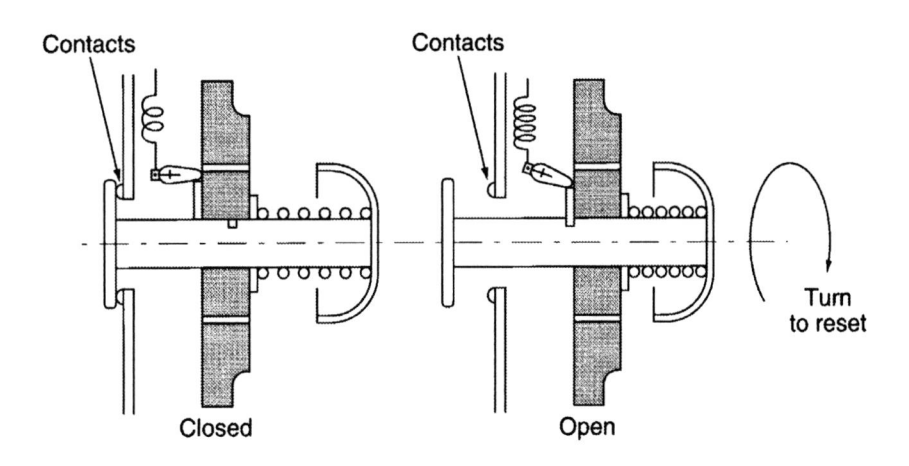

Functional switching (Control) *(537.5)*

The switching of electrically operated equipment in normal service is referred to as 'functional switching'. If this is not provided by a device which allows for switching off for mechanical maintenance or for emergency switching, then it is necessary to provide a switch to interrupt the 'on-load' supply for any circuit or appliance capable for the most onerous duty called upon.

All functional devices must be suitable for the most onerous duty requirements.

Semiconductor switching devices may be used to control the current without necessarily opening the corresponding poles (e.g. remote control electronic dimmers).

Plugs and socket-outlets of a rating not exceeding 16 A may be used as a switching device.

A plug and socket-outlet exceeding 16 A rating may be used as a switching device (a.c. only) provided it has a breaking capacity appropriate to the use intended.

Off-load isolators, fuses or links must **not** be used for functional switching.

Fixed or stationary appliances (connected other than by a plug and socket-outlet) shall have a means of interrupting the supply on-load. The device used may be incorporated in the appliance itself or, if separate, must be in an accessible position that does not put the operator in danger. Where two or more appliances are installed in the same room, e.g. separate oven and hob, one device may be used to control all the appliances.

Fire-fighters' switches *(537.6)*

A fire-fighter's switch * is required for installations operating at more than low voltage (i.e. ≥ 1,000 V) in the case of:

- exterior electrical installations
- interior discharge lighting installations.

* Not applicable to a portable discharge lighting luminaire or to a sign of rating ≤ 100 W and fed from a readily accessible socket-outlet.

Applications

- A covered market, arcade or shopping mall is classed as exterior.
- A temporary installation to an exhibition in a permanent building is classed as interior.
- For an exterior installation the switch should be located outside the building and adjacent to the equipment. Otherwise, a notice should indicate its position with a further notice close to the switch to clearly identify it.
- For interior installations the switch should be in the main entrance to the building or in a position agreed with the local fire authority.

- It should be conspicuous, reasonably accessible to fire fighters and not more than 2.75 m from the ground (except where otherwise agreed with the local fire authority).
- When more than one such switch is used, each switch must be clearly labelled to indicate the installation it controls.

Other requirements are that every fire-fighter's switch should be coloured **red** and have fixed to it (or adjacent) a label with the words '**fire-fighter's switch**' clearly indicated. The label must be a minimum size of 150 mm x 100 mm with letters not less than 36 point, easily legible from a distance.

The **on** and **off** positions must be clearly indicated by lettering which can easily be read by someone standing on the ground. The switch must have its **off** position at the top and be fitted with a lock or catch to prevent it being inadvertently returned to the **on** position.

The switch must be installed so as to be accessible for operation by fire fighters (537.6.4).

Monitoring devices *(538)*

Monitoring devices observe the operation of a system to verify correct operation or detect incorrect operation. In some electrical installations insulation monitoring devices, called IMDs, are used.

When an IMD is used in an IT system it shall be permanently connected and continuously monitor the insulation resistance of the complete system, including the installation and secondary side of the power supply. When an IMD detects a fault this shall be located and rectified as soon as possible.

Installation of Insulation Monitoring Devices (IMDs) *(538.1.2)*

Line terminals of IMDs shall be connected as close as possible to the system origin, either to:

- the power supply neutral point
- the artificial neutral point
- a line or several line conductors.

IMDs shall be suitable to withstand line to line voltages.

In d.c. installations the 'line' terminals of an IMD shall be connected either to the mid-point or to a supply conductor. The IMD 'earth' or 'functional earth' terminal shall be connected to the installation's main earth terminal.

Adjustment of IMDs *(538.1.3)*

IMDs shall be set to a lower value than the normal insulation of the system, with maximum connected loads.

The installation of IMDs shall prevent settings being changed without the use of a key, tool or password, unless the installed location is only accessible to instructed (BA4) or skilled (BA5) persons.

Passive insulation monitoring devices *(538.1.4)*

In some d.c. IT installations a passive IMD may be used if it does not inject current into the system and the insulation of all live distributed conductors are being monitored, all exposed-conductive-parts in the installation are interconnected and the installation methods used to install circuit conductors reduce the risk of faults to earth.

Equipment for insulation fault location in IT systems

When a circuit has safety equipment, arrangements shall be made to ensure the IMD is automatically deactivated whenever the safety equipment is activated.

Residual current monitoring (RCM) *(538.4)*

RCMs permanently monitor leakage current in downstream installations before protective devices are activated.

When RCMs are used in IT systems, directionally discriminating types are recommended to avoid inopportune signalling of leakage current, especially where high leakage capacitance exists downstream of the RCM. When RCMs are used in a.c. systems they shall conform to BS EN 62020.

In situations where an RCD is installed upstream of an RCM, the RCM shall have a rated residual operating current not exceeding a third of that of the RCD. RCMs shall not have a higher residual operating current greater than the first fault current intended to be detected.

For an IT system where the first insulation fault to earth is not required to disconnect the supply, an RCM may be installed to facilitate the location of a fault by installing it at the beginning of outgoing circuits.

Earthing arrangements and protective conductors
(Chapter 54)

Purpose of earthing

By connecting non-current carrying metalwork to earth, a path is provided for any leakage current that can be detected and, if necessary, interrupted by the following devices:

- fuses
- circuit breakers
- residual current devices (RCDs).

General *(541)*

The earth can be considered to be a large conductor which is at zero potential. The purpose of earthing is to connect all metalwork (other than that which is intended to carry current) to earth so that dangerous potential differences cannot exist either between different metal parts, or between metal parts and earth.

Every means of earthing and protective conductor must satisfy the requirements of BS 7671.

Where there is a lightning protection system, account must be taken of the requirements of BS 6651: Code of Practice for protection of structures against lightning.

Earthing arrangements *(542)*

The earthing arrangement of an installation must be such that:

- the value of impedance from the consumer's main earthing terminal to the earthed point of the supply (TN systems) or to earth (TT and IT systems) complies with the protective and functional requirements of the installation and is expected to remain continuously effective

- earth fault currents and protective conductor currents which may occur under fault conditions can be carried without danger, particularly from thermal, thermo mechanical and electromechanical stresses

- they are sufficiently robust or are protected from mechanical damage appropriate to the assessed conditions.

The installation should be so installed as to avoid risk of subsequent damage to any metal parts or structures through electrolysis (542.1.7).

Earth electrodes *(542.2)*

The following items are recognised by BS 7671 as suitable earth electrodes:

- earth rods or pipes
- earth tapes or wires
- earth plates
- underground structural metalwork embedded in foundations
- welded metal reinforcement of concrete embedded in the earth, except pre-stressed concrete
- lead sheaths or other metallic coverings of cables, where not precluded by Regulation 542.2.5
- other suitable underground metalwork.

Earth electrodes must be installed in such a way that their resistance does not increase above the required value through climatic conditions, such as soil drying or freezing, or from corrosion, etc.

The metalwork and pipes of public gas, water or other services must not be used as an earth electrode. This metalwork must still be bonded.

The lead sheaths and other metallic coverings of cables may be used as earth electrodes provided that:

- adequate precautions are taken to prevent excessive deterioration by corrosion
- they are in effective contact with earth
- consent is obtained from the owner of the cables
- the owner of the electrical installation is informed of any proposed changes that will affect the use of the cables as earth electrodes.

For details of types and methods of installing earth electrodes, refer to Technical Data Sheet 5B.

Earthing conductors *(542.3)*

Where earthing conductors are buried in the ground, they must have a cross-sectional area not less than that stated in Table 54.1 of BS 7671 and, where PME conditions apply, they must comply with Regulation 544.1.1 for the cross-sectional area of protective bonding conductors.

All connections of earthing conductors to earth electrodes must be electrically and mechanically sound and fitted with a permanent label to BS 951. The label must be permanently fixed in a visible position with the words, '**Safety electrical connection – do not remove**' (514.13.1).

For tape or strip conductors, their thickness must be suitable to withstand corrosion and mechanical damage.

Main earthing terminals or bars *(542.4)*

A main earthing terminal must be provided in every installation to enable the earthing conductor to connect to:

- circuit protective conductors
- protective bonding conductors
- functional earthing conductors (if required)
- lightning protection system bonding conductor (if present).

Provision must be made in an accessible position for disconnection of the earthing conductor for test measurement of the earthing arrangements.

The method of disconnecting the earthing terminal from the means of earthing must be such that it can only be effected with the use of tools. It must be mechanically strong and ensure the maintenance of electrical continuity. It may conveniently be combined within the main earthing terminal.

Protective conductors *(543)*

Cross-sectional areas

The minimum cross-sectional area (csa) of protective conductors shall be:

- calculated in accordance with Regulation 543.1.3
- selected in accordance with Regulation 543.1.4.

If the protective conductor csa is to be calculated then the formula used is:

$$s = \frac{\sqrt{I^2 t}}{k} \text{ mm}^2$$

Where:
- s = minimum size of protective conductor in mm²
- I = maximum earth fault current in amperes
- t = time in seconds for the protective device to operate
- k = factor for specific protective conductors from Tables 54.2 to 54.7 of BS 7671

Where a protective conductor is common to several circuits, the cross-sectional area must be calculated (543.1.3) for the most onerous values of fault current and operating time in each of the circuits; or selected to correspond with the cross-sectional area of the largest line conductor.

If the protective conductor does not form part of a cable, is not a conduit, ducting or trunking, and is not contained in an enclosure formed by the wiring system, the cross-sectional area must not be less than:

- 2.5 mm² copper equivalent if sheathed, or otherwise provided with mechanical protection
- 4 mm² copper equivalent where mechanical protection is not provided.

When metal enclosures are used as protective conductors, the cross-sectional area must be at least equal to the application of the calculation or selection method described above.

Its electrical continuity must be assured and it must be protected against mechanical, chemical or electrochemical deterioration. It must also permit other protective conductors to be connected at every predetermined tap-off point (543.2.4).

The nominal cross-sectional areas for metal conduits are, as illustrated below, based on the standard sizes given in British Standards.

Size	Nominal cross-sectional area mm²	
	Light gauge	Heavy gauge
16	47	72
20	59	92
25	89	131
32	116	170

Size mm x mm	Nominal cross-sectional area mm²
50 x 37.5	125
50 x 50	150
75 x 50	225
75 x 75	285
100 x 50	260
100 x 75	320
100 x 100	440
150 x 50	380
150 x 75	450
150 x 100	520
150 x 150	750

The values given in these tables should only be used if the joints made in either conduit or trunking systems do not reduce the nominal cross-sectional area.

Note: The value of resistance of steel conduit and trunking manufactured to BS 4568 and BS 4768 respectively must not exceed 5×10^{-3} Ω/m (0.005 Ω/m).

Some manufacturers have published resistance values for their conduit and trunking.

If the protective conductor csa is to be selected then Table 54.7 of BS 7671 is used.

Table 54.7 establishes the minimum cross-sectional area of protective conductor in relation to the cross-sectional area and material of the associated line conductor. For example:

Cross-sectional area of line conductor SIZE (s) mm²	Minimum cross-sectional area of the corresponding protective conductor SIZE (s) mm²	
	If the protective conductor is of the same material as the line conductor	If the protective conductor is not the same material as the line conductor
$S \leq 16$	S	$\frac{k_1}{k_2} \times S$
$16 < S \leq 35$	16	$\frac{k_1}{k_2} \times 16$
$S > 35$	$\frac{S}{2}$	$\frac{k_1}{k_2} \times \frac{S}{2}$

Where:

k_1 is the value of k for the line conductor, selected from Table 43.1 in Chapter 43 according to the materials of both conductor and insulation

k_2 is the value of k for the protective conductor, selected from Tables 54.2, 54.3, 54.4, 54.5 or 54.6, as applicable

Types of protective conductor *(543.2)*

A protective conductor may be one or more of the following:

- a single-core cable (colour **green** and **yellow**)
- a conductor in a cable
- an insulated or bare conductor in a common enclosure (trunking, conduit, etc.) with insulated live conductors
- a fixed bare or insulated conductor
- the metal sheath, screen or armouring of a cable
- metal conduit or other enclosure or electrically continuous support system for conductors
- an extraneous-conductive-part complying with Regulation 543.2.6.

The first four items above, if less than 10 mm² cross sectional area, shall be copper.

When the protective conductor is formed by conduit, trunking or the metal sheath and/or armour of a cable, a separate protective conductor must be installed from the earthing terminal of each accessory (e.g. socket-outlet) to the earthing terminal of its associated box or enclosure (543.2.7).

Preservation of electrical continuity of protective conductors *(543.3)*

Protective conductors must be installed so that they are protected against mechanical and chemical deterioration and electrodynamic effects.

Where protective conductors of cross-sectional areas up to and including 6 mm² are installed, and are not an integral part of a multicore cable or a cable enclosure such as conduit or trunking, they must be protected throughout by insulation at least equivalent to that provided for a single-core non-sheathed cable of appropriate size insulated for at least 450/750 V.

When the sheath of a cable containing an uninsulated protective conductor is removed for the purpose of making terminations or joints, such protective conductors up to and including 6 mm² must be protected by a **green** and **yellow** insulating sleeve complying with the BS EN 60684 series. This requirement does not apply to joints in metal conduit, ducting or trunking or support systems.

Regulation 543.3.3 requires all connections to be accessible for inspection, testing or maintenance apart from:

- a joint designed to be buried in the ground
- a compound filled/encapsulated joint
- a connection between a cold tail and a heating element e.g. underfloor heating
- joints made by welding, soldering, brazing or compression tools
- a joint forming part of the equipment complying with the appropriate product standard.

Switches must not be inserted in protective conductors unless allowed by Regulation 537.1.5.

Multipole linked switching or plug-in devices are used where the protective conductor circuit is not interrupted before the live conductors and not later than the live conductors on reconnection.

All joints in protective conductors must be electrically and mechanically sound. Joints in metal conduit systems must be screwed or mechanically clamped. Plain slip or pin-grip sockets are not suitable as they will not ensure an effective low resistance joint throughout the life of the installation due to the small area of contact achieved with this type of fitting.

Where an installation is connected to an electrical earth monitoring system, the operating coil of that system must be connected in the pilot conductor, not the protective earthing conductor.

For further details of earth monitoring systems, refer to Technical Data Sheet 5C.

Combined protective and neutral (PEN) conductors *(543.4)*

A PEN conductor may only be used on systems where:

- the local distributor has authorised its use in respect of the installation concerned, or
- the installation is fed from a privately owned transformer or converter with no metallic connection with the distributor's network (except for the earthing connection).
- the supply is fed from a private generating plant.

Provided that the part of the installation concerned is not fed through an RCD, the following types of cable may be used as PEN conductors:

- conductors of cables not subject to flexing, such as armoured PVC-insulated cables and MI cable, used on fixed installations having a cross-sectional area of not less than 10 mm² for copper conductors and 16 mm² for aluminium.

The outer conductor of a concentric cable must not be common to more than one circuit. This requirement does not prevent the use of twin or multicore cables to serve a number of points contained within one final circuit.

At every joint and termination point, the continuity of the outer conductor of a concentric cable must be ensured by a conductor additional to any means of sealing and clamping the outer conductor (543.4.5). Illustrated below is a single-core MICC cable where the outer metallic sheath provides the function of both cpc and neutral (PEN).

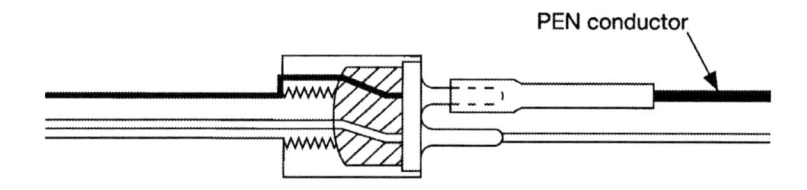

Mineral insulated (MI) cable termination

No isolation or switching devices may be installed in the outer conductor of a concentric cable. The PEN conductor must be insulated so as to be suitable for the highest voltage present.

If, for any reason, the neutral and protective functions are provided by separate conductors run separately from any point in the installation, they must not be connected together anywhere beyond that point.

For further information on PME supply systems, refer to Technical Data Sheet 5D.

Earthing requirements for the installation of equipment having high protective conductor currents *(543.7)*

This section applies to circuits supplying equipment with a protective conductor current exceeding 3.5 mA in normal use and circuits where the total protective conductor current exceeds 10 mA in normal use.

Examples of this type of equipment include:

- IT equipment
- industrial and telecommunication equipment with RF filtering
- heating elements.

All data processing equipment can lose information due to electrical mains disturbance, in the form of voltage spikes or transients. These problems can often be eliminated or reduced by feeding the electrical supplies to the equipment via filter circuits.

These filter circuits usually include resistive and capacitive components which are connected between live conductors and earth. The effect of these components is to increase earth currents which are driven by the voltage through these components. Therefore, equipment fitted with these filters in the mains supply will normally have high levels of earth current, which may rise to even higher levels when the equipment is initially switched on and until the capacitors in the filter circuits are charged up.

General

Except where a number of items of equipment are installed on the same circuit, there are no special requirements for equipment having a protective conductor current not exceeding 3.5 mA.

Any item of stationary equipment with an earth leakage current exceeding 3.5 mA, but not exceeding 10 mA in normal use, shall be connected directly to the fixed wiring circuit by means of such devices as double-pole switches, fused connection units, or plug and socket-outlets manufactured to BS EN 60309-2.

Equipment with protective conductor currents in excess of 10 mA must be connected by one of the following methods:

- Permanent connection to the wiring of the installation. The protective conductor must comply with Section 543.7 and the permanent connection to the wiring may be by means of a flexible cable.

- A flexible cable with a plug and socket-outlet to BS EN 60309-2, provided that either:

    - the protective conductor of the flexible cable is not less than 2.5 mm² for 16 A plugs and not less than 4 mm² for plugs rated above >16 A, or

    - the protective conductor of the flexible cable is of a cross-sectional area not less than that of the line conductor.

- A protective conductor complying with Section 543 with an earth monitoring system to BS 4444 (as Technical Data Sheet 5C) which automatically disconnects the supply if the protective conductor fails.

For final circuits and distribution circuits supplying one or more items of equipment where the total protective conductor current will be in excess of 10 mA, a high integrity protective connection complying with at least one of the following must be provided:

- A protective conductor of not less than 10 mm².

- A single, copper protective conductor of not less than 4 mm², complying with Regulations 543.2 and 543.3, and being enclosed to provide additional protection against mechanical damage.

- Duplicate protective conductors complying with Section 543. They may be of different types (for example, metal conduit together with a cpc installed in the conduit). If the two cpcs are incorporated in a multicore cable, the total cross-sectional area of all the conductors, including the live conductors, must not be less than 10 mm².

    Note: Where two protective conductors are used, they must be terminated independently of each other throughout the circuit. This will require accessories to have two separate earth terminals (543.7.1.4).

- An earth monitoring system to BS 4444 (as Technical Data Sheet 5C) which automatically disconnects the supply if the protective conductor fails.

- Connecting the equipment by the use of a double-wound transformer or equivalent (e.g. motor alternator). The protective conductor of the supply is to be connected to the exposed-conductive-parts of the equipment and to a point of the secondary winding of the transformer or equivalent device. The protective conductors between the equipment and the transformer or equivalent device must comply with Regulations 543.7.1.3.

Socket-outlet final circuits

Ring circuit

When several items of equipment are to be installed and their total protective conductor currents in normal use will exceed 10 mA, a ring final circuit may be used.

The protective conductors must be separately terminated throughout the circuit (this requires accessories to have two separate earth terminals), with the start and finish ends of the protective conductor being separately connected at the earthing terminal of the distribution board. Any spurs off the ring circuit will require a high integrity protective connection as Regulation 543.7.1, e.g. the cpc of the spur being wired as a ring with duplicate cpcs (see diagram on next page).

Radial circuits

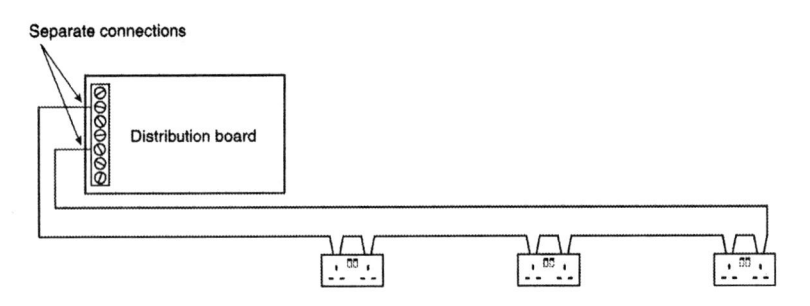

A radial final circuit may be used if one of the following applies:

- The cpc is connected as a ring.
- A separate protective conductor is provided at the final socket-outlet which is connected to the metal conduit or trunking.
- If two or more radial circuits supply socket-outlets in adjacent areas and are fed from the same mains board, have identical means of short circuit and overcurrent protection and cpcs of the same size. Then a second protective conductor may be provided at the final socket-outlet on one circuit by connecting it to the cpc of the adjacent circuit.

The diagrams below show the protective conductor arrangements for a socket-outlet ring final circuit and radial final circuit where the total protective conductor current exceeds 10 mA.

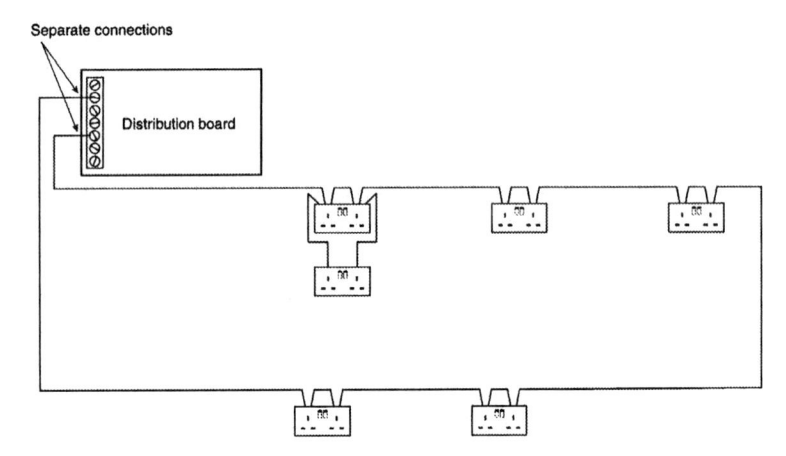

Ring final circuit

Radial final circuit

Circuits that have a high protective conductor current must be identified as such at the distribution board and this information must be visible to a person working on the circuit.

Protective bonding conductors

Where TN-S and TT conditions apply, the main bonding conductors must be not less than half the cross-sectional area of the earthing conductor; the minimum size being 6 mm² with a maximum size of 25 mm².

Where PME conditions apply, the cross-sectional area must be in accordance with Table 54.8 of BS 7671 (see below), which gives the minimum cross-sectional area of main bonding conductors in relation to the neutral conductor of the incoming supply. The conductor sizes are for copper conductors or other conductors affording equivalent conductance.

Table 54.8 of BS 7671

Supply neutral conductor cross-sectional area	Main equipotential bonding conductor cross-sectional area
35 mm² or less	10 mm²
Over 35 mm² up to 50 mm²	16 mm²
Over 50 mm² up to 95 mm²	25 mm²
Over 95 mm² up to 150mm²	35 mm²
Over 150 mm²	50 mm²

Note: The local distributor's network conditions may require larger conductors.

The extraneous-conductive-parts within an installation, such as gas and water services, must be at the same potential as the exposed-conductive-parts, i.e. the metalwork of the electrical installation. This creates an equipotential zone.

This is achieved by connecting all the exposed-conductive-parts of the electrical installation to the main earthing terminal by the circuit protective conductors (cpcs) and by installing main bonding conductors from the main earthing terminal to the gas, water and other services at their point of entry to the premises as illustrated below.

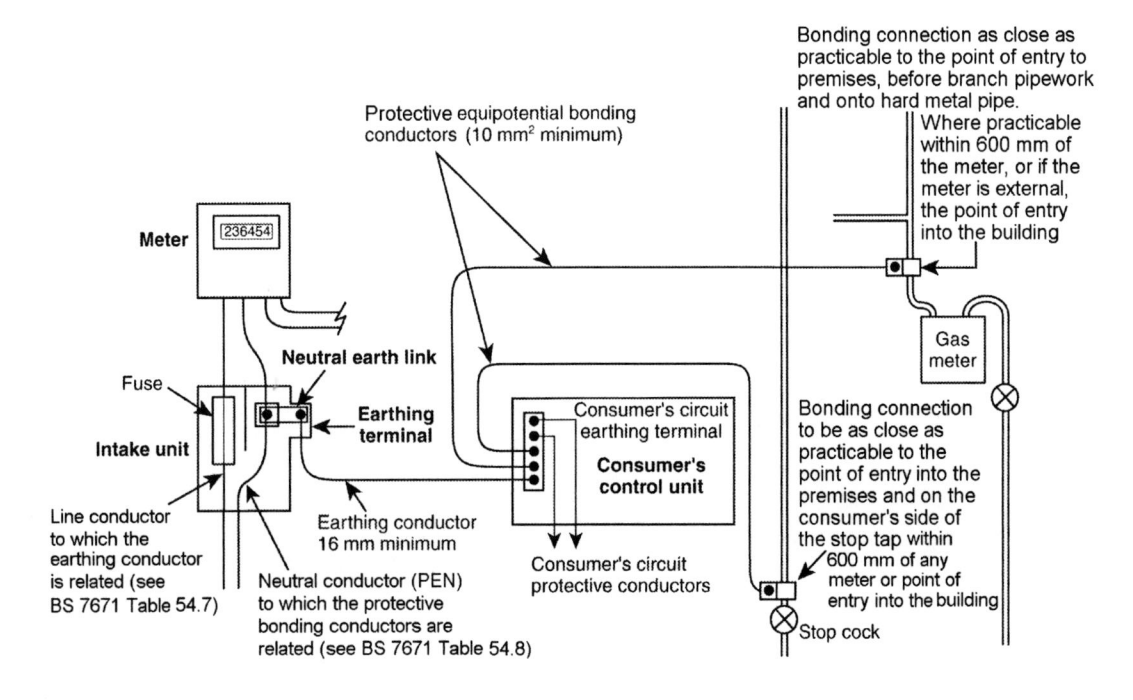

Main equipotential bonding to gas, water and other services must be made as close as practicable to the point of entry of the service to the building. The bonding must be applied on the consumer's side of any meter, stopcock or insulating section and before branch pipework. The connection is made to hard metal pipes, not to soft metal or flexible pipework, using earth clamps to BS 951 that will not be affected by corrosion at the point of connection.

If there is a meter, the bonding connection must be on the consumer's side, preferably within 600 mm of the meter outlet before any branch pipework. If the meter is external to the building, ideally the connection must be made as close as practical to the point of entry of the service to the premises.

The main bonding conductors must be separate (as shown) or a single conductor may be used provided it passes **unbroken** through the connection at one service and directly on to the other.

The use of plastic pipework for installations within buildings can affect equipotential bonding, therefore if the incoming service pipes are plastic:

- and the pipes within the installation are also plastic, they do not require main bonding
- but the pipes within the installation are metal, the main equipotential bonding must be carried out to the metal installation pipes.

BS 951 earth clamps

The connection of a bonding conductor to metal pipework is usually by means of an earth clamp to BS 951. The clamp selected must be suitable for the environmental conditions at the point of connection. This is sometimes identified by the colour of the clamp body or a coloured stripe on the warning label, where:

- a **red** stripe on the label indicates it is only suitable where the conditions are non-corrosive, clean and dry (as a guide, hot pipes only)
- a **blue** or **green** stripe on the label indicates that the clamp is suitable for all conditions (including corrosive and humid).

The clamp must be adequate for the size of conductor being connected and it must have a label attached with the words **'Safety electrical connection – do not remove'**.

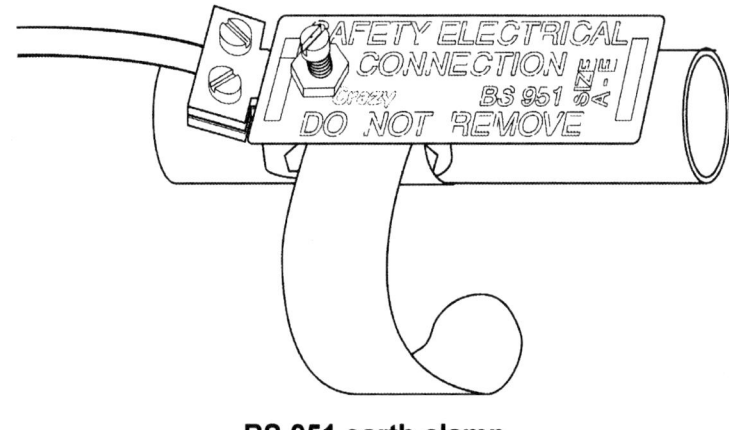

BS 951 earth clamp

If a gas meter is fitted in an external, semi-concealed box and it is not practicable to provide a bonding connection at the point of entry due to the gas installation pipe entering the premises at low level (i.e. below the floorboards or buried in a concrete floor) then it is possible to make the bonding connection within the meter box itself.

The bonding conductor, when returning inside the premises, must pass through a separate hole above the damp course (with the hole sealed on both sides). The Gas Safety (Installation and Use) Regulations 1994 prohibit the bonding conductor passing through the same hole as a gas pipe.

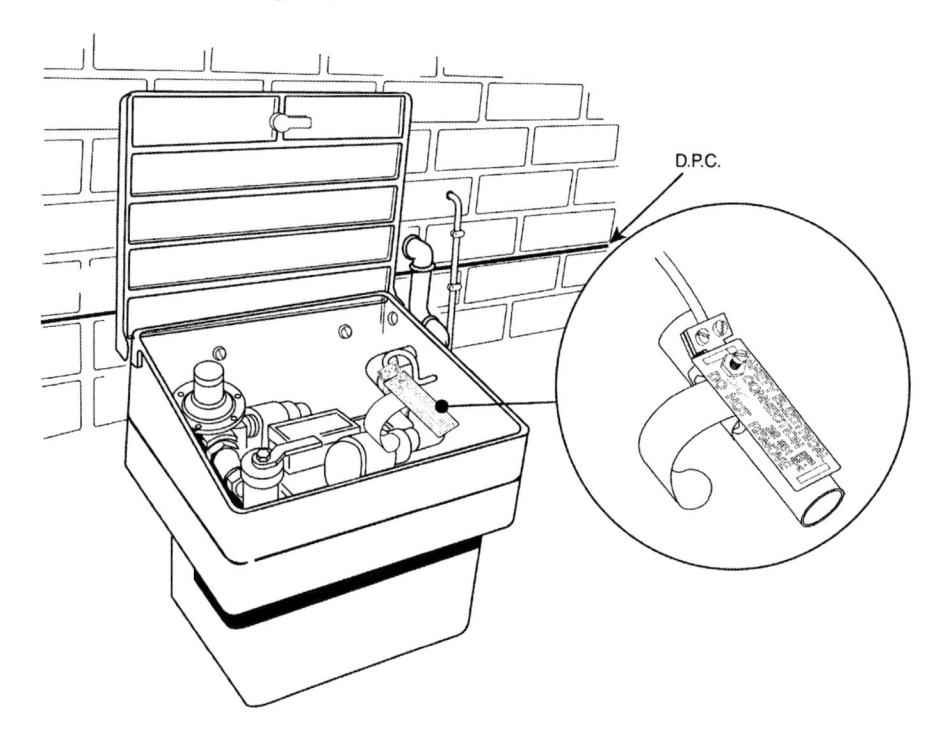

A connection made to the gas installation pipe externally between the meter box and entry point to the building would increase the risk of corrosion and mechanical damage.

If there is a water meter on the consumer's side of, and close to, the stop tap then the main bonding conductor could be connected directly after the meter. Otherwise, the main bonding connection must be as close as possible to the point of entry on the consumer's side of the stop tap as normal.

If a water meter is installed within the premises on the consumer's side of, and some distance from, the stop tap then the water meter must be bridged by a bonding conductor. This is to prevent damage to the working parts of the meter in the event of a fault current flowing through that section of pipework in which the meter is connected.

Supplementary bonding conductors *(544.2)*

If connecting two exposed–conductive-parts, a supplementary bonding conductor (if sheathed, or otherwise mechanically protected) must have a conductance of not less than the smaller protective conductor connected to the exposed-conductive-part, subject to a minimum of:

- 4 mm² if mechanical protection is not provided.

If connecting an exposed-conductive-part to an extraneous-conductive-part, a supplementary bonding conductor (if sheathed or otherwise mechanically protected) must have a conductance of at least half that of the protective conductor to the exposed-conductive-part, subject to a minimum of:

- 4 mm² if mechanical protection is not provided.

In situations where a supplementary bonding conductor connects two extraneous-conductive-parts, neither of which are connected to an exposed-conductive-part, the minimum cross-sectional area of the supplementary bonding conductor must be:

- 2.5 mm² if sheathed or mechanically protected
- 4 mm² if mechanical protection is not provided.

Supplementary bonding can be provided by conductors, or by permanent and reliable conductive parts or by a combination of the two.

When supplementary bonding is applied to fixed appliances, supplied by a short length of flexible cord from an appropriate electrical accessory, the cpc in the flexible cord provides the supplementary bonding to the exposed-conductive-parts of the appliance from the earthing terminal in the electrical accessory.

Other equipment *(Chapter 55)*

Low voltage generating sets *(551)*

This section covers LV and ELV installations, incorporating generating sets intended to supply all or part of an installation either constantly or occasionally, regarding the supply to:

- an installation not connected to the distributor's network
- an installation as an alternative to the distributor's network
- an installation in parallel with the distributor's network
- combinations of the above.

Generators with the following types of power source are covered (551.1.1):

- combustion engine
- turbines
- electric motors
- photovoltaic cells
- electrochemical sources
- other suitable means.

And the following electrical characteristics are considered (551.1.2):

- mains-excited and separately-excited synchronous generators
- mains-excited and self-excited synchronous generators
- mains and self-commutated static inverters with or without bypass facilities.

Generators used for the following supplies are included (551.1.3):

- permanent installations
- temporary installations
- portable equipment that is not connected to a permanent fixed installation.

General requirements *(551.2)*

Means of excitation and commutation must be appropriate for the operation of the generating set and must not affect the correct operation of other sources of supply.

The prospective short-circuit current and prospective earth fault currents must be assessed for every source of supply to ensure that suitable protective devices have been selected.

The capacity and operating characteristics of a generating set must not cause danger or damage to equipment on connection or disconnection of the load. Load shedding must be utilised where the full load is beyond the generator capacity.

Extra-low voltage provided by SELV and PELV *(551.3)*

When it is necessary to maintain the supply of an extra-low voltage system following the loss of one or more supply sources, each supply source, which can be operated independently, shall be capable of supplying the intended load of the extra-low voltage system.

If the loss of a low voltage supply to an extra-low voltage source occurs, provision shall be made to prevent danger or damage to any other extra-low voltage equipment.

Fault protection *(551.4)*

Fault protection shall be provided for the installation in respect of each source of supply.

Automatic disconnection of supply *(551.4.3)*

Protection by automatic disconnection of the supply shall be provided in accordance with Section 411.

Additional requirements for standby systems *(551.4.3.2)*

A generator used for standby purposes on a TN system must not rely on the public means of earthing. A suitable earth electrode must be provided.

Additional requirements for installations with static converters *(551.4.3.3)*

When the requirements for automatic disconnection cannot be met for parts of the installation on the load side of a static converter, supplementary bonding must be provided on the load side in compliance with Section 415.2.

The resistance of the supplementary bonding conductor must comply with the following condition:

$$R \leq \frac{50}{I_a}$$

Where:

- R = resistance of supplementary bonding conductor
- I = maximum earth fault current which can be supplied by the inverter alone for up to 5 seconds

Where a static inverter operates in parallel with the distributor's network, Regulation 551.7 also applies.

The correct operation of protective devices must not be affected by d.c. generated by a static converter or any filters.

Additional requirements for automatic disconnection where both the installation and generating set are not permanently fixed *(551.4.4)*

- Protective conductors between separate items of equipment complying with Table 54.7 must be provided.
- In TN, TT and IT systems an RCD of 30 mA maximum operating current must be installed to meet the automatic disconnection requirements of Regulation 415.1.

Protection against overcurrent *(551.5)*

Where there is protection for overcurrent of the generating set, the device(s) must be located as near as possible to the generator terminals.

For a generator operating in parallel with the distributor's network or another generator, any harmonic currents must be limited so the thermal rating of conductors is not exceeded (551.5.2).

There are five options for limiting harmonic currents and one or more may be used:

- generators with compensated windings
- a suitable impedance in the connection to the generator star points
- interlocking switches to break the circulatory circuit but which do not impair fault protection
- filtering equipment
- other suitable means.

Additional requirements where the generating set provides a switched alternative to the distributor's network (standby systems) *(551.6)*

Precautions complying with Section 537 of BS 7671 for isolation are required to prevent the generator operating in parallel with the distributor's network. One or more of the following methods may be adopted:

- electrical, mechanical or electromechanical interlocks of the changeover devices
- locking system with single transferable key
- three-position break – before – make changeover switch
- automatic changeover switching device with interlock
- other means of equal security of operation.

For TN-S systems where the neutral is not isolated, any RCD shall be sited so that any incorrect operation due to parallel neutral paths is avoided (551.6.2).

Additional requirements where the generating set may operate in parallel with the distributor's network *(551.7)*

This Regulation requires care to be taken to avoid adverse effects to the distributor's network and other installations with regard to:

- power factor
- voltage changes
- harmonics
- imbalance
- starting
- synchronising
- voltage fluctuation.

Where parallel operation is required and synchronisation is necessary, the use of automatic synchronising systems for frequency, phase and voltage is preferred.

If the distributor's network fails or deviates from declared values of voltage or frequency for the supply, there must be means of disconnecting the generator from that supply. This protection must be in accordance with the distributor's requirements (551.7.4).

Also, there must be means to prevent the connection of a generator to the distributor's network if the declared values of voltage or frequency of the network are outside normal limits or in the event of loss of the supply (551.7.5).

Means of isolation accessible to the distributor at all times (so as far is reasonably practicable) must be provided to enable the generating set to be isolated from the distributor's network in accordance with national rules. When the generator output exceeds 16 A the distributors system operator requirements shall also be met (551.7.6).

Where a generating set is to also provide a standby supply to the distributor's network, the requirements of Regulation 551.6 are applicable.

Requirements for installations incorporating stationary batteries *(551.8)*

When installations include stationary batteries they shall be installed so that they are only accessible to skilled or instructed persons, by being placed in a secure location or, for smaller batteries, a secure enclosure. The location or enclosure must be ventilated.

Battery connections shall be provided with basic protection using suitable insulation or an enclosure, or be arranged in such a way that where any two bare conductive parts which are simultaneously accessible and have a potential difference exceeding 120 V cannot be inadvertently touched.

Rotating machines *(552)*

Equipment and cables feeding motors must be rated to carry the full-load current of the motor. On starting, a heavy current will generate extra heat which will be dissipated in the cables for a short period of time. This will not cause overheating and must be ignored unless the motor is subject to frequent starting and stopping, in which case it may be necessary to install a cable of a larger cross-sectional area.

Electric motors with ratings exceeding 0.37 kW must be supplied from a starter which includes overload protection. This requirement does not apply to a motor incorporated in an item of equipment complying with an appropriate British Standard.

Every motor must be provided with a means to prevent automatic restarting after a stoppage due to a drop in voltage or failure of supply, where unexpected restarting of the motor could cause danger. For example, in a machine shop a failure of supply to a lathe could result in the operator making some adjustment or cleaning the machine which, if the machine restarts unexpectedly, could cause injury to the operator. In order to prevent this happening, 'no volt' protection must be provided. This is usually achieved by using a starter of the type illustrated.

The coil is fed through hold-in contacts, which open when the **stop** button is operated or the overload trip contacts open, or the voltage to the coil drops due to a loss or fluctuation of supply voltage. The motor can only be restarted by operating the **start** button.

Note: *The above requirement does not apply to motors where automatic starting is necessary to avoid danger, e.g. pumping systems, refrigeration and air conditioning units.*

When safety depends on the correct rotation of a motor, provision shall be made to prevent reverse operation if a phase is lost. In installations where reverse current braking of a motor is provided, provision shall be made to avoid reverse rotation at the end of a braking sequence.

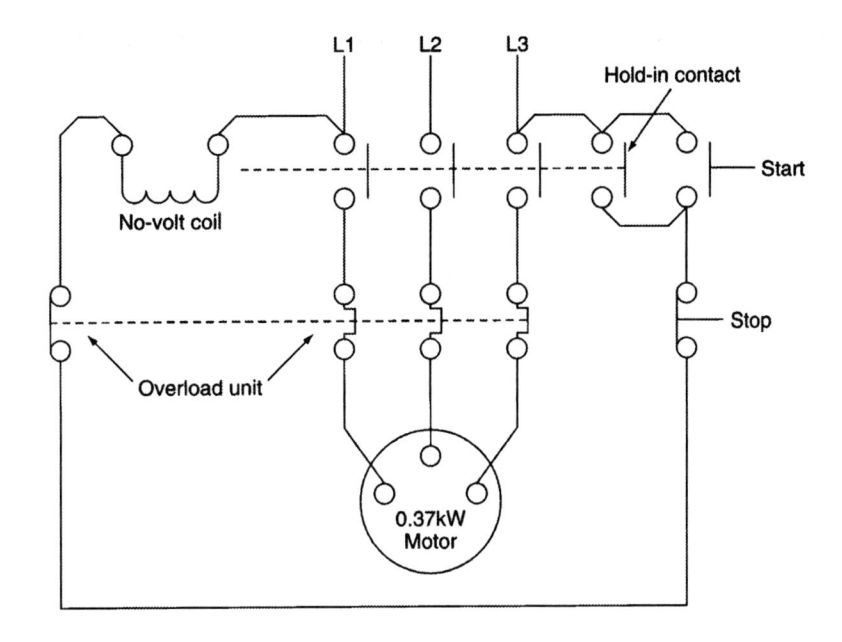

Accessories *(553)*

Plugs and socket-outlets *(553.1)*

The plugs and socket-outlets which are recognised as being suitable for low voltage circuits are listed in Table 55.1 of BS 7671 and are shown below.

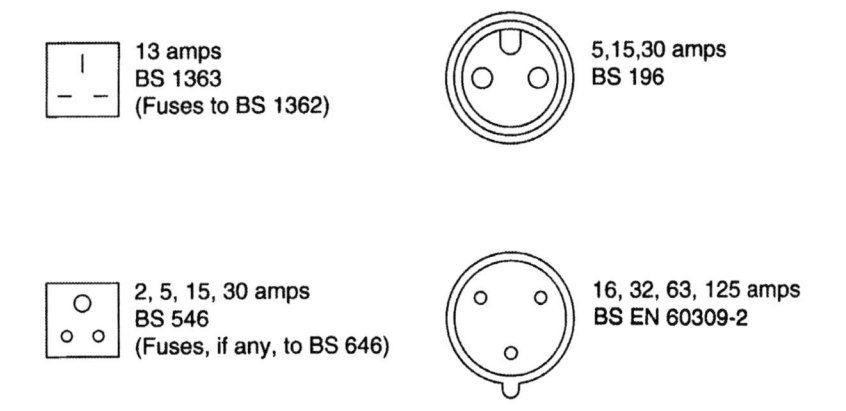

13 amps
BS 1363
(Fuses to BS 1362)

5, 15, 30 amps
BS 196

2, 5, 15, 30 amps
BS 546
(Fuses, if any, to BS 646)

16, 32, 63, 125 amps
BS EN 60309-2

These plugs and socket-outlets are designed so that it is not possible to engage any pin of the plug into a live contact of a socket-outlet whilst any other pin of the plug is exposed (not a requirement for SELV circuits) and the plugs are not capable of being inserted into sockets of systems other than their own.

With the exception of SELV or special circuits having characteristics where danger may arise, all socket-outlets must be of the non-reversible type, with a point for the connection of a protective conductor.

Plugs and socket-outlets other than those shown above may be used on single-phase a.c. or two wire d.c. circuits operating at voltages not exceeding 250 V for the connection of:

Electric clocks - Clock connector unit incorporating a fuse BS 646 or 1362 not exceeding 3 A.

Electric shaver - Shaver supply unit to BS EN 61558-2-5 for use in bath or shower rooms. In other locations a socket to BS 4573 can be used.

On construction sites (but not necessarily in site offices, toilets, etc.) only plugs, socket-outlets and couplers to BS EN 60309-2 must be used.

Socket-outlets must be fixed at a height above floor level or working surface so that the socket-outlet, its associated plug and its flexible cord are not subjected to mechanical damage during insertion, use or withdrawal of the plug.

Where portable equipment is likely to be used, an adequate number of socket-outlets must be provided so that the equipment can be supplied from an adjacent and accessible socket-outlet.

Cable couplers *(553.2)*

Cable couplers may be used in conjunction with the following types of plug and sockets (but these are not to be used on SELV circuits or Class II circuits):

- BS 196
- BS 61535
- BS 6991
- BS EN 60309-2
- BS EN 60320.

Cable couplers must be connected at the end of the cable remote from the supply. They must be non-reversible and have provision for the connection of the protective conductor.

Current-using equipment *(554)*

Electrode water heater and boilers *(554.1)*

Electrode boilers and water heaters must be connected to a.c. systems only and be controlled by a multipole linked circuit breaker, which when operated disconnects the supply from all electrodes simultaneously. Overcurrent protective devices must be installed in each conductor feeding an electrode.

The shell of any electrode boiler or heater must be bonded to the metallic sheath or armour of the incoming supply cable with a protective conductor complying with Regulation 543.1.1.

Electrode boilers or water heaters supplied directly from the supply at voltages in excess of low voltage must be protected by an RCD which will disconnect the supply from the electrodes in the event of a sustained protective conductor current of more than 10% of the rated current of the boiler or heater. If this arrangement does not allow stable operation then the setting of the RCD for tripping may be increased to a maximum of 15%, or a time delay may be included to prevent tripping due to imbalance of short duration.

For electrode boilers and water heaters connected to a three-phase low voltage supply, the shell of the boiler or heater must be connected to the neutral conductor of the supply and also to the earthing conductor. The cross-sectional area of the neutral conductor must not be less than that of the largest line conductor connected to the equipment (see below).

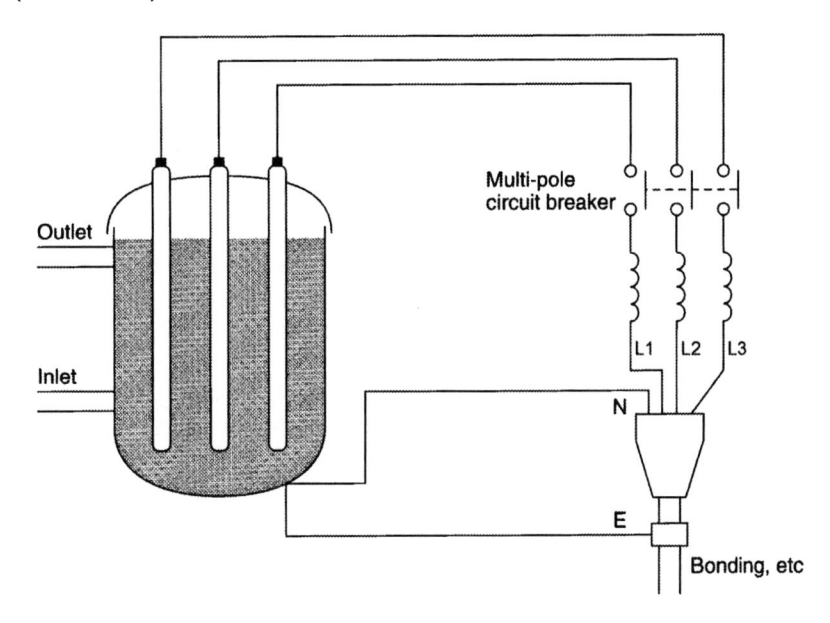

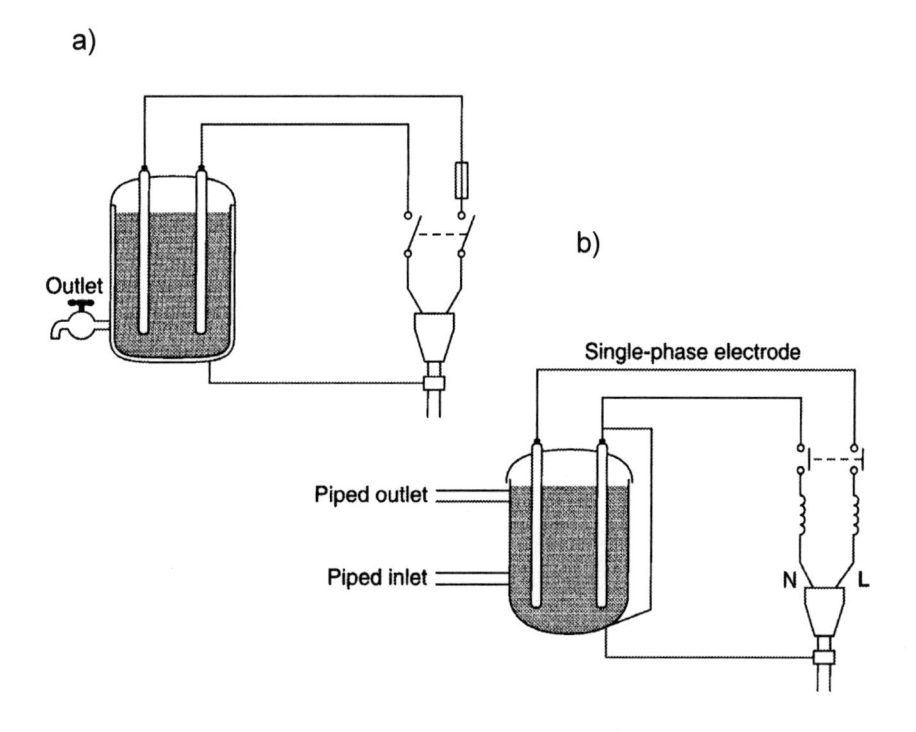

a)

b)

For single-phase electrode boilers and heaters (which have one electrode connected to the neutral conductor earthed by the distributor), the shell of the boiler or heater must be connected to the neutral of supply as well as the earthing conductor (see previous page).

Where the electrode boiler or heater is not piped to a water supply or in physical contact with any earthed metal, and the electrodes and water in contact with the electrodes are shielded in insulating material so that it is impossible to touch either the electrodes or the water whilst the electrodes are live, a fuse in the line conductor may be substituted for the circuit breaker required by Regulation 554.1.2 and the shell of the boiler need not be connected to the neutral of the supply.

Every type of heater for liquids or other substances must incorporate or be provided with an automatic device (such as thermostats or thermal cut-outs) to prevent a dangerous rise in temperature.

Water heaters with immersed and uninsulated heating elements *(554.3)*

All the metal parts of heaters or boilers that are in contact with water but do not carry current must be solidly and metallically connected to a metal water pipe through which the water supply to the boiler is provided. The water pipe must also be effectively bonded to the main earth terminal by means independent of the circuit protective conductor.

The supply to such heaters or boilers must be run through a double-pole linked switch, incorporated in or separate from (but within easy reach of) the heater. The wiring from the heater or boiler being connected directly to the switch without the use of any plug or socket-outlet. If the heater or boiler is installed in a room with a fixed bath, the switch must comply with Section 701.

The installer must confirm that no single-pole protective devices are connected in the neutral conductor in any part of the circuit between the heater or boiler and the origin of the supply.

Heating conductors and cables *(554.4)*

Soil, floor and ceiling heating systems are used in a number of installations, such as:

- installations under sports grounds to enable sports to be played even when frost and snow exists

- installations in roads and ramps of car parks to prevent icing and dangerous road conditions in winter

- installations under the soil in horticulture installations to enable plants and vegetables to establish early and healthy root systems

- installations in the floors or ceilings of buildings.

Where heating cables pass through or are in close proximity to materials which may present a fire hazard, the cables must be installed in fire-resistant material with ignitability characteristic 'P' (BS 476 Part 12) and be protected from mechanical damage (554.4.1).

Heating cables intended for laying directly in soil, concrete, cement screeds or other materials used for the construction of roads and buildings should be:

- capable of withstanding any mechanical damage that is likely to occur during installation
- constructed of materials that will be resistant to the effects of damp and corrosion under normal service condition.

Heating cables installed in soil or roads or the structure of buildings must be:

- completely embedded in the material they are to heat
- installed so as to prevent damage in the event of normal movements expected in the cables or in the substances in which they are embedded
- installed in compliance with the manufacturer's instructions.

Electric surface heating systems *(554.5)*

- Equipment for surface heating systems (ESH) must comply with BS 6351.
- Installation and testing of the system must be carried out as provided for in BS 6351.

Transformers *(555)*

Autotransformers and step-up transformers *(555.1)*

- Where an autotransformer is connected to a circuit having a neutral conductor, the common terminal of the winding must be connected to the neutral.
- A step-up transformer must not be used on IT systems.
- A step-up transformer must be provided with a linked switch to disconnect it from all the live conductors of the supply.

Luminaires and lighting installations *(559)*

This section applies to luminaires and fixed lighting installations, and highway power supplies and street furniture. Requirements are given for:

- fixed outdoor lighting
- extra-low voltage lighting supplies from a source not exceeding 50 V a.c. or 120 V d.c.
- display stand lighting.

The following are not covered:

- high voltage signs supplied at low voltage, e.g. neon signs
- signs and luminous discharge tubes operating at no-load output voltages from 1 kV TO 10 kV.

Outdoor installations *(559.3)*

The following are classed as outdoor lighting installation locations or places:

- **Locations:**
    roads
    parks
    car parks
    gardens
    places open to the public
    sporting areas
    illumination and floodlighting of monuments.

- **Places:**
    telephone kiosks
    bus shelters
    advertising panels
    town plans
    road signs
    road traffic and signal systems.

Temporary festoon lighting is excluded, as is distributors' equipment.

General requirements for installations *(559.4)*

All luminaires shall comply with the relevant standard of manufacture and shall be installed in accordance with manufacturer's instructions.

When luminaires are installed in a pelmet, care must be taken to ensure that curtains, blinds and their operation are not adversely affected, especially regarding proximity and fire risk.

If track systems are used, these shall comply with BS EN 60570.

When luminaires without transformers or converters are fitted with extra-low voltage lamps, which are connected in series, these shall be treated as low voltage equipment for the purposes of this section of the Regulations.

For details of symbols used to identify the different types of luminaires and control gear, reference should be made to Table 55.2 of BS 7671.

Protection against fire *(559.5)*

When selecting and installing luminaires, the thermal effects of radiant and convected energy from the luminaires shall be considered, especially the maximum permissible power dissipated by the lamps, the fire resistance of adjacent materials and the minimum distance allowed to combustible material, including any materials in the path of a spotlight beam.

Wiring systems *(559.6)*

Connections to fixed wiring

The wiring of lighting circuits shall be terminated using one of the accessories in the table below.

Accessory	Standard
Ceiling rose	BS 67
Luminaire supporting coupler	BS 6972 or BS 7001
Batten lampholder	BS EN 60598
Luminaire	BS EN 60598
Suitable socket-outlet	BS 1363-2, BS 546 or BS EN 60309-2
Plug-in lighting distribution unit	BS 5733
Connection unit	BS 1363-4
Box	BS EN 60670 series or BS 4662
Device for connecting a luminaire (DCL) outlet	BS EN 61995-1

Ceiling roses or lampholders used for filament lamps shall not be used on circuits exceeding 250 V. A ceiling rose shall only be used with one flexible cord unless specifically designed for more.

Fixing luminaires *(559.6.1.4)*

All luminaires shall be adequately fixed using screws, hooks or other suitable means. When pendant luminaires are to be installed the fixings shall be capable of supporting a mass of not less than 5 kg. All fixing of luminaires shall be in accordance with manufacturer's instructions. Any ceilings used to fix luminaires shall be capable of supporting them.

Cables and flexible cords used to connect luminaires shall be installed so they place no stress on them or their terminations.

Lighting circuits using lampholders of the following types shall be protected using an overcurrent device not greater than 16 A.

Small bayonet cap	B15
Bayonet cap	B22
Small Edison screw	E14
Edison screw	E27
Giant Edison screw	E40

Bayonet lampholders B15 and B22 shall have temperature rating T2 and comply with BS EN 61184.

For circuits connected to TN and TT systems the outer contact of Edison screws or single centre bayonet cap lampholders shall always be connected to the neutral conductor, unless they are Type E14 or E27, complying with BS EN 60238.

Through wiring *(559.6.2)*

Through circuit wiring in luminaires is permitted if the luminaire is designed for this purpose. When through wiring is allowed, the cable used shall be capable of withstanding the temperature within the luminaire, as specified in the manufacturer's instructions. For luminaires complying with BS EN 60598 but with no temperature markings heat-resistant cables are not required, unless specified in the manufacturer's instructions. In cases of no information, heat-resistant cables and/or conductors of types HO5S-U, HO5S-K, HO5SJ-K or HO5SS-K (BS 6007), or other suitable equivalent, shall be used. Lighting installations shall be controlled by switches complying with BS 3676/BS EN 60669-1 and/or BS EN 60669-2-1, or a suitable automatic control system.

Independent lamp control gear *(559.7)*

Only control gear which is marked as suitable and to the relevant standard may be used external to a luminaire.

Stroboscopic effect *(559.9)*

When lighting installations involve premises where machines have moving parts, which may result in stroboscopic effect giving the impression that the moving machinery is stationary, luminaires with lamp and control gear designed to avoid this effect shall be selected. Alternatively, lighting loads may be distributed across the phases of a three-phase supply.

Outdoor lighting, highway power supplies and street furniture *(559.10)*

Live parts of electrical equipment shall be protected by insulation, barriers or enclosures, providing basic protection.

A door used for access to electrical equipment in street furniture, etc. must not be used as a barrier or enclosure for compliance with the Regulation.

A door that is less than 2.5 m above ground level must be locked with a key or secured with a tool. Also there must be an intermediate barrier for basic protection when the door is open. The barrier must achieve protection to IP2X or IPXXB.

Luminaires installed less than 2.8 m above ground level must permit access to the light source only by removing a barrier or enclosure with the use of a tool.

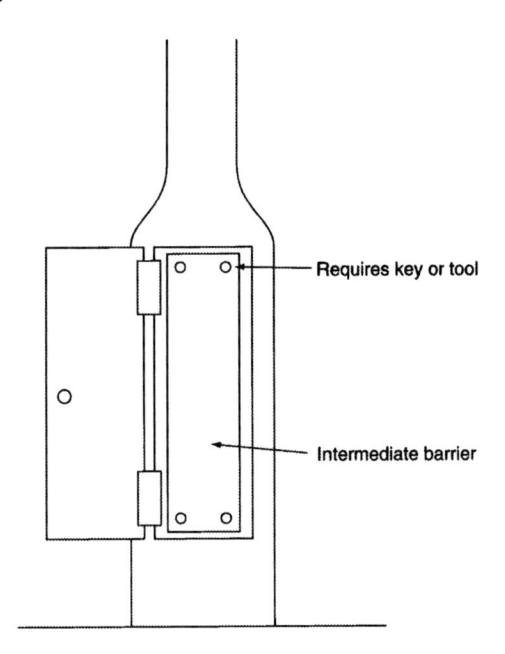

Any metal structures (e.g. a fence or pedestrian barrier), which are not part of or not connected to street furniture or street located equipment, need not be connected to the main earthing terminal.

For lighting arrangements in places such as bus shelters and telephone kiosks, it is recommended that additional protection using 30 mA RCDs is provided.

Double or reinforced insulation *(559.10.4)*

There is no need to provide a protective conductor or to connect conductive parts of lighting columns to the earthing system.

Fire risk due to short circuit *(559.11.4)*

When a device providing protection against fire risk is used it shall continuously monitor the power demand of the luminaires. It shall automatically disconnect the supply circuit within 0.3 s in the case of a short circuit or failure causing a power increase greater than 60 W.

Wiring systems *(559.11.5)*

No metal structural parts of buildings, for example parts of street furniture or pipe systems, shall be used as a live conductor.

Minimum cross-sectional area of conductors for extra-low voltage are:

- 1.5 mm² copper but for flexible cables with a maximum length of 3 m, 1 mm² copper may be used
- 4 mm² copper for flexible suspended cables
- 4 mm² copper for composite cables.

Suspended systems *(559.11.6)*

When suspension devices are used for extra-low voltage luminaires, they shall be fixed by insulated distance cleats and be accessible throughout the route. They shall be capable of carrying five times the mass of the luminaires, which include the lamps and cables, but must be capable of supporting at least 5 kg.

Connection of conductors shall be by screw terminals or screwless clamps to the relevant BS EN standards.

Table 55.2 of BS 7671 provides a table of symbols used in luminaires, control gear and their installation.

Safety services *(Chapter 56)*

This chapter gives the general requirements for safety services and the electrical supply systems. Typical examples of such systems are:

- emergency lighting
- fire detection and alarm
- CO-detection
- fire pumps
- fire rescue service lifts
- fire evacuation systems
- smoke ventilation
- fire service communication
- essential medical
- industrial safety.

Classification *(560.4)*

Types of supply:

- operator initiated non-automatic
- independent automatic supply. These types of supply are classified by the maximum time to change over from the original power source:

- no break. An automatic supply ensuring continuous supply within specified conditions
- very short break. Supply available within 5 s
- short break. Supply available between 0.15 s and 0.5 s
- lighting break. Supply available between 0.5 s and 5 s
- medium break. Supply available between 5 s and 15 s
- long break. Supply available after 15 s.

Reference shall be made to relevant British Standards of the specific safety system for the rated operating details of the safety source. When a British Standard is not available the requirements shall be determined by a risk assessment.

General *(560.5)*

When a safety service is required to be operational during a fire, the source of supply shall provide fire resistance and maintain the supply for an adequate duration.

Electrical sources for safety services *(560.6)*

Electrical supply sources for safety services are to be installed as fixed equipment, which is not affected by failure of the normal supply source. They shall be installed in locations accessible to only skilled persons (BA5) or instructed persons (BA4) and be adequately ventilated.

Central power supply sources *(560.6.9)*

Batteries shall be vented or valve-regulated types with a minimum life of 10 years.

Low power supply sources *(560.6.10)*

The power output of low power supply systems is limited to:

- 500 w for 3 hours
- 1,500 w for 1 hour.

The minimum battery life in accordance with BS EN 50171 is five years.

Circuits *(560.7)*

Safety services circuits shall be independent of any other circuits and shall not pass through any explosive risk zones (BE3) or fire risk zones (BE2), unless for the fire risk zone they are risk-resistant and this is the only route available.

Switchgear and control gear shall be identified clearly and grouped in locations only accessible to skilled or instructed persons. Cables for safety circuits shall be adequately separated from cables of other circuits (including cables of other safety circuits) by distance or barriers, unless metallic screened fire-resistant cables are used.

Full details and drawings of safety source installations shall be displayed at the origin of the installation. Details of all current-using equipment and operating instructions shall be made available.

Wiring systems *(560.8)*

The following types of cables can be used for safety services, required to operate in fire conditions:

- fire-resistant cables complying with:

    BS EN 50362

    BS EN 50200

- the British Standard for the fire alarm systems 5839-1 specifies cables to:

    BS EN 60702-1

    BS 7629

    BS 7846

TECHNICAL DATA SHEET 5A

FIRST NUMERAL		SECOND NUMERAL	
(a) PROTECTION OF PERSONS AGAINST CONTACT WITH LIVE OR MOVING PARTS INSIDE ENCLOSURE (b) PROTECTION OF EQUIPMENT AGAINST INGRESS OF SOLID BODIES		PROTECTION OF EQUIPMENT AGAINST INGRESS OF LIQUID	
No./SYMBOL	DEGREE OF PROTECTION	No./SYMBOL	DEGREE OF PROTECTION
0	(a) No protection. (b) No protection.	0	No protection.
1	(a) Protection against accidental or inadvertent contact by a large surface of the body, e.g. hand, but not against deliberate access. (b) Protection against ingress of large solid objects <50mm diameter.	1	Protection against drops of water. Drops of water falling on enclosure shall have no harmful effect.
2	(a) Protection against contact by standard finger. (b) Protection against ingress of medium size bodies <12mm diameter <80mm length.	2	Drip Proof:- Protection against drops of liquid. Drops of falling liquid shall have no harmful effect when the enclosure is tilted at any angle up to 15° from the vertical.
3	(a) Protection against contact by tools, wires or suchlike more than 2.5mm thick. (b) Protection against ingress of small solid bodies.	3	Rain Proof:- Water falling as rain at any angle up to 60° from vertical shall have no harmful effect.
4	(a) As 3 above but against contact by tools, wires, or the like, more than 1.0mm thick. (b) Protection against ingress of small foreign bodies.	4	Splash Proof:- Liquid splashed from any direction shall have no harmful effect.
5	(a) Complete protection against contact. (b) DUSTPROOF:- Protection against harmful deposits of dust, dust may enter but not in amount sufficient to interfere with satisfactory operation.	5	Jet Proof:- Water projected by a nozzle from any direction (under stated conditions) shall have no harmful effect.
6	(a) Complete protection against contact. (b) DUST-TIGHT Protection against ingress of dust.	6	Watertight Equipment:- Protection against conditions on ship's decks, etc. Water from heavy seas or power jets shall not enter the enclosures under prescribed conditions.
IP CODE NOTES -Degree of protection is stated in form IPXX. -Protection against contact or ingress of water respectively is specified by replacing first or second X by digit number tabled e.g. IP2X defines an enclosure giving protection against finger contact but without any specific protection against ingress of water or liquid.		7	Protection Against Immersion in Water:- It shall not be possible for water to enter the enclosure under stated conditions of pressure and time.
		8	Protection Against Indefinite Immersion in Water Under Specified Pressure:- It shall not be possible for water to enter the enclosure.

IPXXB:- Means a finger could be inserted safely to a depth of 80mm.
IPXXC:- Tools or wire 2.5mm diameter can be inserted safely for 100mm.
IPXXD:- 1mm diameter wire can be inserted 100mm safely.

Index of protection (IP) code for general guidance

TECHNICAL DATA SHEET   5B

Types and methods of installing earth electrodes

Note: The information in this section has been obtained from manufacturers and the Code of Practice for Earthing, BS 7430: 1998.

Earth plates

These are used where high fault currents are possible and normal soil conditions exist.

Typical manufacturer's type and sizes are (imperial measurements supplied by manufacturers):

- Solid copper   2' x 2' or 3' x 3'
  $1/16$" or $1/8$" thick

- Lattice   2' x 2' or 3' x 3'
  $1/8$" thick

Solid copper earthplate

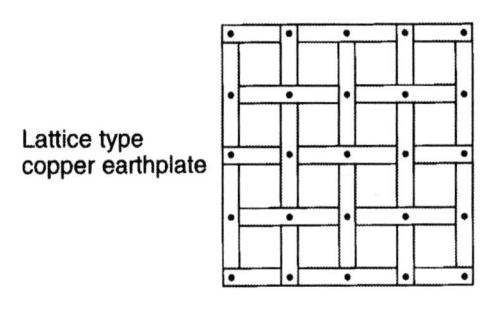

Lattice type copper earthplate

Method of installing earth plates

Earth plates are buried vertically and must be covered with a minimum of 600 mm of soil to reduce the effects of climatic conditions and extensive voltage gradients appearing on the ground surface under fault conditions. Care must be taken to ensure the electrode connection is protected against corrosion and mechanical damage.

Method of terminating at earth plates

The following methods are used to connect earthing conductors to electrodes:

- brazing
- tape connectors
- mechanical clamping devices
- aluminothermic welding or explosive welding techniques.

Protection of terminations

Earth electrode connections must be protected at the point of termination by the use of grease, bitumastic paint or bitumastic compounds.

Earth electrodes for areas of rocky soil structure

In areas where there is rock at or near to the surface of the soil, copper tape, stranded conductors or wire mesh electrodes may be used.

These types of electrodes must be buried to a depth suitable to minimise the risk of their becoming damaged and to protect them from climatic conditions such as frost. A suitable depth may be 457 mm.

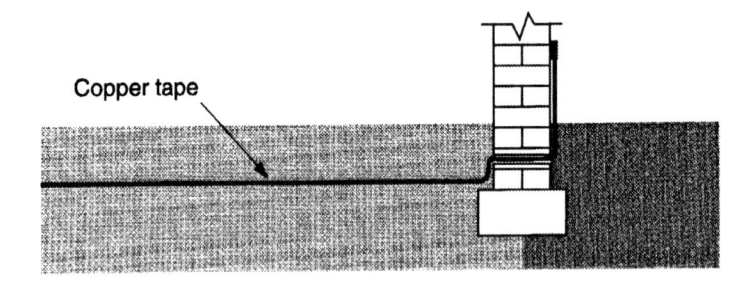

Methods of installing copper tapes or stranded conductors

Copper tapes and stranded conductors can be arranged in single lengths, or as parallel or radial groups. The usual size of copper tape is 25 mm wide by 3 mm thick.

Earth rods

Earth rods are used where low earth fault currents exist and the soil resistivity is high.

Types of earth rods

The following types of earth rod are available:

- **Copper earth rods**
  Non-extensible 0.61 m and 1.22 m long with ribbed construction to provide maximum surface contact with soil. Extensible 1.22 m long, 15 mm diameter.

- **'Biclad' earth rods**
  Extensible 1.22 m long, 14.3 mm diameter made from copper-sheathed mild steel bar.

- **'Bimetal' earth rods**
  Non-extensible 1.22 m long, 9 mm diameter complete with integral clamp for both strip and stranded conductors. Extensible 1.22 m. 1.83 m and 2.44 m rods, 15.9 mm diameter. 'Bimetal' rods consist of a corrosion-resistant thick copper exterior, permanently molten-welded to a high tensile steel core.

- **Stainless steel and galvanised rods**

Method of installing earth rods

Earth rods are driven completely below ground using a hand-held hammer or power hammer. They are set out at a distance of **not less** than their own length apart.

Earth electrode resistance area

Every earth electrode has a definite electrical resistance to earth. Current flowing from the electrode to the general mass of earth has to traverse the concentric layers of soil immediately surrounding the electrode. Since the soil is a relatively poor conductor of electricity and as the cross-sectional areas of the layers of soil nearest to the electrode are small, the result is that of a graded resistance concentrated mainly in the area of soil surrounding the electrode. Moreover, the surface of the soil near the electrode will become 'live' under fault conditions.

Surface voltage gradients

The illustration shows a typical surface voltage distribution near an earth electrode. The cow standing on the ground near the 'live' electrode may receive a considerable voltage between its fore and hind feet resulting in a dangerous and possibly lethal shock since voltages of around 25 V are dangerous to livestock.

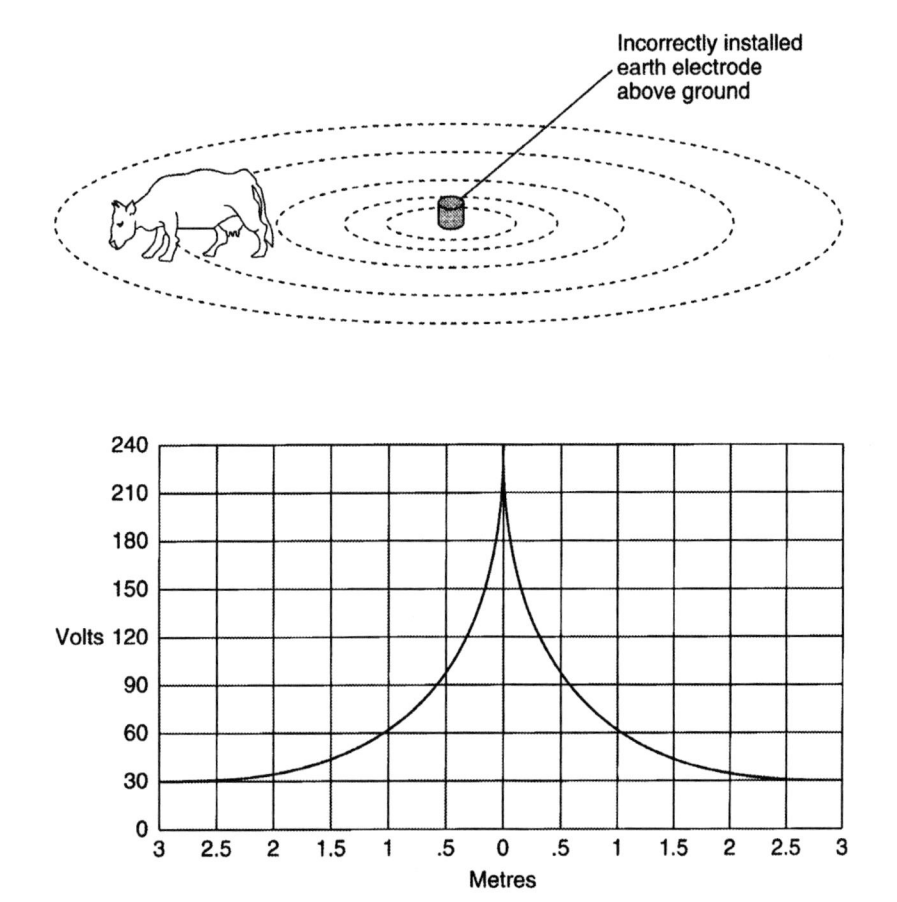

Surface voltage gradients – danger to livestock

Termination to earth electrodes

The connection of earthing conductors to electrodes requires adequate insulation where they enter the ground, to avoid possible dangerous voltage gradients at the surface. All electrode connections must be thoroughly protected against corrosion and mechanical failure.

It is important that the electrode is made accessible for inspection purposes, and a label must be fitted at or near the point of connection.

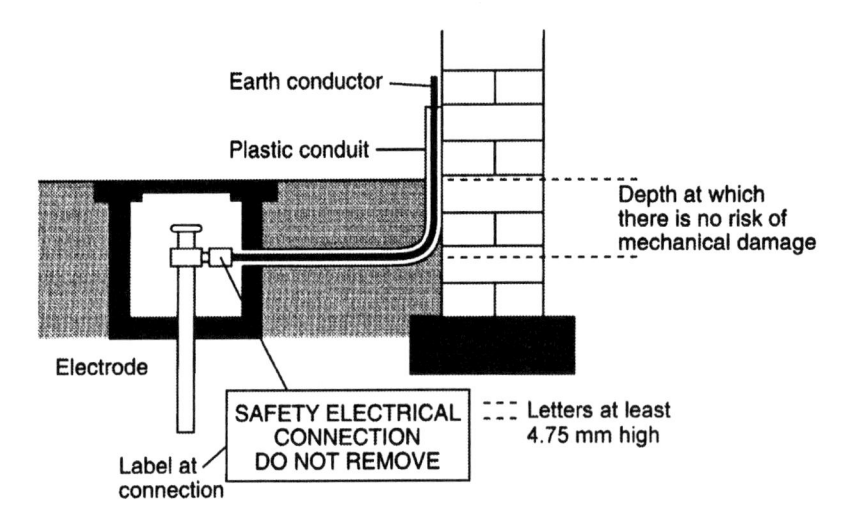

Inspection covers

Constructed of concrete with a galvanised steel lifting handle, the underside is recessed to enclose and protect the electrode connection.

Standard pathway covers with cast-iron frames with a concrete filling may also be used.

Glazed earthenware and fibreglass covers are also available.

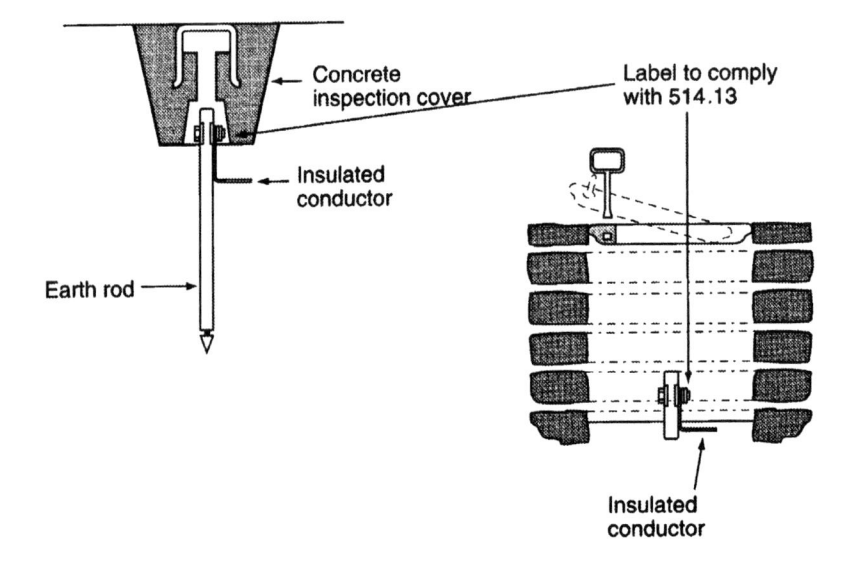

TECHNICAL DATA SHEET 5C

Earth monitoring equipment

Basic earth monitoring units

The primary objective of a basic earth monitoring unit is to guard against the failure of the circuit protective conductor in a flexible or trailing cable. See BS 4444 1995.

Basic circuit

This is as illustrated below and consists of the following:

- portable tool
- protective circuit breaker
- transformer
- relay.

The portable tool is supplied by a four-core flexible cable consisting of the:

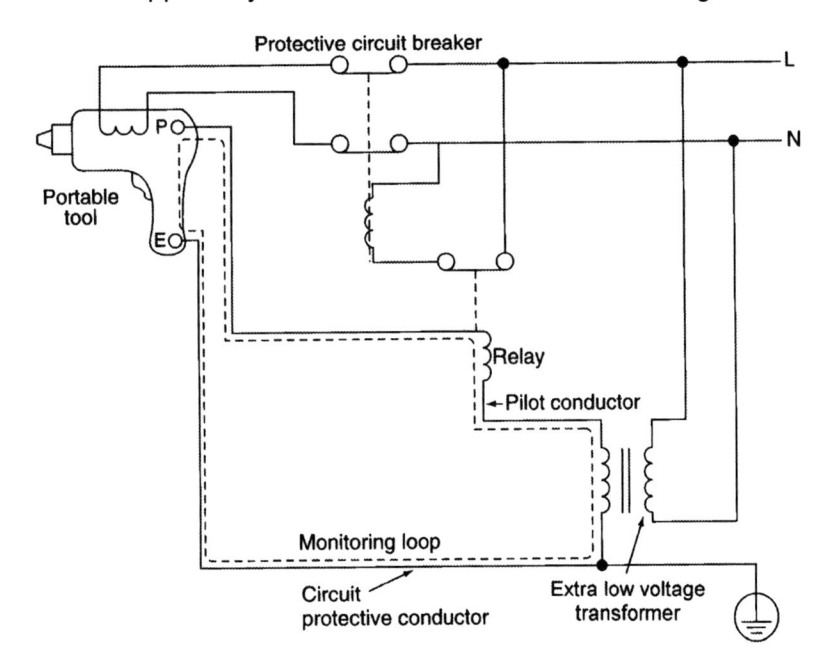

- primary circuit protective conductor which is bonded in the normal manner
- pilot conductor
- line conductor
- neutral conductor.

The transformer is used to provide an extra-low voltage supply of 12 V and a current of not more than 3 A around the loop circuit, consisting of:

- the circuit protective conductor
- a section of the metallic housing of the portable tool
- the pilot conductor
- the relay.

Operation of the relay and circuit breaker

The circulating current holds in the relay and so energises the hold-on coil of the circuit breaker for as long as the earth monitoring loop remains intact.

Use of earthed screened cables

Cable incorporating an earthed screening must be used in situations where vehicles or other equipment could possibly damage the cable. The line conductor will become short circuited to the earthed screening in the event of the cable being cut. This fault must then be cleared by an independent overcurrent device.

TECHNICAL DATA SHEET 5D

Protective multiple earthing (PME)

With this system the transformer neutral conductor is earthed and, in addition, the neutral is earthed at selected points in the distribution system. This has the effect of changing a phase-to-earth fault to a phase-to-neutral fault. When a PME installation is functioning correctly the degree of protection afforded is the same as that provided on a TN-S system.

Methods of providing TN-C-S systems (PME supplies)

The following diagrams illustrate the methods used by local distributors on PME distribution systems to provide supplies to consumers.

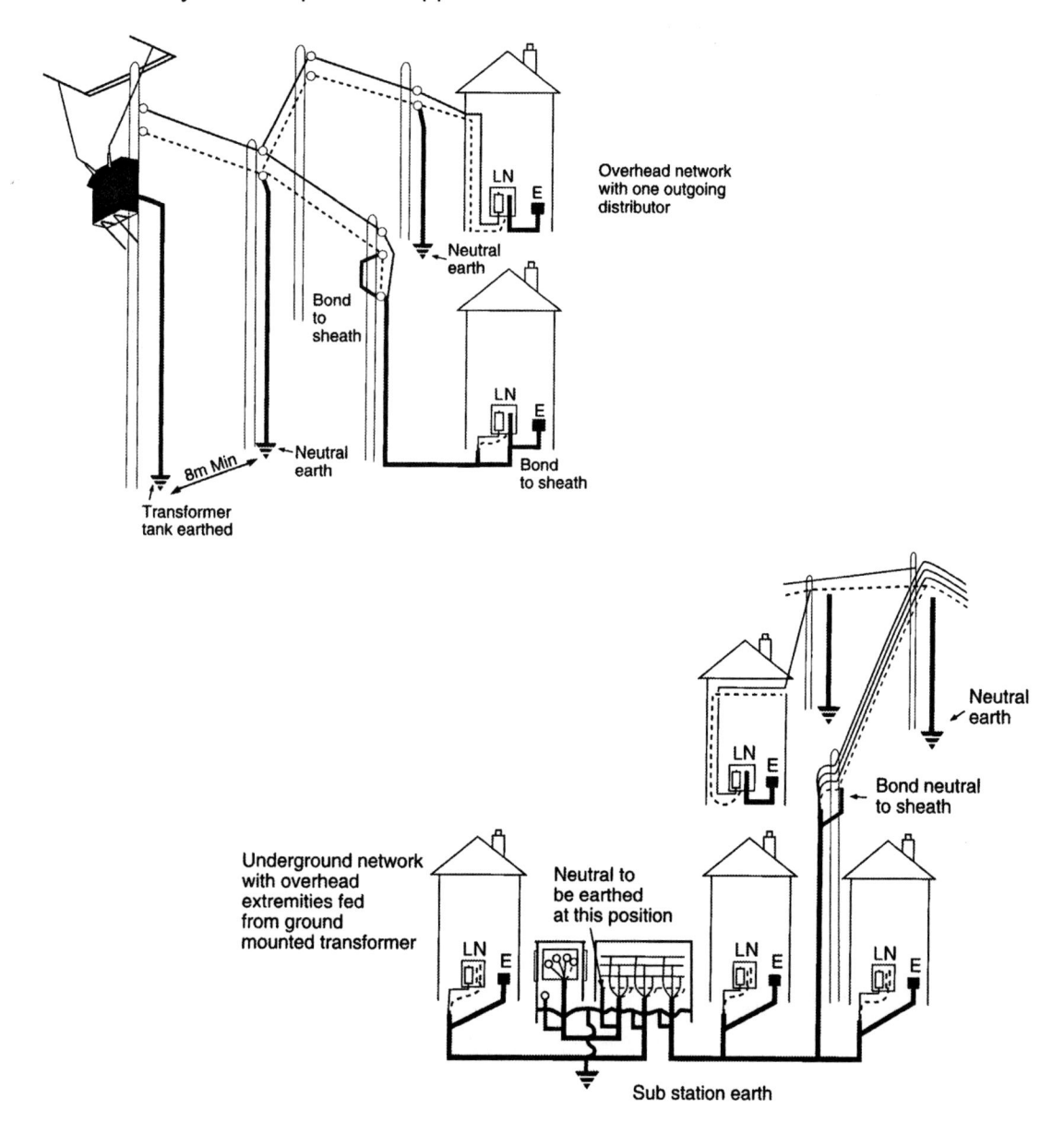

PNB supply system

This is used where only one consumer is fed from a supply transformer on the network.

On construction sites and some agricultural installations, it may not be possible to comply with the bonding requirements for PME approval. In these circumstances, where a transformer supplies only one consumer, the local distributor will earth the neutral at one point only (at the service point) to provide an earthing terminal. This is known as protective neutral bonding (PNB).

The method used to provide a PNB supply to the consumer is illustrated below.

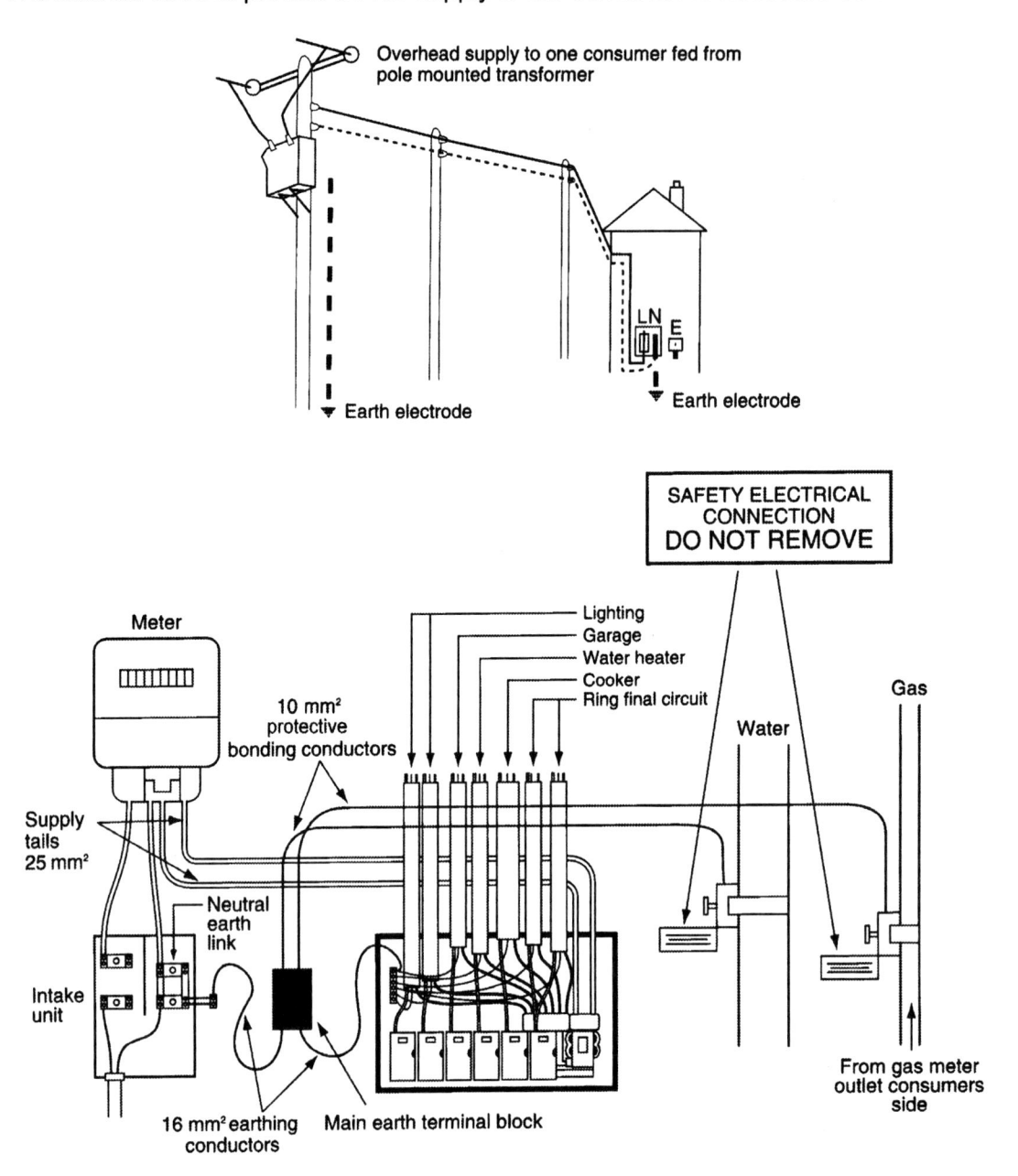

Typical TN-C-S installation

INSPECTION AND TESTING

BS 7671 Part 6

Initial verification *(Chapter 61)*

General *(610)*

During installation and on completion, every installation must be inspected and tested before being connected to the supply to establish that the requirements of the Regulations have been met. Precautions must be taken to avoid dangers to persons and damage to property and equipment during inspection and testing.

The following information must be made available to the persons carrying out the inspection and testing of an installation (Section 311, 312 and 313, Section 131 and the information required by Regulation 514.9.1).

- Assessment of general characteristics:
    - maximum demand (after diversity)
    - the number and type of live conductors of the supply (and of the circuits of the installation), and the earthing arrangements of the installation
    - nominal voltage(s), nature of the load current and frequency*
    - PFC value at the origin of the installation*
    - value of Z_e*
    - suitability for the requirements of the installation (including maximum demand)
    - type and rating of the protective device at the origin of the installation*.

* Regulation 28 of the Electricity Safety, Quality and Continuity Regulations 2002 requires distribution network operators to provide this information to relevant persons free of charge.

- Diagrams, charts or tables indicating:
    - the type of circuits
    - the number of points installed
    - the number and size of conductor
    - the type of wiring system.
- Details of the characteristics of the protection devices for automatic disconnection.
- The location and types of devices used for:
    - protection

- isolation and switching.
- Details of circuits or equipment sensitive or vulnerable to tests – e.g. equipment with voltage-sensitive electronics, such as central heating controls with electronic timers and displays, also intruder alarm equipment, certain types of RCD, etc.

Note: Information may be given in a schedule for simple installations. See table below for a domestic installation. A durable copy of the schedule relating to distribution board must be provided inside or adjacent to the distribution board.

Schedule for modified consumer unit arrangement

CCT No.	Type of circuit	Points served	Line conductor mm²	Protective conductor mm²	Protective devices	Type of wiring
1	Fire alarm	5	1.5	1	6 A MCB	PVC/PVC
2	D/S lights	7	1.5	1	6 A MCB	PVC/PVC
3	U/S sockets	8	2.5	1	32 A MCB	PVC/PVC
4	Cooker	1	10	4	40 A MCB	PVC/PVC
5	Spare					
6	U/S lights	6	1.5	1	6 A MCB	PVC/PVC
7	Bathroom light	1	1.5	1	6 A MCB	PVC/PVC
8	D/S sockets	9	2.5	1	32 A MCB	PVC/PVC
9	Kitchen sockets	8	2.5	1	32 A MCB	PVC/PVC
10	Shower	1	10	4	40 A MCB	PVC/PVC

See Technical Data Sheet 4B for the accompanying consumer unit configuration (Figure 2).

Inspection *(611)*

A detailed inspection must be made of installed electrical equipment, usually with the part of the installation being inspected disconnected from the supply. A Schedule of Inspection must be completed and signed and appended to the Electrical Installation Certificate or Periodic Inspection Report (along with a Schedule of Test Results) as appropriate.

The purpose of the inspection is to verify that the electrical installation:

- complies with the British Standards or harmonised European Standards (this may be ascertained by mark or by certificate from the installer or manufacturer)
- is correctly selected and erected in accordance with these Regulations
- is not visibly damaged so as to impair safety.

The inspection must include the following where relevant:

- connections of conductors
- identification of conductors

- routing of cables in safe zones or mechanical protection methods meeting the requirements of Section 522, in relation to external influences
- selection of conductors for current-carrying capacity and voltage drop
- connection of single-pole devices for protection or switching in line conductors only
- correct connection of accessories and equipment
- presence of fire barriers, suitable seals, and protection against thermal effects
- methods of protection against electric shock

a) protection against both basic and fault protection, i.e.:
   - SELV
   - PELV
   - double insulation
   - reinforced insulation

b) basic protection:
   - protection by insulation of live parts
   - protection by barriers or enclosures
   - protection by obstacles
   - protection by placing out of reach

c) fault protection:
   - automatic disconnection of supply
   - presence of earthing conductors
   - presence of circuit protective conductors
   - presence of protective bonding conductors
   - presence of supplementary bonding conductors
   - earthing arrangements for combined protective and functional purposes
   - presence of adequate arrangements for alternative sources
   - FELV
   - choice and setting of protective and monitoring devices for fault and/or overcurrent protection
   - non-conducting location (measurement of distances, where appropriate)
   - absence of protective conductors
   - presence of earth-free local equipotential bonding conductors
   - electrical separation
   - additional protection

- prevention of mutual detrimental influence
- presence of appropriate devices for isolation and switching

- presence of undervoltage protective devices
- labelling of circuits, fuses, switches and terminals
- selection of equipment and protective measures appropriate to external influences
- adequacy of access to switchgear and equipment
- presence of danger notices and other warning notices
- presence of diagrams, instructions and similar information
- erection methods.

Note: *During any re-inspection of an installation, all relevant items in the checklist must be checked.*

Testing *(612)*

For initial verification, the following items (where relevant) must be tested in the following sequence and the results compared with all relevant criteria.

Before the supply is connected (or with the supply safely isolated), test:

- continuity of protective conductors including main and supplementary bonding
- continuity of ring final circuit conductors
- insulation resistance
- protection by SELV, PELV or electrical separation
- insulation, resistance/impedance of floors and walls
- polarity
- earth electrode resistance.

With the electrical supply reconnected, carry out the following live tests:

- confirmation of polarity
- protection by automatic disconnection of the supply
- earth fault loop impedance
- additional protection
- prospective fault current
- check of phase sequence
- functional testing
- verification of voltage drop.

Suitable reference methods of testing are described in IEE Guidance Note No. 3 (Inspection and Testing). The use of other methods of testing is not precluded provided that they will give results which are no less effective.

The applicable tests must be carried out and the results compared with relevant criteria. The required tests up to, and including, polarity must be carried out in the stated order prior to the installation being energised. Where the installation has an earth electrode, it must also be tested before the installation is energised. After the installation has been connected to the supply, the polarity should be confirmed.

If a test indicates failure to comply, that test and the preceding tests (whose results may have been affected by the fault) must be repeated after rectification of the fault.

Continuity of protective conductors *(612.2)*

The initial tests applied to protective conductors, including main bonding conductors and supplementary bonding conductors, are intended to verify that the conductors are both correctly connected and electrically sound. They also verify that the resistance of circuit protective conductors is such that the overall earth fault loop impedance of the circuits is of a suitable value to allow the circuit to be disconnected from the supply in the event of an earth fault (within the disconnection times selected to meet the requirements of Section 411).

Test methods 1 or 2 are used to confirm the continuity of protective and bonding conductors.

Test method 1, which measures the $R_1 + R_2$ value for the circuit being tested, may be the more convenient method to use as it will also confirm polarity as each test is completed. Test method 2 measures R_2 only and often has the inconvenience of a long trailing test lead.

Only test method 2 can be used to test the continuity of main and supplementary bonding conductors.

Test methods 1 and 2 can only be simply carried out on 'all insulated' installations. For installations where parallel paths are present through steel conduit, trunking or pipework, etc. a degree of disconnection must be carried out (where practicable) to remove these conductive parts. In some instances, where it is not practical to remove all parallel paths, a protective conductor can only be verified by visual inspection in addition to testing during installation and on completion (e.g. high level lighting fixed to steelwork and wired without a plug and socket arrangement).

Test instrument

This should be a low-resistance ohmmeter with a recommended supply that has a no load voltage of between 4 V and 24 V d.c. or a.c. with a short-circuit current of not less than 200 mA.

Note: Remember to subtract the value of the instrument's test leads from the test results. Some instruments have an in-built facility to zero out or 'null' the test leads.

Test method 1 ($R_1 + R_2$ value)

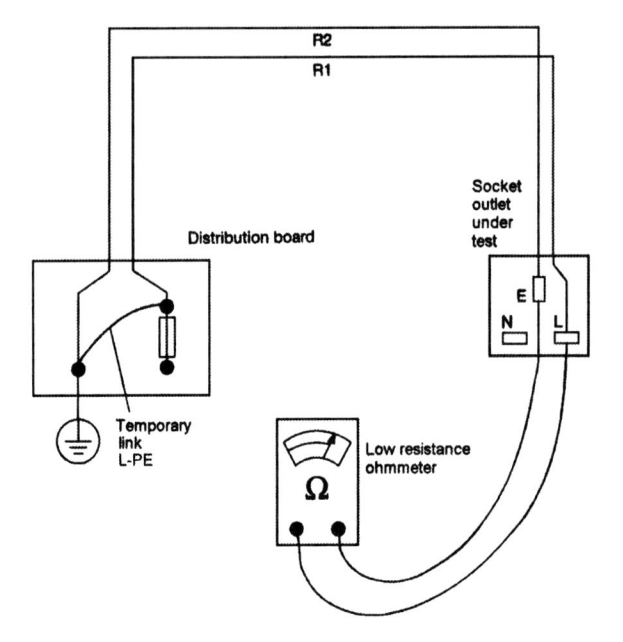

To carry out the test:

- safely isolate the supply

- disconnect any parallel paths (where practicable) or bonding connections that could influence the test readings

- temporarily link the line and protective conductor at the supply end of the circuit, e.g. consumer unit or distribution board

- prepare the low-resistance ohmmeter for use, not forgetting to short the leads together and zero (null) the display or note the resistance of the test leads

- measure between the line and protective conductor at each outlet, point or accessory on the circuit – this will also confirm polarity. Record the value obtained at the end of the circuit (max $R_1 + R_2$) on the Schedule of Test Results

    Note: Remember to deduct any test lead resistance if null facility not available

- reconnect any parallel paths and bonding connections removed for the test

- remove the temporary (L-PE) link.

Using this method will tell you a number of things about the circuit under test. It will:

- check the continuity of the line and protective conductor

- verify that the polarity is correct at each test position

- provide a measure of the $R_1 + R_2$ value which, when added to the known value of impedance for the circuit distribution board (which would be Z_e if the distribution board is at the origin), will give a calculated figure of Z_s for the circuit which could be compared to the design value.

Test method 2 (R_2 value)

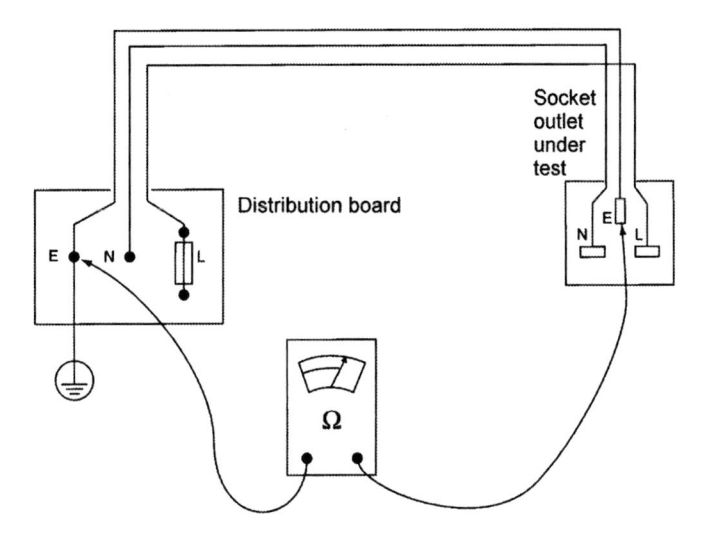

If the two ends of the protective conductor are within the span of the meter test leads, then the resistance of the protective conductor, R_2, may be measured directly. If the distance between the two ends of the protective conductor exceeds the span of the instrument test leads, then they must be extended with a suitable length of cable.

Test method 2 is carried out by connecting one lead of the instrument to the main earthing terminal and using the other lead to make contact with all the protective conductors under test at the various points, accessories, outlets, exposed and extraneous-conductive-parts.

Note: *The resistance of the extended test lead and instrument test leads must be deducted from the test results by either the 'null' feature on the instrument (if fitted) or measuring the overall test lead resistance and subtracting it from the results.*

To carry out the test:

- safely isolate the supply

- disconnect any parallel paths (where practicable) or bonding connections that could influence the test results. A main or supplementary bonding conductor can be simply tested by disconnecting one end before testing

- prepare the low-resistance ohmmeter for use

- measure the resistance of the protective or bonding conductor(s) by connecting one lead from the instrument to the main earthing terminal (MET) and connecting the other lead to the various points under test

- remember to deduct or null the test lead resistance from your test results, and record the value obtained (R_2) on the Schedule of Test Results

- reconnect any parallel paths or bonding connections that were disconnected for the test.

Testing of protective conductors comprising steel enclosures

When the protective conductor of the installation is a steel conduit, trunking or steel wire armouring, etc. the following test procedure must be observed:

- visually inspect the enclosure to verify it is complete and in good order throughout the installation
- carry out a continuity test with a low-resistance ohmmeter using the more suitable of the two methods previously described.

Continuity of ring final circuit conductors *(612.2.2)*

This test must be made to verify the continuity of the line, neutral and protective conductors (unless the cpc is formed by conduit or trunking, etc.) of every ring final circuit.

The test result must also establish that the ring is complete and has not been interconnected, creating an apparent continuous ring circuit that is actually broken. See illustration.

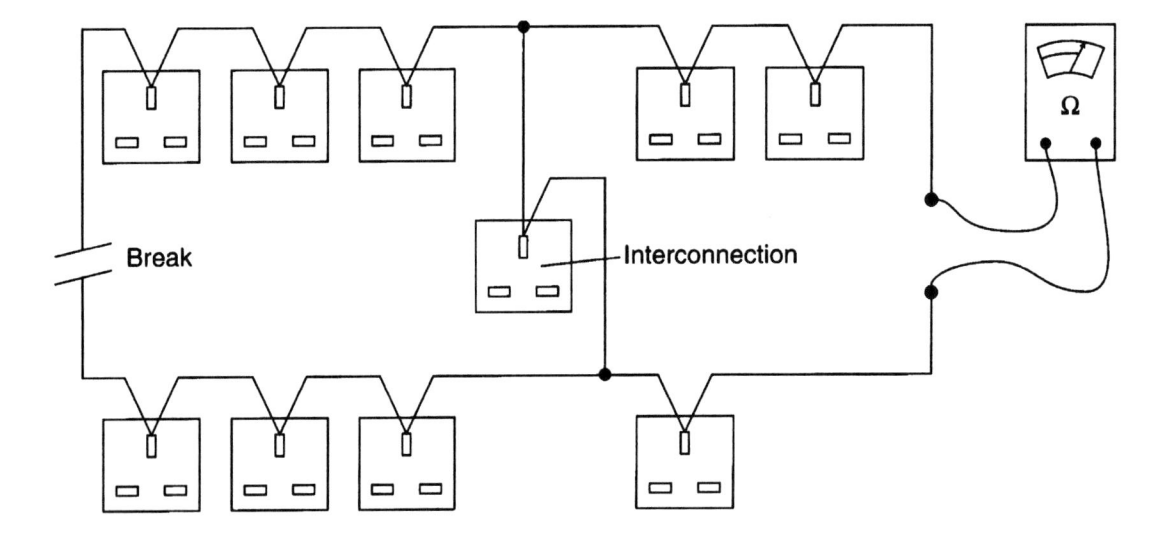

Test instrument

Low-resistance ohmmeter.

Step 1: Ring continuity

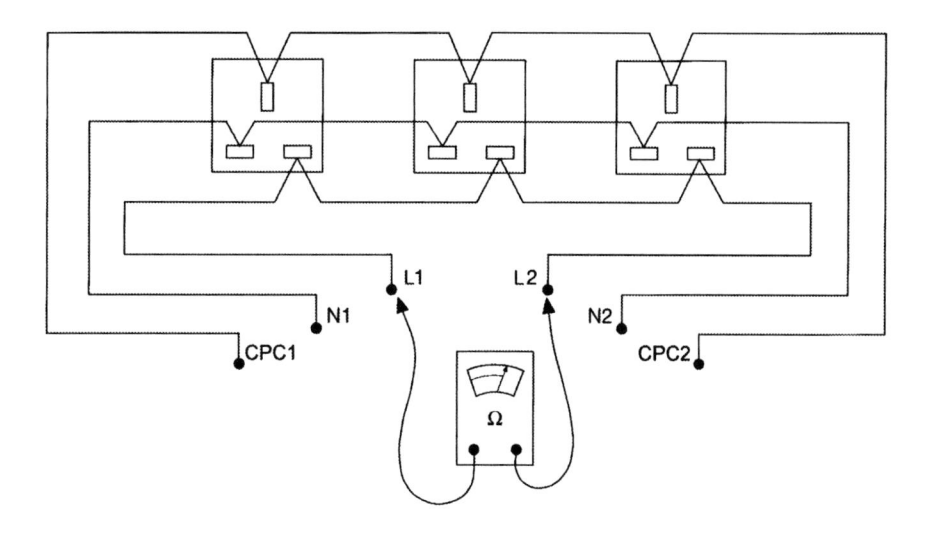

To carry out the test:

- safely isolate the supply

- identify and disconnect the line, neutral and the circuit protective conductors at the distribution board or consumer unit, etc.

- prepare the low-resistance ohmmeter for use

- measure the end-to-end resistance of each conductor. If the conductor size is the same, the instrument readings must be too (within 0.05Ω). Any variation beyond this tolerance could be due to a wiring defect, loose connection or incorrect identification of the ring conductors.

Note: The values of the cpc ring will differ if the cpc has a smaller cross-sectional area conductor than the line or neutral, for example, 2.5/1.5 mm². In this case, the value of the cpc ring would be 1.67 times that of the line or neutral ring, i.e. if the line and neutral are both 1Ω, the cpc should be approximately 1.67Ω.

Step 2: Socket-outlets, phase-neutral

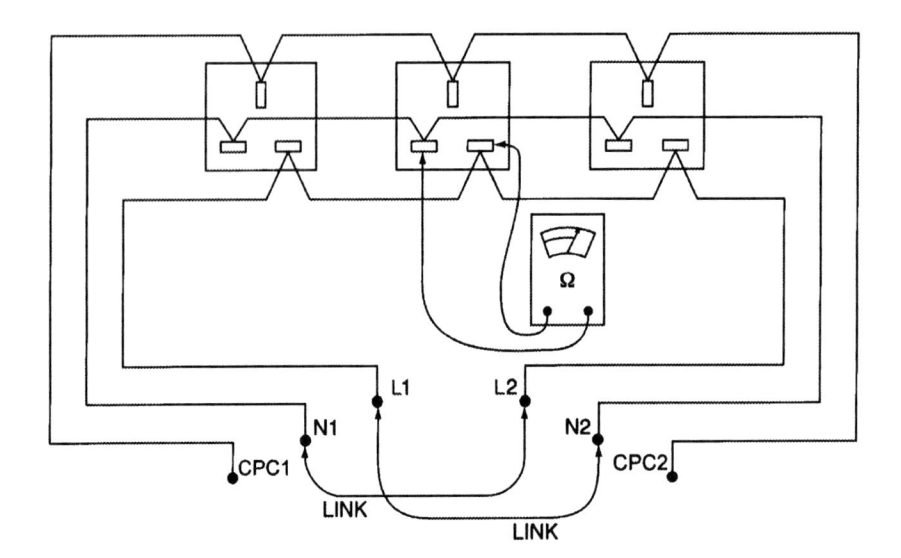

- Join the opposite ends of phase and neutral at the distribution board and measure between phase and neutral at every outlet or point on the circuit.

- The values obtained must be substantially the same – approximately ¼ of the phase ring plus ¼ of the neutral ring resistance. Any sockets wired as spurs will have a higher reading in proportion to the length of the spur.

Step 3: Socket-outlets, phase-cpc ($R_1 + R_2$ value)

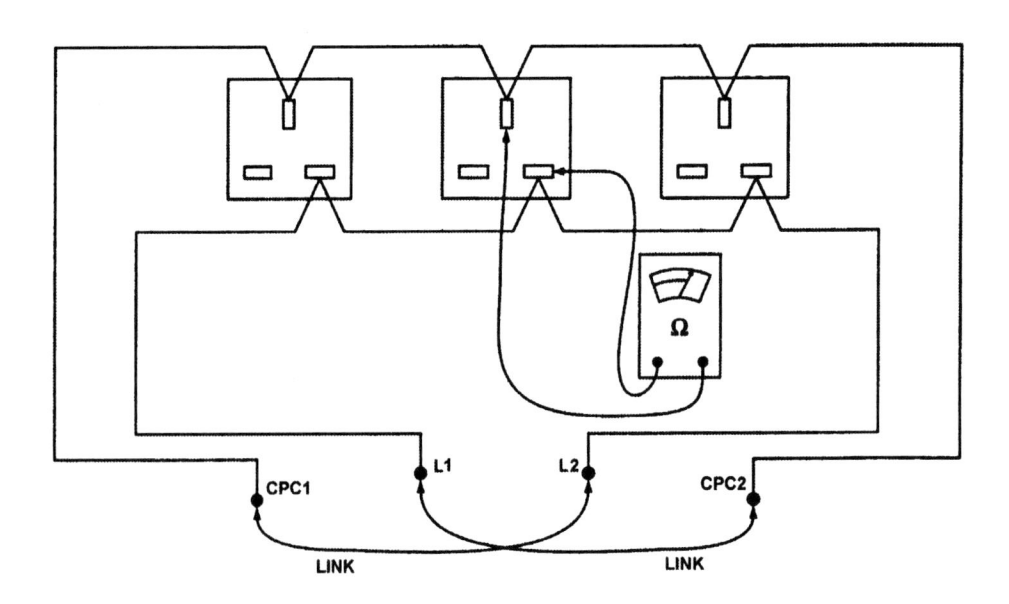

- Join the opposite ends of phase and cpc at the distribution board and measure between phase and cpc at every outlet or point on the circuit.

- The values obtained must be substantially the same – approximately ¼ of the phase ring plus ¼ of the cpc ring. Any sockets wired as spurs will have a higher reading in proportion to the length of spur.

- Record the results – the phase-phase (end-to-end) measured value is known as r_1, the neutral-neutral is r_n and cpc-cpc is r_2. Usually, the maximum value of $R_1 + R_2$ is the only one recorded on the Schedule of Test Results.

- Remember to reconnect all the conductors.

Note: When testing at the outlets for Step 2 or 3, if the readings at the outlets increase towards the centre of the ring and decrease back towards the distribution board, then it is likely that the opposite ends of the ring conductors have not been crossed as intended.

Where the cpc is formed by metal conduit or trunking, this metalwork must still be tested for continuity. However, circuit polarity will need to be confirmed separately.

Example

A 60 m end-to-end ring circuit wired in 2.5 mm² conductors with 1.5 mm² cpc. The circuit has no spurs.

The results of the test will be approximately as follows:

Step 1	Phase-phase (r_1) reading	= 0.45 Ω
	Neutral-neutral (r_n) reading	= 0.45 Ω
	Cpc-cpc (r_2) reading	= 0.72 Ω
Step 2	Phase-neutral reading at outlets	= 0.23 Ω
Step 3	Phase-cpc ($R_1 + R_2$) reading at outlets	= 0.29 Ω

It can therefore be seen that, for socket-outlets wired on a true ring circuit, the readings at the socket-outlets in Step 2 are approximately ½ the readings in Step 1 (i.e. ¼ phase + ¼ neutral) and the values of Step 3 are approximately a ¼ of the phase + ¼ of the cpc values of Step 1.

If this test is carried out correctly, the results are satisfactory and the circuit has a separate cpc, then polarity will also be confirmed at each outlet. Also, the need for a separate cpc continuity test will be removed.

Insulation resistance *(612.3)*

These tests are to verify that the insulation of conductors, electrical accessories and equipment is satisfactory and that there is no unwanted path between live conductors or between live conductors and earth.

Test instrument

An insulation resistance tester must be used which is capable of providing a d.c. voltage of not less than twice the nominal voltage of the circuit to be tested (rms value for an a.c. supply). The instrument must provide the following test voltages of the table below, which gives the minimum permissible values of insulation resistance. Insulation resistance values are usually much higher.

(From Table 61 of BS 7671)

	Circuit nominal voltage	Test voltages	Minimum value of insulation resistance
1	SELV and PELV	250 V d.c.	$\geq 0.5\ M\Omega$
2	All circuits (other than SELV or PELV) up to and including 500 V	500 V d.c.	$\geq 1.0\ M\Omega$
3	In excess of 500 V	1,000 V d.c.	$\geq 1.0\ M\Omega$

Test procedure

- Safely isolate the supply, including neutral.

- Ensure that all current-using equipment is disconnected (including neons, capacitors, discharge lighting, etc.) and all filament lamps are removed. Where it is impracticable to disconnect or remove current-consuming equipment, the local switches controlling this equipment must be left open. This will avoid any damage to equipment and prevent misleading results being obtained due to component resistances.

- Disconnect control equipment or apparatus constructed with voltage sensitive devices (semiconductors) e.g. Surge Protective Devices (SPD), dimmer switches, timers, touch switches, electronic control gear for lighting, and will include certain types of RCD. The devices will be liable to damage if exposed to the high test voltages used in insulation resistance tests.

    Apart from disconnecting the above, the installation must be as complete as possible with accessories fitted, trunking lids replaced, etc.

- Ensure switches and circuit breakers are closed, and all fuses are in place.

- Prepare the test instrument for use, checking that the leads are in good order and the instrument is working correctly on the voltage range for the circuit or installation under test.

- Carry out the tests at the distribution board/consumer unit as required for the installation or circuit under test and record the readings.

Note: *If a reading of less than 2 MΩ is obtained then the reason for this low reading must be investigated.*

Insulation resistance tests between live conductors

Single-phase installations

Test:

- line to neutral (any two-way switches to be operated).

Three-phase installations

Test:

- L1 to L2
- L1 to L3
- L2 to L3
- L1 to neutral
- L2 to neutral
- L3 to neutral.

Record the lowest reading on the Test Result Schedule.

Insulation resistance tests to earth

Where the circuit includes electronic devices (that are impracticable to disconnect), a measurement to earth can be made with the phase and neutral joined together. In addition other precautions, such as disconnection, may be required to prevent damage to electronic devices (612.3.3).

Single-phase installations

Test:

- line and neutral (connected together) to earth
  (any two-way switches to be operated).

Three-phase installations

Test:

- all live conductors (including neutral) connected together to earth.

Note: *Where circuits or equipment are not sensitive to insulation tests, phases and neutrals may be tested **separately** to earth.*

Equipment

Fixed equipment, which is likely to influence the insulation resistance test results, should be disconnected before testing. Where this is impractical, the test voltage for the circuit involved may be reduced to 250 V d.c. but the insulation resistance must be at least 1 MΩ.

Protection by SELV, PELV or electrical separation *(612.4)*

The resistance values obtained for the following tests shall be for the circuit with the highest voltage present and should be in accordance with Table 61 of BS 7671.

Protection by SELV *(612.4.1)*

The supply source should be inspected to ensure conformity with Section 414. This may involve measuring the supply source voltage to ensure that it does not exceed 50 V a.c. or 120 V d.c.

The separation of live parts from those of other circuits and from earth shall be confirmed by measuring the insulation resistance. This can be done by testing between the live conductors of each SELV circuit connected together, and those of any adjacent circuits operating at higher voltages which are also connected together.

When the circuit is supplied from a safety source complying with BS 3535 an insulation resistance test at 250 V d.c. should be made, with the minimum acceptable value being 0.5 MΩ.

In order to confirm compliance with BS 7671, an insulation resistance test of 500 d.c., between live conductors of SELV circuits and protective conductors of circuits operating at higher voltages shall be carried out. The value of insulation resistance must not be less than 1 MΩ.

Protection by PELV *(612.4.2)*

Installations should be inspected and tested, as described for SELV installations, with the exception that an insulation resistance test is not made between live conductors of the PELV circuits and earth.

Protection by electrical separation *(612.4.3)*

For electrical separation with more than one item of current-using equipment, verification of the protection can be made by measurement or calculation. If two coincidental faults of negligible impedance occur between different line conductors, any protective bonding conductors or exposed-conductive-parts, it is necessary to verify that at least one faulty circuit is disconnected in accordance with the protective measure disconnection times for a TN system.

Inspection and testing for electrical separation

The supply source should be inspected to confirm compliance with BS 7671, Section 413. When the supply source does not comply with Section 413, compliance with Section 418 must be verified.

Functional extra-low voltage circuits *(612.4.4)*

Extra-low voltage circuits that are not SELV or PELV should be inspected and tested as low voltage circuits.

Basic protection by a barrier or an enclosure provided during erection *(612.4.5)*

Where barriers and enclosures in accordance with Regulation 416.2 have been provided during erection it must be confirmed that the minimum degree of protection of IP2X, IPXXB, IP4X or IPXXD (as applicable) has been achieved (see IEE Guidance Note No. 3).

IP2X or IPXXB is the required degree of protection to prevent contact with live parts. For accessible top surfaces of an enclosure, IP4X is required to prevent entry of foreign objects greater than 1 mm diameter or width.

Insulation resistance/impedance of floors and walls *(612.5)*

In order to comply with the requirements of Regulation 418.1 in a non-conducting location, such as an all-insulated room in a testing laboratory of a works, the resistance of the floors and walls of the room to the main protective conductor of the electrical installation must be measured at not less than three positions in the same location.

One of the measured points must be around 1 m from any accessible extraneous-conductive-part. The other two measurements shall be made at greater distances. These measurements must be repeated for each relevant surface within the location.

The insulation or insulating arrangements must be able to withstand a test voltage of at least 2 kV a.c. and not pass a leakage current exceeding 1 mA in normal conditions of use.

When tested for insulation resistance at 500 V d.c., the value must be at least 1 MΩ.

Such tests should be regarded as being special in character requiring the advice of those experienced in this field, i.e. probably the person who will use the non-conductive location. Further information and examples of measuring insulation resistance/impedance are given in Appendix 13 of BS 7671.

Polarity (612.6)

This test must be carried out to verify that:

- all fuses, circuit breakers and single-pole control devices such as switches are connected in the line conductor only
- except for E14 and E27 lampholders to BS EN 60238, all centre contact bayonet or Edison screw lampholders must have their outer screwed contact connected to neutral
- socket-outlets and similar accessories have been correctly connected.

The installation must be tested with all switches in the 'on' position and all lamps and power consuming equipment removed.

Test instrument

Low-resistance ohmmeter.

Test procedure

- Note that this test is the same as the cpc continuity test method 1 – the $(R_1 + R_2)$ version.
- If correct polarity is verified using this test, a tick is usually placed in the polarity column of the Test Results Schedule.

A test of polarity can be carried out as illustrated on the next page.

Polarity test ($R_1 + R_2$ method)

Polarity test: lighting

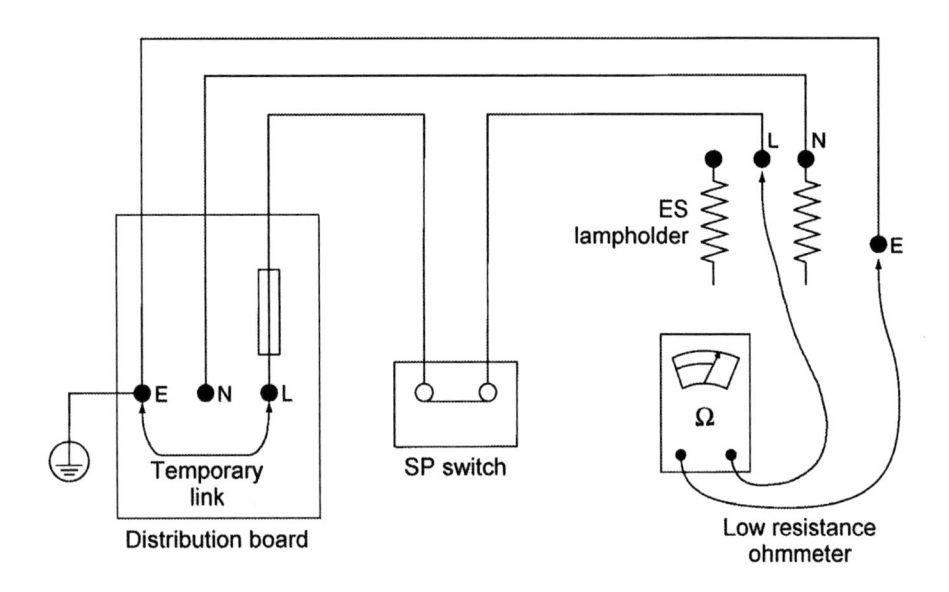

Polarity test: socket-outlet

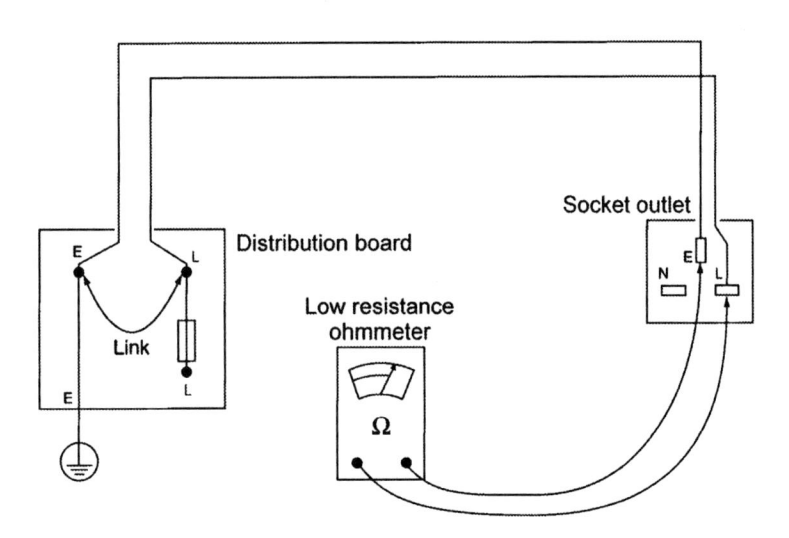

For ring final circuits that contain a separate cpc (e.g. not using conduit as a cpc), if the circuit has been tested to 612.2.2 then polarity will have been confirmed at each outlet during the test.

After connection of the supply, polarity must be reconfirmed with a voltage indicator. This will establish the correct reconnection of any conductors previously disconnected during the testing process.

Earth electrode resistance *(612.7)*

After an earth electrode has been installed, it is necessary to verify that the resistance of the electrode meets the conditions of BS 7671.

There are two common methods of testing earth electrode resistance:

- Test method 1: Proprietary earth testers
- Test method 2: Phase earth loop impedance tester.

Test method 1

Test instrument

Earth electrode resistance tester.

Most three-wire earth testers today pass a current through the resistance under test from terminals E to C1, and the resultant potential is measured between terminals E and P1.

Earth electrode resistance testers generally come with test spikes, coloured connecting leads and coloured terminals on the meter. The test leads are also different lengths. Always connect the test leads to the earth electrode under test, test spikes and meter in accordance with the manufacturer's instructions. Remember, always to start on the highest resistance range and work down until a satisfactory reading has been obtained. This will avoid damage to the meter movement of the test meter.

Test procedure

- **Safely isolate the supply** (this must be carried out before disconnecting the earthing conductor).
- Disconnect the earthing conductor (this is to ensure that all the test current flows through only the earth electrode).

  The distance between the test spikes is important. If they are too close, their resistance areas will overlap. The distance between the electrode under test and the current spike must be approximately 10 times the electrode length, e.g. 20 m for a 2 m earth electrode.

- The test is carried out with P in the mid position, then two further readings are carried out with P either side of the mid-point (approximately 10% of the total distance apart).
- The average value of these three readings must be calculated and compared to the individual readings. None of the individual readings should vary from the average by more than 5%.
- **Reconnect the earthing conductor correctly.**

Connections for a three-wire earth electrode resistance tester are illustrated below.

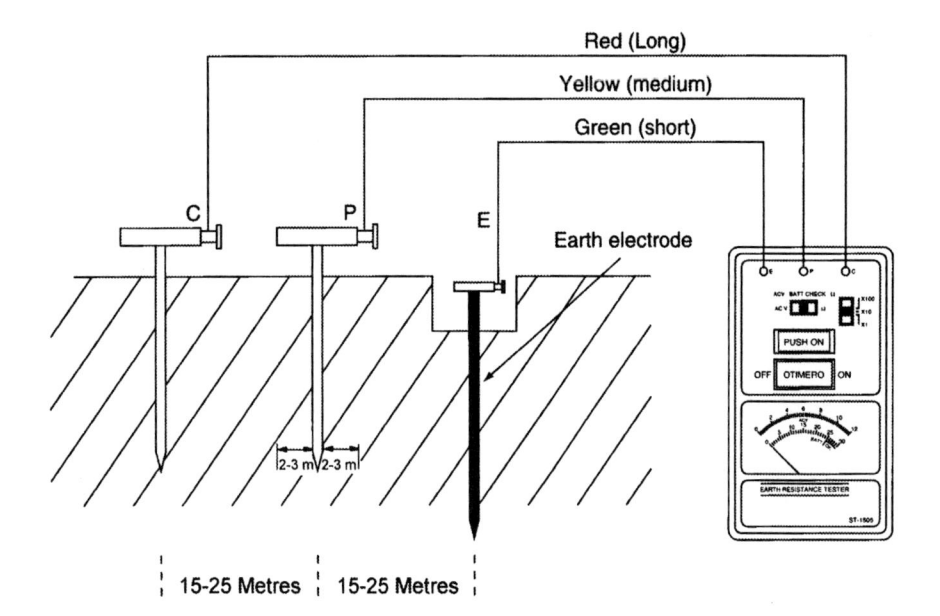

Test method 2 (RCD-protected TT installations)

Where the electrode to be tested is in use with an RCD, this method may be used as an alternative to test method 1.

Test instrument

Earth loop impedance tester.

Test procedure

- **Safely isolate the installation.**

- Disconnect the earthing conductor from the main earthing terminal (this allows the test current to flow through the earth electrode and remove any problems with parallel earth return paths e.g. service pipes, etc.).

- Prepare the instrument for use, checking that the instrument and leads, probes and clips are in good order and suitable for purpose.

- Connect one lead of the earth loop impedance tester to the earthing conductor then, using a probe, connect the other lead to the line conductor at the incoming side of the main switch for the installation (assuming the use of a two-lead instrument), carry out the test.

- The reading obtained is taken to be the electrode resistance.

Note: *After the test, ensure that you re-connect the earthing conductor BEFORE restoring the supply.*

Table 41.5 gives the maximum earth fault loop impedance (Z_s) and therefore the maximum earth electrode resistance (R_A) to ensure satisfactory RCD operation in TT systems. Part of this table is shown below.

RCD residual operating current ($I_{\Delta n}$)	Maximum earth fault loop impedance (Z_s)	
	Normal, dry locations	Agricultural, horticultural or construction sites
	230 V	400 V
30 mA	1,667 Ω	1,533 Ω
100 mA	500 Ω	460 Ω

In practice, any value obtained above 200 Ω would require further investigation.

Protection by automatic disconnection of the supply *(612.8)*

When RCDs are installed to provide protection against fire, verification of the conditions for protection by automatic disconnection of the supply may be considered as satisfying the requirements of Chapter 42.

Methods of verification of the effectiveness of fault protection methods can be achieved for TN, TT and IT systems as follows.

TN systems

In order to ensure compliance with Regulation 411.4:

- measure the earth fault loop impedance
- verify the characteristics/effectiveness of protective devices by:
    - overcurrent devices. These should be visually inspected for short-time or instantaneous tripping settings for circuit breakers, and the current rating and type for fuses
    - RCDs. These should be visually inspected and tested using a suitable test instrument to confirm the requirements of Chapter 41 are met. see Regulation 612.1 or 612.2 for testing methods.

TT systems

In order to ensure compliance with Regulation 411.5:

- measure the earth electrode resistance R_A for exposed-conductive-parts of the installation
- verify the characteristics/effectiveness of the associated protective device by:
    - overcurrent devices. These should be visually inspected for short time or instantaneous tripping settings for circuit breakers, and the current rating and type for fuses
    - RCDs. These should be visually inspected and tested to confirm the requirements of Chapter 41 are met.

IT systems

In order to ensure compliance with Regulation 411.6, verification can be made by calculation or measurement of the current (Id) for the first fault at the line or neutral conductor.

When conditions are similar to that of a TT system, in the event of a second fault in another circuit (411.6.4(i)), verification is carried out as for a TT system.

When conditions are similar to that of a TN system, in the event of a second fault in another circuit (411.6.4(i)), verification is carried out as for a TN system.

Phase earth fault loop impedance *(612.9)*

The earth fault loop impedance comprises the following parts:

- the point of fault
- circuit protective conductor
- the main earthing terminal
- earthing conductor
- for TN systems, the metallic return path – or, with TT systems, the earth return path (through the mass of earth)
- the earthed neutral point of the transformer
- the transformer winding
- the line conductor from the transformer to the point of fault.

The earth fault loop path of a TN-S system is illustrated on the next page.

Note: The impedance of the earth fault loop is denoted by the symbol Z_s, where Z = impedance and S = system (a supply of energy and an installation).

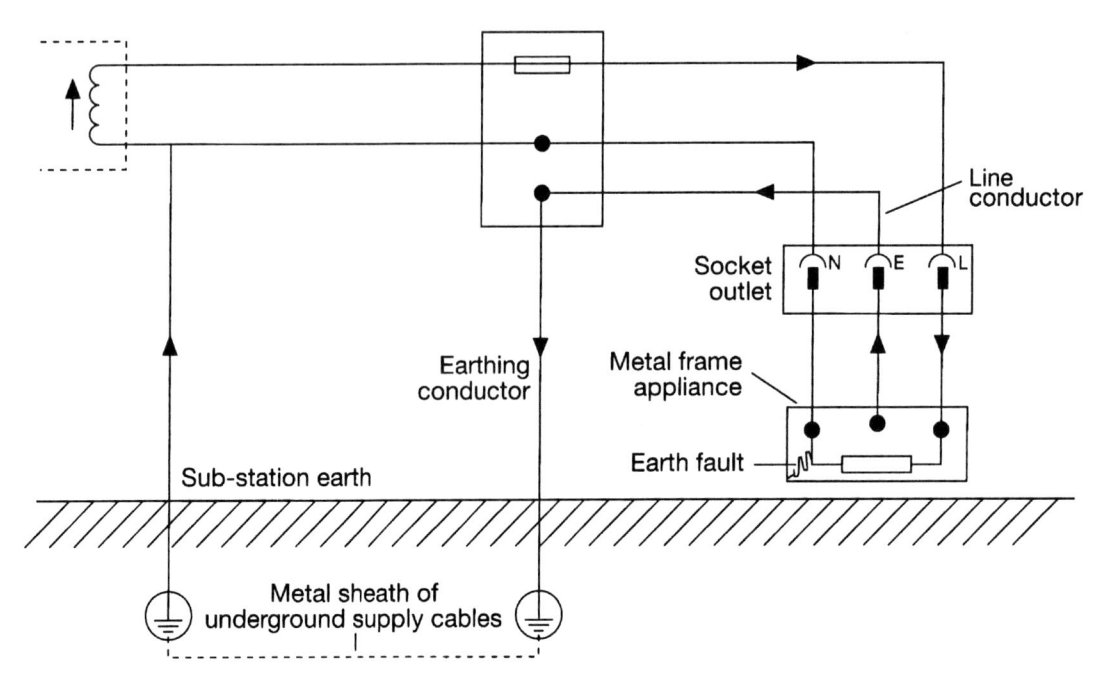

Diagram: Earth fault loop path showing sub-station earth, earthing conductor, line conductor, socket outlet (N, E, L), metal frame appliance with earth fault, and general mass of earth with metal sheath of underground supply cables.

Test instrument

An earth loop impedance tester.

The earth loop impedance tester uses the circuit voltage to pass a test current of around 20 A through the phase-earth path for a duration of approximately 20-30 milliseconds. The current passes through a known resistor within the instrument and the instrument compares the circuit when loaded and not loaded with this resistance and displays the impedance value in ohms.

When using this test instrument, care must be taken to ensure that no hazardous effects can arise in the event of any defect in the earthing circuit, such as would arise if there was a break in the protective conductor of the system under test. This would prevent the test current from flowing and the whole of the protective conductor system would be connected directly to the line conductor.

Important: Live polarity must be re-confirmed before carrying out this test.

Test procedure

- This is a live test so be aware of the safety precautions to prevent danger.

- Prepare the instrument for use, checking that the instrument and leads, probes and clips are in good order and suitable for purpose.

- If using the probes or clips, care must be taken to minimise the risk of electric shock or burns due to working on or near exposed live terminals. Always connect the instrument securely before turning on the supply; then switch off the supply after carrying out the test and before disconnecting the instrument.

- Before carrying out a test, check the status of the LEDs or neons on the instrument to see if they indicate it is safe to proceed with the test. If they indicate problems with the wiring then, for safety reasons, these problems must be rectified before carrying out the test.

- With socket-outlets, test the furthest one on each circuit as a minimum. Otherwise, test all the socket-outlets in the circuit and record the highest reading (testing at socket-outlets requires the instrument manufacturer's lead with fitted plug).

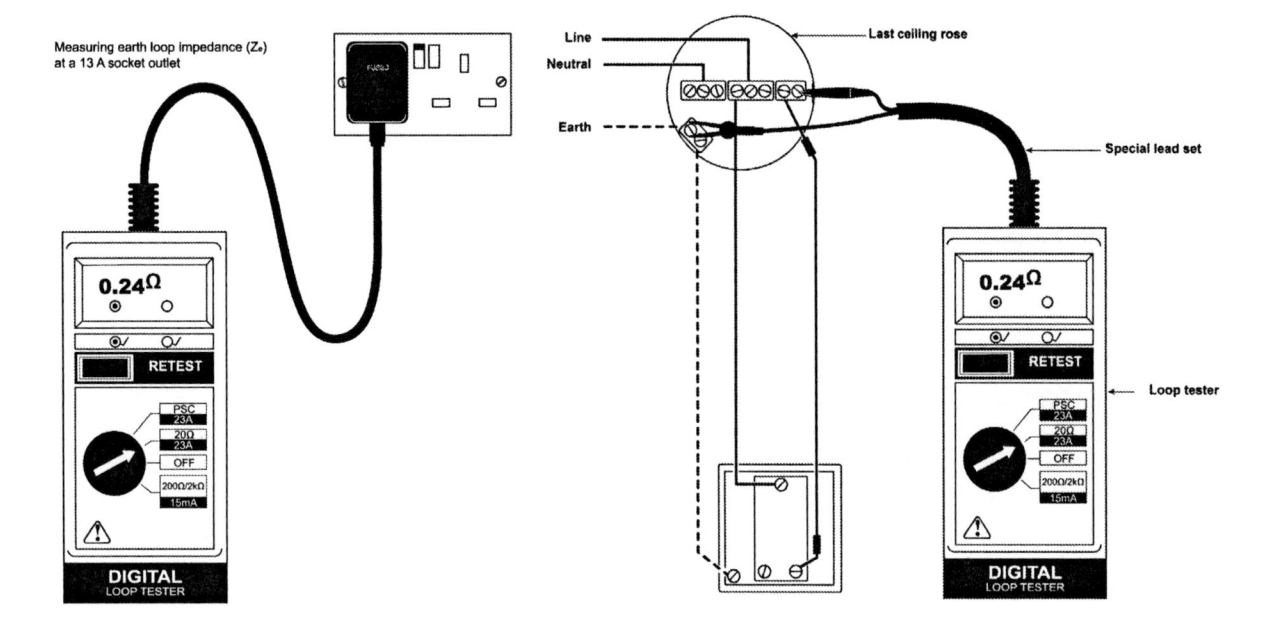

- When testing lighting circuits, the test (as a minimum) must be carried out at the electrically furthest point of the circuit. This could be at a luminaire or switch.

- Earth fault loop impedance (Z_s) must be verified at the furthest point of each circuit, e.g. distribution lighting and socket-outlet circuits and any fixed equipment.

- For three-phase circuits, a test must be carried out for each of the line conductors separately.

- The test result value of Z_s must be checked for compliance by being compared to:
    - the relevant values in Appendix 2 of the IEE On-site Guide or IEE Guidance Note No. 3, or
    - values provided by the electrical designer, or
    - values in BS 7671 when corrected for temperature (see the tables in Section 411.4 of BS 7671 and Appendix 14).

    Each of these methods is described in the following sections.

IEE On-site Guide, Appendix 2: maximum measured earth loop impedance

The values of maximum earth loop impedance in Tables 2A–D of the On-site Guide and IEE Guidance Note No. 3 are maximum **measured** values at a testing temperature of 10°–20°C, but these will need correcting if the ambient temperature at the time of test falls outside this range (using the correction factors from Table 2E).

For example, if the ambient temperature is 25°C then the maximum measured phase earth loop impedance for a socket-outlet circuit protected by a 32 A Type B circuit breaker would be:

$$1.20 \times 1.06 = \mathbf{1.27\ \Omega}$$

Values provided by the electrical designer

To verify compliance, a designer would need to give values for R_1 and R_2 at the ambient temperature expected during circuit testing.

BS 7671 values, corrected for temperature

Tables 41.2, 41.3 and 41.4 in BS 7671 give maximum values of phase-earth loop impedance to achieve maximum disconnection times should an earth fault occur. The note below these tables states that if the conductors are tested at a temperature which is different to the normal operating temperature of the cables, which will usually be the case, then the relevant maximum Z_s value must be adjusted accordingly. Assuming the cables are at 20°C when tested the Z_s value can be corrected by multiplying it by 0.8.

Measured values of phase earth loop impedance must be less than those in BS 7671, Tables 41.2 or 41.3 and 41.4. The values found in these tables can be corrected by multiplying by a correction factor for ambient temperature at the time of testing, as indicated above.

The test reading must then be compared to the corrected value of the appropriate table – BS 7671 Tables 41.2 or 41.3 and 41.4.

External earth fault loop impedance (Z_e)

Z_e is the earth fault loop impedance of the supply, i.e. that part of the system which is external to the installation.

Since the purpose of this test is to establish that the means of earthing is present and it is of an acceptable ohmic value, any parallel paths must be disconnected. This requires the disconnection of the earthing conductor, otherwise misleading readings would be obtained and defects, or even the lack of earthing, could be concealed.

For safety reasons, the whole installation must be safely isolated from the supply before the earthing conductor is disconnected and must stay isolated while the test is carried out and until the earthing conductor is reconnected correctly. Only then can the supply be safely restored.

Test instrument

An earth loop impedance tester.

Some earth loop impedance testers have two leads, while others have three leads. With the two-lead set, the connection is to the incoming phase and earth; with the three-lead set, a neutral connection is required. Where the installation under test is three-phase and no neutral is present, normally the neutral lead of the instrument is connected to earth. Always refer to the instrument manufacturer's instructions for the correct connections.

Test procedure

- **Safely isolate the whole installation from the supply.**
- Disconnect the earthing conductor.
- Prepare the instrument for use, checking the instrument and leads, probes and clips are in good order and suitable for purpose.
- Remember to take great care during the test as it is at the origin where the highest levels of fault conditions can be found.
- Following the safety procedures, connect one lead of the instrument to the earthing conductor then, using a probe, connect the other lead to the incoming live line conductor (checking any instrument polarity indication).
- Carry out the test and record the reading obtained.
- Reconnect the earthing conductor correctly.
- Restore the supply.

Note: *For three-phase installations, a test must be carried out separately on each supply line conductor.*

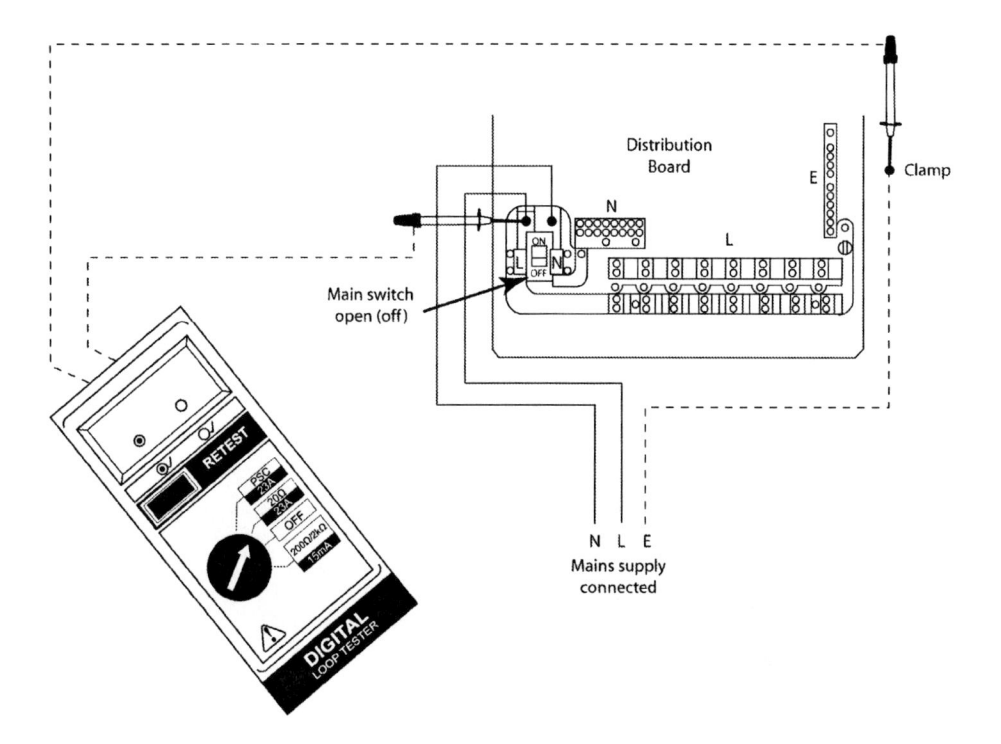

The reasons for carrying out the test are to verify there is an earth connection and to establish that the actual Z_e value is equal to or less than the value used by the electrical designer in his/her calculations.

Values for Z_e can also be obtained by enquiry to the local distributor or by calculation. Values of Z_e provided by network operators for single-phase supplies are typically:

- TN-C-S: 0.35 Ω
- TN-S: 0.8 Ω
- TT: 20 Ω

Sensitive protective devices

Carrying out an earth fault loop impedance test will cause sensitive protective devices (such as RCDs, and possibly 6 A Type B circuit breakers and any 6 A Type 1 circuit breaker) of existing installations to operate. In these circumstances, the following options are available.

- Instrument manufacturers can supply instruments that:
    - limit the test current (typically to 15 mA, as this means an RCD with a 30 mA or greater operating current should not trip), or
    - d.c. bias the RCD (this method momentarily stuns the RCD by saturating the core of the device so the test current is not detected).
- An earth loop test can be carried out on the incoming supply to the RCD with the circuit's $R_1 + R_2$ value being added to the test result to give an approximate value of Z_s.

Testing an RCD-protected socket-outlet by limiting the test current is shown below – note the 15 mA setting.

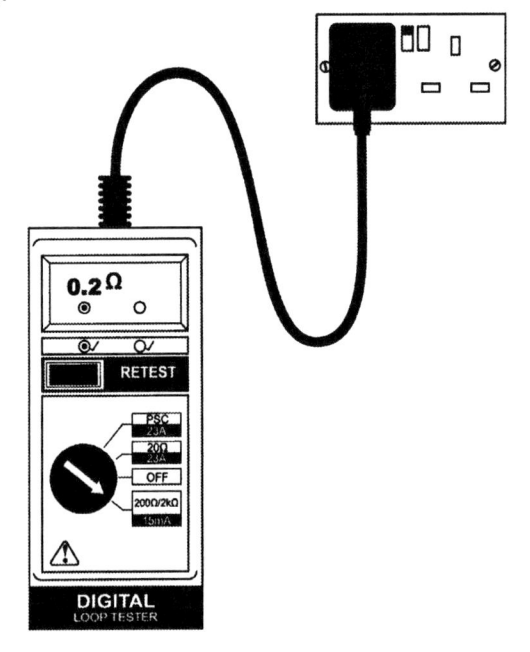

Additional protection *(612.10)*

The effectiveness of the measures applies to an electrical installation for additional protection shall be verified by visual inspection and test.

Additional protection as included in Section 415 could involve:

- supplementary equipotential bonding
- RCDs.

Supplementary equipotential bonding effectiveness can be confirmed by inspection and a continuity test to ensure effective connection between any simultaneously accessible exposed-conductive-parts of fixed equipment and extraneous-conductive-parts. The bonding is connected to the protective conductors of equipment, including those of socket-outlets.

If any doubt exists as to the effectiveness of the supplementary equipotential bonding, it must be confirmed that the resistance between simultaneously accessible exposed-conductive-parts and extraneous-conductive-parts satisfy the following condition:

$$R \leq \frac{50\ V}{I_a} \text{ for a.c. systems}$$

$$R \leq \frac{120\ V}{I_a} \text{ for d.c. systems}$$

Note: I_a is the operating current of the protective devices in amps.

For RCDs it is $I_{\Delta n}$. For overcurrent devices it is the current causing automatic disconnection within 5 s.

The use of RCDs with residual operating currents ($I_{\Delta n}$) not exceeding 30 mA and operating times not exceeding 40 ms at a residual current of 5 $I_{\Delta n}$ is a recognised system of additional protection. The test required is that described for RCDs under functional testing (612.13).

Prospective fault current *(612.11)*

Both the prospective short-circuit current and the prospective earth fault current (PFC) must be measured, calculated or determined by enquiry at the origin and other relevant points in the installation (typically other distribution boards throughout the installation).

The value of PFC must be determined to establish that the type of protective device at the origin would interrupt a fault current up to and including the level of PFC that could flow if a short circuit or earth fault occurs.

The short-circuit (duty) rating of any protective device must be equal to or greater than the value of PFC that could occur at the point where the device is installed. The exception to this rule is where a local device with a lower duty rating is backed up by a device with a higher duty rating.

An example would be a domestic installation with a 100 A supply and 16KA PFC value, with a 100 A 80KA BS 88 service fuse in the distributor's cut-out, and circuit breakers with a 6KA duty rating in the installation consumer unit. Therefore, this value of PFC (prospective fault current covers both prospective short-circuit current and prospective earth fault current) must be determined and compared to the short-circuit duty rating of the relevant protective device.

Test instrument

An earth loop impedance/PFC tester.

Some earth loop impedance/PFC testers use three leads, others use only two leads. Always refer to the instrument manufacturer's instructions for the correct connections.

Test procedure

- Great care is required during the testing procedure. Fault currents are at their highest values at the origin of an installation where this live test will be carried out.
- Prepare the instrument for use, checking that the instrument and leads, probes and clips are in good order and suitable for purpose.
- Select the correct function and range on the instrument (KA).
- The tests must be conducted at the main switch or wherever the tails from the distributor's metering equipment are connected.

Being aware of all the safety requirements, connect the instrument (in accordance with the manufacturer's instructions) to the incoming live supply to measure the phase to neutral value of PFC. See diagram A on the next page.

- Observe the polarity indication on the instrument for correct connection.
- Carry out the test and record the readings.

Now repeat the process, but with the instrument being connected (in accordance with the manufacturer's instructions) to measure the phase to earth value of PFC. See diagram B on the next page.

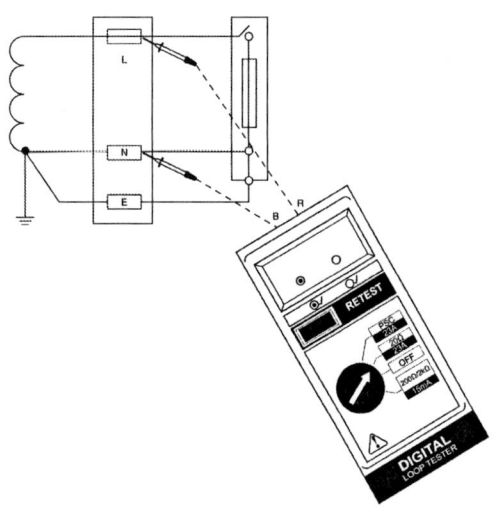

**A: Measurement of prospective short-circuit current
(using a two-lead instrument)**

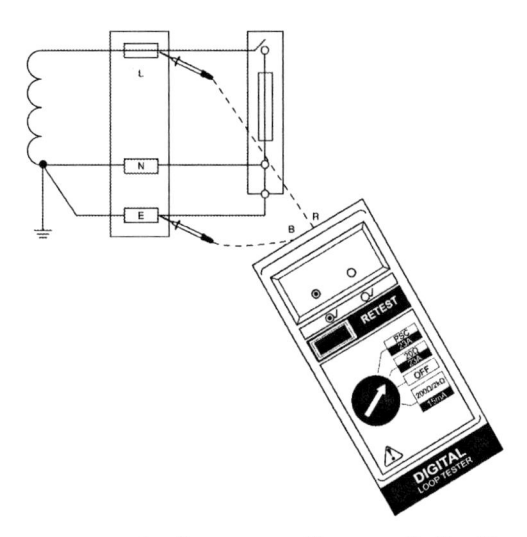

**B: Measurement of prospective earth fault current
(using a two-lead instrument)**

The result to be recorded is the highest of the phase to neutral or phase to earth value.

Note: For three-phase installations, the highest value of PFC will be phase to phase.
Do not attempt this test with a 230 V instrument.

The value can be found by calculation; it will be approximately double the phase to neutral fault current.

Fault current protective devices installed at the origin of an installation must have a short-circuit duty rating equal to or greater than the level of fault current that would flow at their point of installation.

Check of phase sequence *(612.12)*

Verification to ensure phase sequence is maintained shall be made on multiphase circuits. There are several instruments available that carry out this test.

Functional testing *(612.13)*

RCDs manufactured to BS 4293 or BS EN 61008, RCBOs to BS EN 61009 and RCD socket-outlets (SRCDs) must be tested by simulating appropriate fault conditions, using a test instrument.

The test is made on the load side of the RCD, between the line conductor of the circuit protected and its associated circuit protective conductor. All loads supplied through an RCD are disconnected during the test.

The test must not be carried out until it has been established that the circuit earth loop impedance value is sufficiently low to allow the testing of the RCD (as required by BS 7671) i.e. the impedance value multiplied by the RCD operating current must not exceed 50 V (or 25 V for a construction site or agricultural installation). (See Table 41.5 of BS 7671.)

Test instrument

RCD tester.

This is a live test so it is important to check the status of the instrument LEDs or neons to see if they indicate it is safe to carry out the tests.

The test is carried out on the load side of the RCD (with the load disconnected). The connection can be made at any suitable point in the circuit, e.g. a socket-outlet or at the output terminals of the RCD using the probe lead. However, make sure you have correct polarity by checking the correct polarity indication LEDs or neons on the instrument. limit the test current (typically to 15 mA, as this means an RCD with a 30 mA or greater operating current should not trip), or

Test procedure

Testing a 30 mA $I_{\Delta n}$ RCD-protected socket-outlet:

- plug in the RCD tester and check that the polarity indication on the instrument shows it is safe to carry out the test
- if it is safe to proceed, carry out the three instrument tests required
- with a fault current of 50% of the trip current of the RCD flowing for a period of 2 s, the RCD must **not** open (sometimes called the x½ test)
- with 100% of the rated tripping current flowing, the circuit breaker must open within 200 ms (x1 test)
- additional protection:
  - when the RCD has an operating current $I_{\Delta n}$ not exceeding 30 mA and has been installed to Section 612.10, a test current of 5 x $I_{\Delta n}$ should cause the circuit breaker to open within 40 ms, which is the maximum test time (x5 test)
- the tests must be carried out on both the positive and negative half cycles of the supply and the longest operating time must be recorded.

The effectiveness of the integral test facility of the RCD must also be verified by pressing the test button. Users are advised to carry out this test quarterly.

Note: This should be carried out after the instrument tests so as not to affect their results.

If users have not operated the RCD by means of the test button and, depending also on the environmental conditions where the RCD is installed, the device may not operate on the first test (50%) or for the 100% test. After manual operation, however, the device may trip as required and, for this reason, some test engineers would carry out the 50% test after the other instrument tests have been carried out.

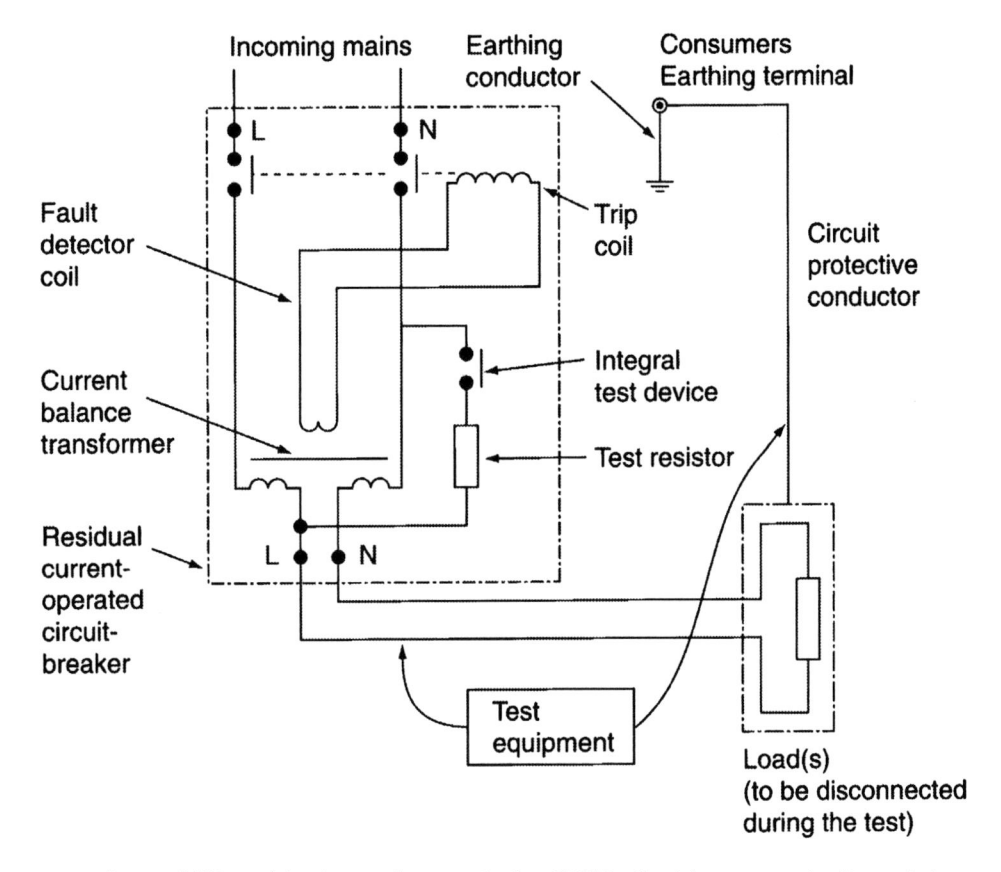

There are also additional test requirements for RCDs that incorporate time delays, etc. See IEE Guidance Note No. 3 for further information.

Functional checks

When switchgear, control gear assemblies, drives, controls and interlocks are installed, they shall be functionally tested to verify they are correctly mounted, adjusted and installed to meet the relevant requirements of the Regulations.

Verification of voltage drop *(612.14)*

Note: Verification of the voltage drop is not usually required during initial verification.

Where it is necessary to verify compliance with Section 525, the following methods may be used:

- evaluate the voltage drop by measuring the circuit impedance
- use of diagrams or graphs, which enable the voltage drop to be evaluated.

Periodic inspection and testing *(Chapter 62)*

General *(621)*

Periodic inspection and testing of an installation must be carried out to establish whether the installation is in a satisfactory condition for continued use.

The same level of testing as for an initial installation is not necessarily required and may not be possible.

Periodic inspection and testing comprises a detailed examination of an installation, without dismantling or with partial dismantling as required, supplemented by the appropriate tests from Chapter 61 of BS 7671.

The scope of the periodic inspection and testing shall be decided by a competent person based upon the availability of records, the condition, use and type of installation.

The inspection and test should provide for:

- the safety of persons and livestock against electric shock and burns
- protection of property from damage by fire and heat due to an installation defect
- confirmation that the installation is not damaged or defective so as to impair safety
- identify defects and departures from BS 7671 which could give rise to danger.

The inspection and testing must not cause danger to persons or livestock, or damage to property and equipment even if the circuit is defective.

Periodic inspection and testing must be undertaken only by competent persons. The results of the periodic inspection and testing, and the extent, must be recorded using appropriate certification.

Frequency of inspection and testing *(622)*

When deciding the frequency of periodic inspection and testing, the following factors must be considered:

- the type of installation
- its use and operation
- the external influences to which it is exposed
- the frequency and quality of maintenance.

A system of continuous maintenance and monitoring of an installation by skilled persons (where records are kept) can replace periodic inspection and testing on an installation that is effectively supervised (622.2).

The Periodic Inspection Report for an electrical installation must be used to report on the condition of an existing installation. The maximum period between inspections is:

	Not exceeding
Domestic installations	10 years or change of occupancy
Commercial installations	5 years or change of occupancy
Industrial installations	3 years
Leisure complexes	3 years
Temporary installations on construction sites	3 months

Note: Shorter intervals between inspections may be necessary depending on the nature of the installation, its use and operation, the external influences to which it is exposed, and the frequency and quality of maintenance. Refer to Section 3 of IEE Guidance Note No. 3 and/or the designer's specifications.

Certification and reporting *(Chapter 63)*

General *(631)*

On completion of the verification of a new installation or alterations to an existing one, an Electrical Installation Certificate must be provided. It must include details of the extent of the installation covered, along with schedule(s) of inspection, and schedule(s) of test results.

A Periodic Inspection Report that includes details of the extent of the installation and any limitations of the inspection and testing, along with schedule(s) of inspection and schedule(s) of test results must be provided for a periodic inspection and test.

If minor installation works do not include a new circuit, but relate to an addition or alteration to an existing circuit, a Minor Electrical Installation Works Certificate must be provided.

An Electrical Installation Certificate, Minor Electrical Installation Works Certificate or Periodic Inspection Report must be signed by a competent person(s). Examples of NAPIT certificates and reports can be found at the end of this section on Technical Data Sheet 6B.

Initial verification *(632)*

After the completion of an initial verification, an Electrical Installation Certificate with schedule(s) of inspection and schedule(s) of test results must be given to the person who ordered the work. The certificate must identify those responsible for the design, construction and inspection and testing of the installation.

The identification of every circuit, including its protective devices, and the record of relevant test results must appear on the schedule(s) of test results.

Any defects or omissions must be made good before the certificate is issued.

Alterations and additions *(633)*

Sections 631 and 632 apply to all the work of an alteration or addition. Any defects found in the existing installation must be recorded on the Electrical Installation Certificate or Minor Electrical Installation Works Certificate.

Periodic inspection and testing *(644)*

A Periodic Inspection Report, together with schedule(s) of inspection and schedule(s) of test results, must on completion of the inspection and testing be given to the person who ordered the inspection.

Any deterioration, damage, defects, dangerous conditions and non-compliances with the Regulations, which may present a risk of danger, must be recorded along with the extent and limitations of the inspection, including the reasons.

Test instruments

Test instruments must be regularly checked and re-calibrated to ensure accuracy. The serial number of the instrument used must be recorded with the test results, to avoid unnecessary re-testing if one of a number of instruments is found to be inaccurate.

For operation, use and care of test instruments, refer to the manufacturer's handbook.

Note: Attention is drawn to HSE Guidance Note GS38 'Electrical test equipment for use by electricians' published by HMSO, which advises on the selection and safe use of suitable test probes, leads, lamps, voltage indicating devices and other measuring equipment.

TECHNICAL DATA SHEET 6A

SAFELY ISOLATE THE SUPPLY before starting the test.

Disconnect all loads, including:

- Lamps and neons
- Capacitors
- Surge Protective Devices (SPD)
- Electronic devices
- Fluorescent lamps
- Current-using equipment

Connect:

- Switches closed
- Circuit breakers closed
- Fuses in place
- Two-way lighting switches operated

REMEMBER: THE SUPPLY MUST BE SAFELY ISOLATED.

For a single-phase installation, Tests Nos. 4, 7 and 10 are required.

For a three-phase and earth installation, Tests Nos. 1, 2, 3, 7, 8 and 9 are required.

For a three-phase, neutral and earth installation, Tests Nos. 1 to 10 are required.

1. L1 – L2
2. L1 – L3
3. L2 – L3
4. L1 – N
5. L2 – N
6. L3 – N
7. L1 – E
8. L2 – E
9. L3 – E
10. N – E

Once the source of supply has been isolated, the installation or circuit must be tested to verify that it is dead before work is carried out.

Test instruments, such as test lamps and voltage indicators, should be checked on a known live source before and after the testing procedure to ensure they are working properly.

TECHNICAL DATA SHEET 6B

NAPIT certificates and reports

NAPIT *Electrical Certificate* Installation/Modification

NA/EC

Requirements for Electrical Installations – BS 7671 [IEE Wiring Regulations 17th Edition] and for compliance with Building Regulations Part P

Can be used for new installations, additions or alterations. Please complete all the unshaded areas.

Page 1 of

1. Details of the Installation

Owner/Occupier
Address
Postcode

Installation *(if different from owner/occupier)*
Address
Postcode

2. Extent and limitations of the inspection (note 6)

Installation is: New ☐ Addition ☐ Alteration ☐   Records available: Yes ☐ No ☐   Date of original installation ☐

Extent of electrical installation covered by this report

Comments

This inspection has been carried out in accordance with BS 7671: _____ (IEE Wiring Regulations), amended to _____ (date)
Details of departures from BS:7671 (Regulations 120-3, 120-4) *See page(s)*
Comments on the existing installation (in the case of alteration or addition) *See page(s)*
(For additions or alterations) cables concealed within trunking and conduits, or cables or conduits concealed under floors, in roof spaces and generally within the fabric of the building or underground may not have been inspected.

3. Next inspection (note 7)

We recommend that this installation is further inspected and tested after an interval of not more than _____ months/years, or on change of occupancy.

DECLARATION: For the Design, Construction & the Inspection and Testing of the Installation as described above

Company name
Inspector name
Company address
Postcode

Signature
Position
Date
NAPIT Membership No.

4. Supply characteristics and earthing arrangements

Supply systems   TN-S ☐   TN-C-S ☐   TT ☐   **Number & type of live conductors**   1-phase, 2 wire ☐   3-phase, 4 wire ☐

Nature of Supply Parameters *(Note: (1) by enquiry, (2) by enquiry or by measurement)*   Supply conductor CSA _____

Nominal voltage, U/U_o [1] _____ V   Nominal frequency, f [1] _____ Hz   Phase sequence _____

Prospective fault current, I_{pf} [2] (note 5) _____ kA   External loop impedance, Z_e [2] _____ Ω

Supply Protective Device Characteristics   BS _____   Type _____   Nominal Current Rating _____ A   Max Demand _____ A

Means of Earthing   Distributor's facility ☐   Installation earth electrode ☐

Details of Installation Earth Electrode *(where applicable)*   Type (e.g. rod(s), tape etc) _____

Location _____   Electrode resistance to earth _____ Ω

Main Protective Conductors   Material   Csa (mm2)   Connection verified
Earthing Conductor
Protective Bonding Conductor

Water CSA _____   Gas CSA _____   Oil CSA _____   Other _____   CSA _____

Main Switch or Circuit Breaker

BS _____   Type _____   Location _____   No. of Poles _____
Fuse or Trip Setting _____   Current rating _____   Voltage rating _____
Rated residual operating current $I_{\Delta n}$ = _____ mA   measured operating time of _____ ms (at $I_{\Delta n}$)
(applicable only where an RCD is suitable and is used as a main circuit-breaker)

NAPIT *Electrical Certificate* Installation/Modification

Requirements for Electrical Installations – BS 7671 [IEE Wiring Regulations 17th Edition] and for compliance with Building Regulations Part P

NA/EC

Can be used for new installations, additions or alterations. Please complete all the unshaded areas.

Page 2 of ____

5. Inspector to record their observations in the 1st column below during the 'first fix' visual check and any ommisions or corrected non-conformances, recorded by the Electrical Inspector in the 2nd column below, during the final inspection.

= Optional 1st Fix

1st Fix		2nd Fix		Schedule of Inspections
Inspected	Rectified	Inspected	Rectified	
				Installation Design Specification is available for the Installer and the Inspector
				Earthing Conductor is present, securely connected and a warning label fitted
				Earthing Conductor of the correct size
				Protective Bonding Conductor verified at: Gas ☐ Water ☐ Other ☐
				Protective Bonding Conductors correctly sized
				Protective Bonding Conductors securely connected and a warning label fitted
				Consumer Unit position accessible and where specified on the design
				Correct Circuit Protection Devices fitted and identified for each circuit
				Correct Cable type and size used, allowing for external influences and volt drop
				Cable run in 'safe' zones and adequately protected
				Cables securely fastened or in appropriate wiring systems
				All Cable cores correctly identified at joints and in accessories
				All cable joints correctly terminated, secure and accessible
				Modifications to the Building Fabric appropriate and safe (Structure)
				Modifications to the Building Fabric appropriate and safe (Fire)
				All Accessories correctly placed as appropriate
				Appropriate Supplementary Bonding present and adequately sized
				Supplementary Bonding securely connected and a warning label fitted if required
				Additional protection provided by RCD where required
				All Accessories have environmental protection appropriate for external influences
				All covers replaced, Accessories secure and neatly aligned
				The number of points and their location agree with the original design
				Circuit details correct on the installation schedule
				Periodic Label, RCD label and other Safety Labels fitted
				Schedule of Test
				External earth loop impedance, (Ze)
				Installation earth electrode
				Prospective fault current, (IPF)
				Continuity of Earth Conductors
				Continuity of Circuit Protective Conductors
				Continuity of Supplementary Bonding Conductors
				Insulation Resistance between Live conductors
				Insulation Resistance between Live conductors & earth
				Polarity (Prior to Energisation)
				Polarity (After Energisation) including Phase Sequencing
				Earth fault loop impedance
				2CDs / RCBOs including discrimination
				Functional testing of devices

The sections above are – Satisfactory (✓), Not Satisfactory (X), Not Checked (N/C) or Not Applicable (N/A)

Observations (if any, if none please put 'none' below)

Inspector's Name – First Fix :
Signature   *For additional report see page(s)*
Inspector's Name – Second Fix:
Signature   *For additional report see page(s)*

SCHEDULE(S)

The attached Schedule(s) are part of this document and this Report is valid only when they are attached to it.

☐ Schedule of Test Results are attached.

NAPIT Electrical Test Sheet

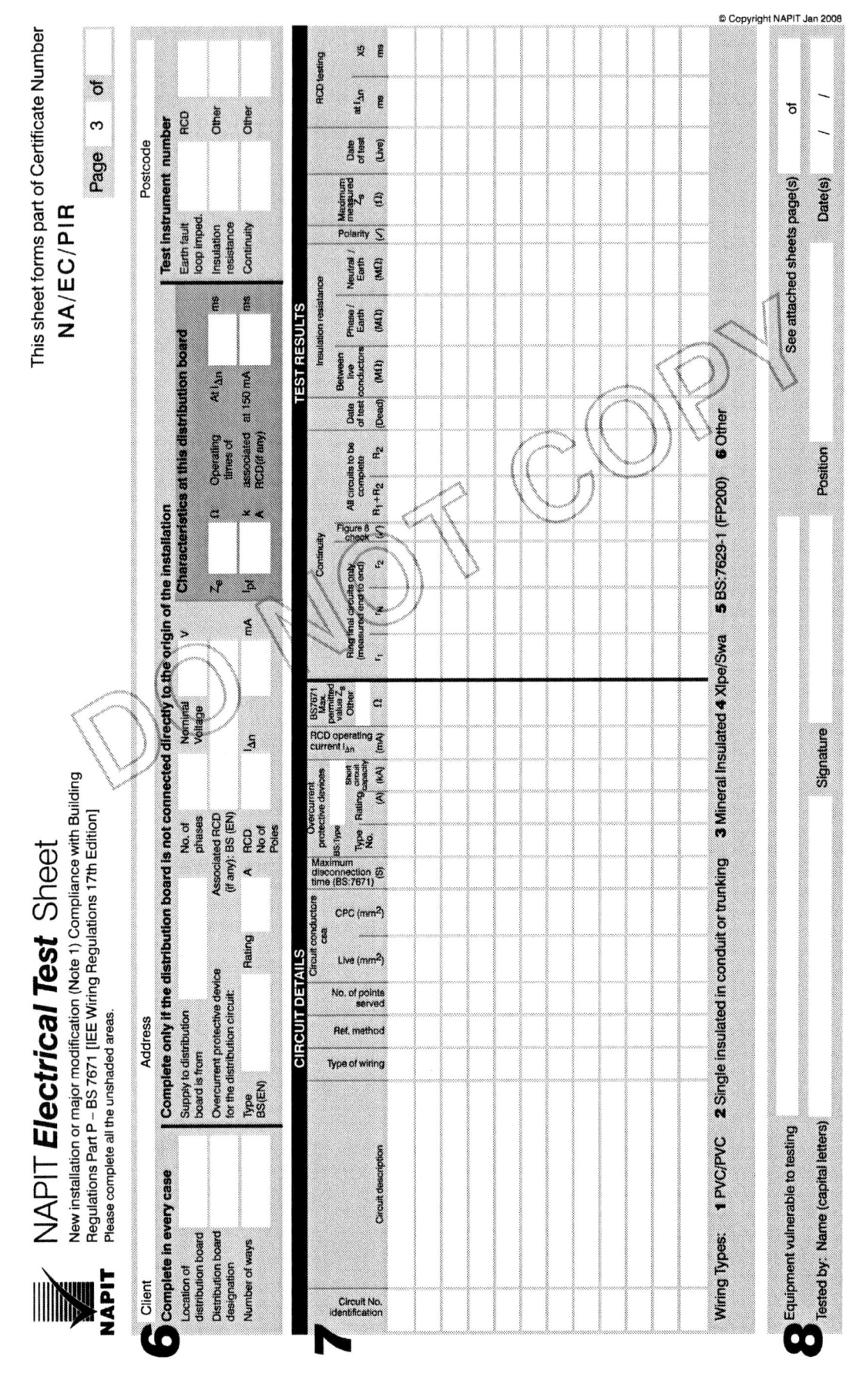

New installation or major modification (Note 1) Compliance with Building Regulations Part P – BS 7671 [IEE Wiring Regulations 17th Edition]
Please complete all the unshaded areas.

This sheet forms part of Certificate Number
NA/EC/PIR

Page 3 of ___

Client
Address
Postcode

6 Complete in every case

Location of distribution board		
Distribution board designation		
Number of ways		

Complete only if the distribution board is not connected directly to the origin of the installation

Supply to distribution board is from	No. of phases	Nominal Voltage (V)
Overcurrent protective device for the distribution circuit:	Associated RCD (if any); BS (EN)	
Type BS(EN)	Rating (A)	RCD No of Poles	$I_{\Delta n}$ (mA)

Characteristics at this distribution board

| Z_e (Ω) | I_{pf} (k A) | Operating times of associated RCD (if any) | At $I_{\Delta n}$ (ms) | at 150 mA (ms) |

Test instrument number

Earth fault loop imped.	RCD
Insulation resistance	Other
Continuity	Other

7 CIRCUIT DETAILS / TEST RESULTS

Circuit description | Circuit No. identification | Maximum disconnection time (BS:7671) (S) | Circuit conductors csa: Live (mm²), CPC (mm²) | Ref. method | No. of points served | Type of wiring | Overcurrent protective devices: BS:Type, Type No., Rating (A), Short circuit capacity (kA) | RCD operating current $I_{\Delta n}$ (mA) | BS7671 Max. permitted value Z_s Other (Ω)

TEST RESULTS

Continuity: Ring final circuits only (measured at each end) r_1, r_n, r_2 | Figure 8 check (Ω) | All circuits to be complete R_1+R_2, R_2 (Ω) | Date of test (Dead) | Insulation resistance: Between live conductors (MΩ), Phase / Earth (MΩ), Neutral / Earth (MΩ) | Polarity ($\checkmark$) | Maximum measured Z_s (Ω) | Date of test (Live) | RCD testing: at $I_{\Delta n}$ (ms), X5 (ms)

Wiring Types: **1** PVC/PVC **2** Single insulated in conduit or trunking **3** Mineral Insulated **4** Xlpe/Swa **5** BS:7629-1 (FP200) **6** Other

8

Equipment vulnerable to testing _____ See attached sheets page(s) ___ of ___

Tested by: Name (capital letters) _____ Signature _____ Position _____ Date(s) __/__/__

© Copyright NAPIT Jan 2008
Sheet 3 of 3 NA/EC/PIR001 (V1)

NAPIT Minor Works/Single Circuit Electrical Certificate

Requirements for Electrical Installations – BS 7671 [IEE Wiring Regulations 17th Edition] and for compliance with Building Regulations Part P.
Can be used for alterations, additions or installation of a new circuit.
Please complete all the unshaded areas.

NA/MW/SCC 0000000000

Part 1: Description of the addition to or the additional circuit

1. Brief description and location
2. Address owner/occupier   Postcode
3. Date the works were completed   /   /

Part 1: Initial assessment of the existing system to be extended

1. BS Number _____ and Type _____ of the Circuit Protective Device
2. Number of existing accessories/circuits already connected to this system _____ New Circuit/Accessory number _____
   (delete as appropriate)   *(delete as appropriate)*
3. Is the addition to or the additional circuit in an a special location? Yes [ ] No [ ]
   If 'Yes', is it a Kitchen [ ] Bathroom [ ] Outdoors [ ] Other [ ] If 'Other', please state _____
4. Supply parameters TN-C-S [ ] TN-S [ ] TT [ ] Ze _____ Ω PFC _____ kA 400v and/or 230v
5. Method of fault protection: ADOS [ ] If 'Other', please state _____
6. Is there enough spare load capacity on the existing system to allow it to be extended? Yes [ ] No [ ]
7. Have tests been carried out prior to installation work to ensure the system or circuit to be extended is adequate? Yes [ ] No [ ]
8. Are the existing protective Earthing and Bonding arrangements adequate? Yes [ ] No [ ]

If you answered 'No' to items 6, 7, 8 or any other comments regarding the Installation, see additional *Report* pages _____ to _____

Part 3: Visual Check (1st Fix) (1st column) Inspection Results (2nd Fix) (2nd column) *Please tick, cross, N/C or N/A*

	Protective Bonding Conductor present and adequate	All Cable Cores correctly identified in joints & accessories
	Correct Circuit Protection Device fitted and identified	Appropriate Supplementary Bonding present & verified
	Correct Cable type and size used, allowing for external influences and volt drop	Modifications to the Building Fabric appropriate and safe
	Cable run in 'safe' zones and adequately protected	All Accessories correctly placed as per Approved Document M and BS8300
	Cables securely fastened or in appropriate wiring systems	All covers replaced, accessories secure & neatly aligned
	All cables joints correctly terminated, secure & accessible	Circuit details updated on the installation schedule (or schedule produced)

Name	Signature	Position	Date

Part 4: Essential Tests _____ %

Ct No.	P/Device BS No.	Type	Rcd A	Max mA	Zs	Ring Circuit Only $r_1 \Omega$	$r_N \Omega$	$r_2 \Omega$	R_1+R_2 or $R_2 \Omega$	L/L L/N MΩ	L/E MΩ	N/E MΩ	Zs Ω	$I_{\Delta n}$ ms	$5I_{\Delta n}$ ms

Satisfactory (1st Column) **Unsatisfactory** (2nd Column) *Please tick, cross, N/C or N/A*

	1. Earth Continuity			4. Earth Fault Loop Impedance
	2. Insulation resistance			5. RCD (If applicable)
	3. Polarity (Dead & Live)			6. Instrument No(s).

Part 5: Declaration

1. I/We certify that the said works do not impair the safety of the existing installation, that the said works, as far as it is reasonably practical to determine, have been designed, constructed, inspected and tested in accordance with BS 7671: _____ , (IEE Wiring Regulations) amended to _____ and that the said works, to the best of my/our knowledge and belief, at the time of my/our inspection, complied with Chapter 13 of BS 7671: _____ .

2. Company name	3. Signature
Inspector name	
Company address	Position
	Date / /
Postcode	4. An inspection and test of this installation is recommended after an interval of not more than _____ years/months.
NAPIT Membership No.	

NAPIT Administration Centre, 4th Floor, Mill 3, Pleasley Vale Business Park, Mansfield, Nottinghamshire NG19 8RL   Sheet 1 of 1   NA/MW/SCC/001 (V1)

© Copyright NAPIT Jan 2008

© Construction Industry Training Board   6B/5   E1: BS 7671 (February 2008)

NAPIT **Periodic Report** on an existing Electrical Installation

Requirements for Electrical Installations –
BS 7671 [IEE Wiring Regulations 17th Edition]
Only for the reporting on the condition of an existing installation

NA/PIR

Page 1 of ___

A. Details of the Installation

Owner/Occupier *(Delete as necessary)*
Address
Postcode

Installation *(If different from owner/occupier)*
Address
Postcode

B. Purpose of the report
This form to be used only for reporting on the condition of an existing installation.
Purpose for which this report is required

C. Details of the Installation

Description of premises: Domestic ☐ Commercial ☐ Industrial ☐ Other *(please state)* ☐
Estimated age of the electrical installation _____ years
Evidence of alterations or addition Yes ☐ No ☐   If 'Yes', estimated _____ years
Date of previous inspection ___/___/___   Electrical Installation Certificate No. or previous Periodic Inspection Report No. _____
Records of installation available Yes ☐ No ☐   Records held by _____
Extent of electrical installation covered by this report *(note 4)*

_____
_____
_____
_____

Limitations

_____
_____
_____
_____

This inspection has been carried out in accordance with BS 7671: _____ (IEE Wiring Regulations), amended to _____ (date)
Cables concealed within the trunkings and conduits, and/or cables and conduits concealed under floors, in roof spaces and generally within the fabric of the building or underground have not been inspected.

D. Periodic Inspection Summary

Date(s) of the inspection from ___/___/___ to ___/___/___
General conditions of the installation
_____
_____

Overall assessment   Satisfactory ☐   Unsatisfactory ☐
We recommend that this installation is further inspected and tested after an interval of not more than _____ months/years, provided that any observations 'requiring urgent attention' are attended to without delay.

E. Declaration

To the best of our knowledge and belief we confirm that the details recorded in this report, including any attached schedules and the recommendations are an accurate assessment of the condition of the Electrical Installation within the limitations described in Section C.
DECLARATION: To the best of our knowledge the details recorded in this report are an accurate indication of the Electrical Installation with Inspection findings listed in the attached schedules.

Company name
Inspector name
Company address

Postcode

Signature

Position
Date
NAPIT Membership No.

NAPIT Administration Centre, 4th Floor, Mill 3, Pleasley Vale Business Park, Mansfield, Nottinghamshire NG19 8RL   Sheet 1 of 4 NA/PIR/001 (V1)

NAPIT Periodic Report on an existing Electrical Installation

**Requirements for Electrical Installations –
BS 7671 [IEE Wiring Regulations 17th Edition]**

Can be used for the reporting on the condition of an existing installation only.

NA/PIR

Page 2 of ___

F. Supply characteristics and earthing arrangements

Supply systems TN-S ☐ TN-C-S ☐ TT ☐ **Number & type of live conductors** 1-phase, 2 wire ☐ 3-phase, 4 wire ☐

Nature of Supply Parameters *(Note: (1) by enquiry, (2) by enquiry or by measurement)* Supply conductor CSA ___

Nominal voltage, U/U_o [1] ___ V   Nominal frequency, f [1] ___ Hz   Phase sequence ___

Prospective fault current, I_{pf} [2] (note 5) ___ kA   External loop impedance, Z_e [2] ___ Ω

Supply Protective Device Characteristics BS ___ Type ___ Nominal Current Rating ___ A **Max Demand** ___ A

Means of Earthing Distributor's facility ☐ Installation earth electrode ☐

Details of Installation Earth Electrode *(where applicable)* Type (e.g. rod(s), tape etc) ___

Location ___ Electrode resistance to earth ___ Ω

Main Protective Conductors   Material   Csa (mm2)   Connection verified
Earthing Conductor
Protective Bonding Conductor

Water CSA ___ Gas CSA ___ Oil CSA ___ Other ___ CSA ___

Main Switch or Circuit Breaker

BS ___ Type ___   Location ___   No. of Poles ___

Fuse or Trip Setting ___   Current rating ___   Voltage rating ___

Rated residual operating current $I_{\Delta n}$ = ___ mA measured operating time of ___ ms (at $I_{\Delta n}$)
(applicable only where an RCD is suitable and is used as a main circuit-breaker)

G. Observations and recommendations for actions to be taken

Referring to the attached schedule of inspection and test results, and subject to the limitations at Section C.

Explanation of codes
1. Requires urgent attention
2. Requires improvement
3. Requires further investigation
4. Does not comply with BS 7671 (as amended)

No remedial work required ☐   OR   The following observations and recommendations are made ☐

Item No.	Defect details	Code

Note: For additional report pages use the continuation report form with the relevant serial number and page numbers detailed on each page.

Urgent remedial work recommended for items ___

Corrective action(s) recommended for items ___

NAPIT *Periodic Report* on an existing Electrical Installation

Requirements for Electrical Installations –
BS 7671 [IEE Wiring Regulations 17th Edition]

NA/PIR

Can be used for the reporting on the condition of an existing installation only.

Page 3 of ____

H Schedule of Inspections

Schedule of Inspection Reports: Page no(s) ____ to ____

Additional pages, including additional source(s) data sheets: Page no(s) ____ to ____

Schedule of Circuit and Test Results for the Periodic Inspection Page no(s) ____ to ____

The attached Schedules are part of this document and this Report is valid only when they are attached to it.

Inspected | **Schedule of Inspections**

- If previous certificate or reports exists have there been any alterations
- Inspection of Incoming Supply and equipment
- Earthing Conductor is present, securely connected and a warning label fitted
- Earthing Conductor of the correct size
- Protective Bonding Conductors verified at: Gas ____ Water ____ Other ____
- Protective Bonding Conductors correctly sized
- Protective Bonding Conductors securely connected and a warning label fitted
- Distribution Board position accessible
- Correct Circuit Protection Devices fitted and identified for each circuit
- Correct Cable type and size used, allowing for external influences and volt drop
- Cable run in 'safe' zones and adequately protected (where readily accessible to inspect)
- Cables securely fastened or in appropriate wiring systems (where readily accessible to inspect)
- All Cable cores correctly identified at joints and in accessories (at samples inspected)
- All cable joints correctly terminated, secure and accessible (at samples inspected)
- Modifications to the Building Fabric appropriate and safe (Structure) (where readily accessible to inspect)
- Modifications to the Building Fabric appropriate and safe (Fire) (where readily accessible to inspect)
- All Accessories correctly placed as per Approved Document M and BS 8300 (where applicable)
- Appropriate Supplementary Bonding present and adequately sized
- Protective Bonding securely connected and a warning label fitted if required
- Additional protection provided by RCD where required
- All Accessories have environmental protection appropriate for external influences
- All covers replaced, accessories secure and neatly aligned
- The number of points and their location agree with the original design (if available)
- Original circuit details correct on the installation schedule
- Periodic Label, RCD label and other Safety Labels fitted
- If installation has wiring colours of two versions to BS 7671, has warning label been fitted?
- Appropriate measures taken in Special Locations
- Fire Detection units installed
- Other *(please state)* ____
- Other *(please state)* ____
- Other *(please state)* ____
- Other *(please state)* ____
- Other *(please state)* ____
- Other *(please state)* ____

Schedule of Test

External earth loop impedance, Ze	____	Insulation Resistance between Live conductors	____
Installation earth electrode	____	Insulation Resistance between Live conductors & earth	____
Prospective fault current IPF	____	Polarity and phase sequence	____
Continuity of Earth Conductors	____	Earth fault loop impedance	____
Continuity of Circuit Protective Conductors	____	RCSs / RCBOs including discrimination	____
Continuity of Bonding Conductors	____	Functional testing of devices	____

The sections above are –
Yes (Y), No (N), Not Known (N/K), Satisfactory (✓), Not Satisfactory (X), Not Checked (N/C), Not Applicable (N/A) or Limitations (Lim)

Inspector's Name ____

Signature ____ Date ____

NAPIT Administration Centre, 4th Floor, Mill 3, Pleasley Vale Business Park, Mansfield, Nottinghamshire NG19 8RL

Sheet 3 of 4 NA/PIR/001 (V1)

© Copyright NAPIT Jan 2008

© Construction Industry Training Board   6B/8   E1: BS 7671 (February 2008)

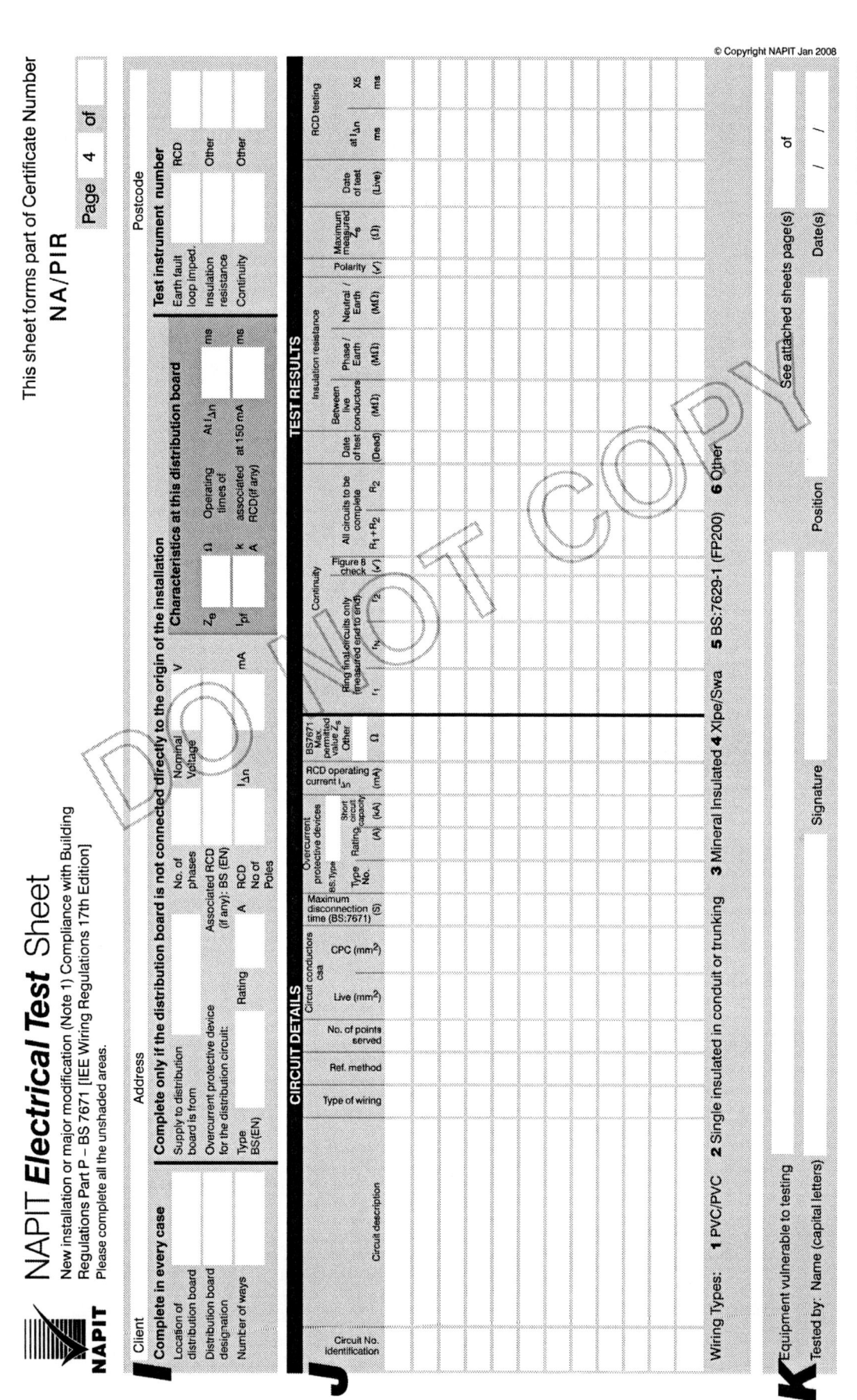

NAPIT *Electrical Test* Sheet

New installation or major modification (Note 1) Compliance with Building Regulations Part P – BS 7671 [IEE Wiring Regulations 17th Edition]
Please complete all the unshaded areas.

This sheet forms part of Certificate Number **NA/PIR**

Page 4 of ☐

Client	Address		Postcode	

I Complete in every case

		Complete only if the distribution board is not connected directly to the origin of the installation		Characteristics at this distribution board		Test instrument number	
Location of distribution board		Supply to distribution board is from		No. of phases		Earth fault loop imped.	RCD
Distribution board designation		Overcurrent protective device for the distribution circuit:		Z_s Ω / I_{pf} kA	Nominal Voltage V / $I_{\Delta n}$ mA	Insulation resistance	Other
Number of ways		Type BS(EN) / Rating A	Associated RCD (if any): BS (EN) / RCD No of Poles	Operating times of associated RCD (if any) At $I_{\Delta n}$ ms / at 150 mA ms		Continuity	Other

J CIRCUIT DETAILS / TEST RESULTS

Circuit No. identification	Circuit description	Type of wiring	Ref. method	No. of points served	Circuit conductors csa Live (mm²)	CPC (mm²)	Maximum disconnection time (BS:7671) (S)	Overcurrent protective devices BS:Type / Type No. / Rating (A) / Short circuit capacity (kA)	RCD operating current $I_{\Delta n}$ (mA)	BS7671 Max permitted value Z_s Other Ω	Continuity Ring final circuits only (measured end to end) r_1 / r_n / r_2	Figure 8 check (✓)	All circuits to be complete R_1+R_2 / R_2	Date of test (Dead)	Insulation resistance Between live conductors (MΩ)	Phase / Earth (MΩ)	Neutral / Earth (MΩ)	Polarity (✓)	Maximum measured Z_s (Ω)	Date of test (Live)	RCD testing at $I_{\Delta n}$ ms	X5 ms

Wiring Types: 1 PVC/PVC   2 Single insulated in conduit or trunking   3 Mineral Insulated   4 Xlpe/Swa   5 BS:7629-1 (FP200)   6 Other

K

Equipment vulnerable to testing			See attached sheets page(s) of
Tested by: Name (capital letters)	Signature	Position	Date(s) / /

© Copyright NAPIT Jan 2008
Sheet 4 of 4 NA/PIR/001 (V1)

NAPIT *Electrical Report* sheet

Requirements for Electrical Installations –
BS 7671 [IEE Wiring Regulations 17th Edition]
Can be used for appending to Electrical Certificate for reporting existing work or for Continuation sheet for a Periodic Inspection Report.

This sheet forms part of *Inspection Report Number/*Certificate Number

NA/*EC/*PIR *Delete as applicable

Page ___ of ___

Site: _____
Inspector: _____   Date: _____

Observations and recommendations for actions to be taken
Referring to the attached schedule of inspection and test results, and subject to the limitations at 4:

☐ No remedial work required   OR   ☐ The following observations and recommendations are made

Explanation of codes
1. Requires urgent attention
2. Requires improvement
3. Requires further investigation
4. Does not comply with BS 7671 (as amended)

Item No.	Defect details	Code

Urgent remedial work recommended for items: _____
Corrective action(s) recommended for items: _____

NAPIT Administration Centre, 4th Floor, Mill 3, Pleasley Vale Business Park, Mansfield, Nottinghamshire NG19 8RL

Sheet 1 of 1   NA/IR/001 (V1)

© Copyright NAPIT Jan 2008

© Construction Industry Training Board   6B/10   E1: BS 7671 (February 2008)

SPECIAL INSTALLATIONS OR LOCATIONS — BS 7671 Part 7

Introduction

Special installations or locations are covered in Part 7 of BS 7671. The particular requirements for each of them supplement or modify the general requirements of BS 7671.

The absence of reference to the exclusion of a chapter, section or clause means the relevant general Regulations apply.

The number quoted after a section number refers to a chapter, section or Regulation within Parts 1 to 6. This means that the numbering does not necessarily follow sequentially.

The special installations or locations are contained in separate sections, as follows:

Locations containing a bath or shower	Section 701
Swimming pools and other basins	Section 702
Rooms and cabins containing sauna heaters	Section 703
Construction and demolition site installations	Section 704
Agricultural and horticultural installations	Section 705
Conducting locations with restricted movement	Section 706
Electrical installations in caravan/camping parks and similar locations	Section 708
Marinas and similar locations	Section 709
Medical locations (reserved for future use)	Section 710
Exhibitions, shows and stands	Section 711
Solar photovoltaic (PV) power supply systems	Section 712
Mobile or transportable units	Section 717
Electrical installations in caravans and motor caravans	Section 721
Temporary electrical installations for structures, amusement devices and booths at fairgrounds, amusement parks and circuses	Section 740
Floor and ceiling heating systems	Section 753

Locations containing a bath or shower *(701)*

This section covers baths, showers (any associated cubicles) and their surroundings. It does not apply to emergency facilities in laboratories or industrial areas. Areas that contain baths or showers for medical treatment or disabled persons may require special consideration.

Zone classification

a)   Bath tub

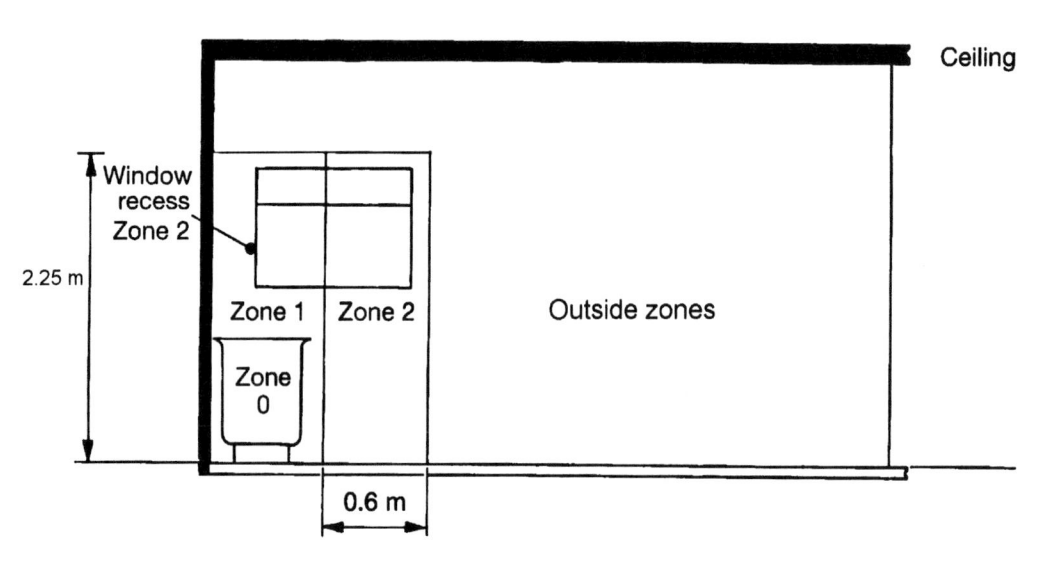

b)   Shower basin

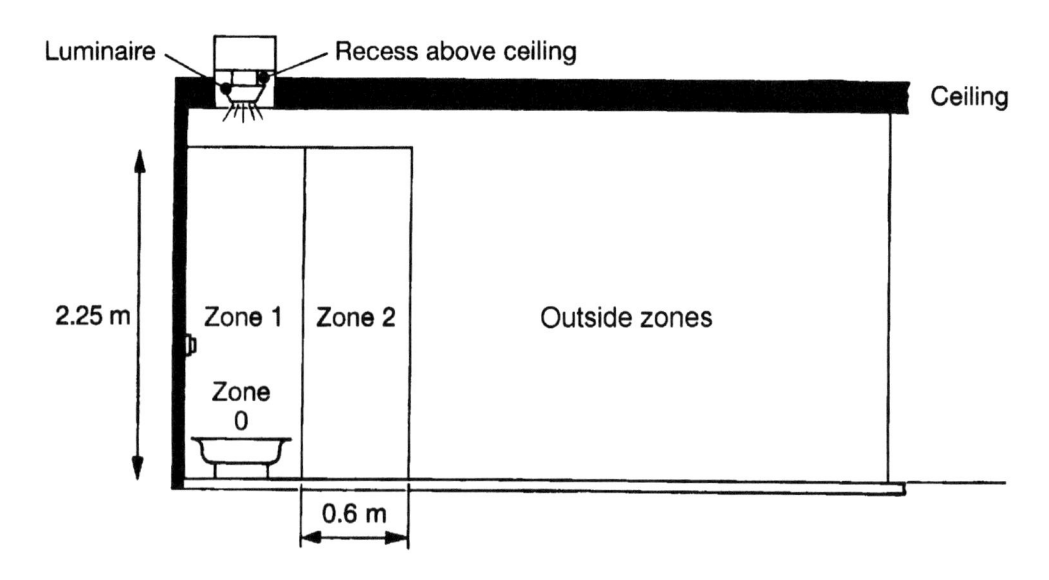

c) Shower without basin, with permanent fixed partition

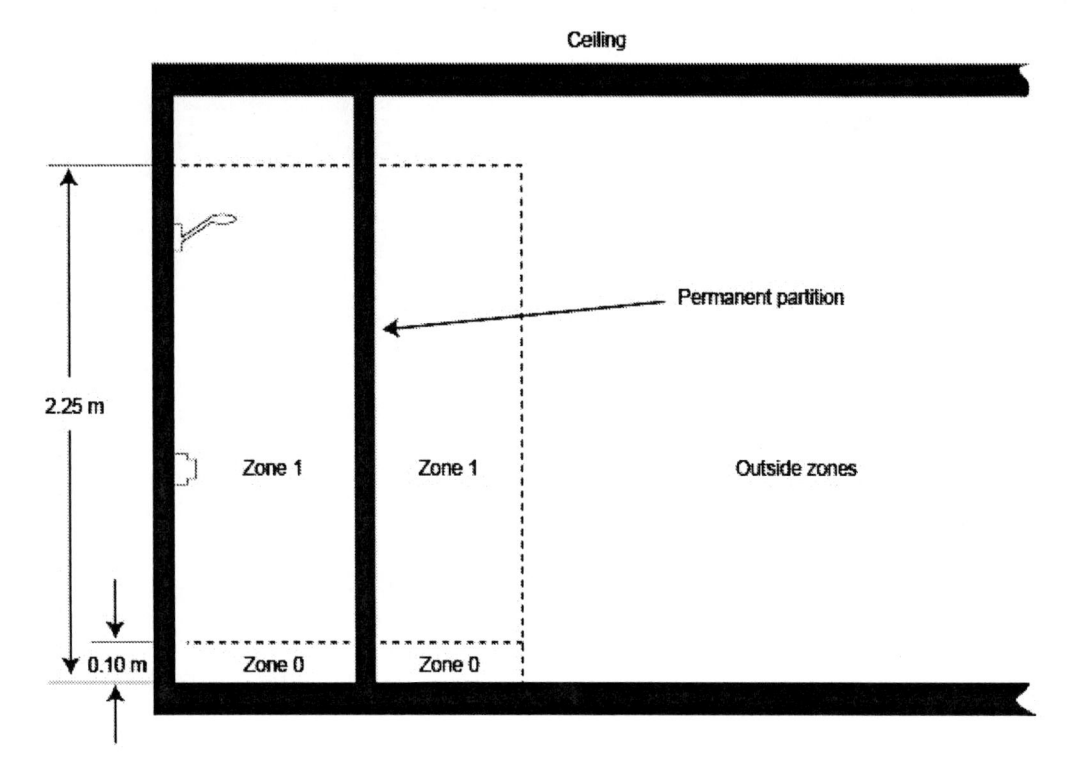

- Ceiling
- Permanent partition
- 2.25 m
- 0.10 m
- Zone 1
- Zone 1
- Outside zones
- Zone 0
- Zone 0

d) Bath tub

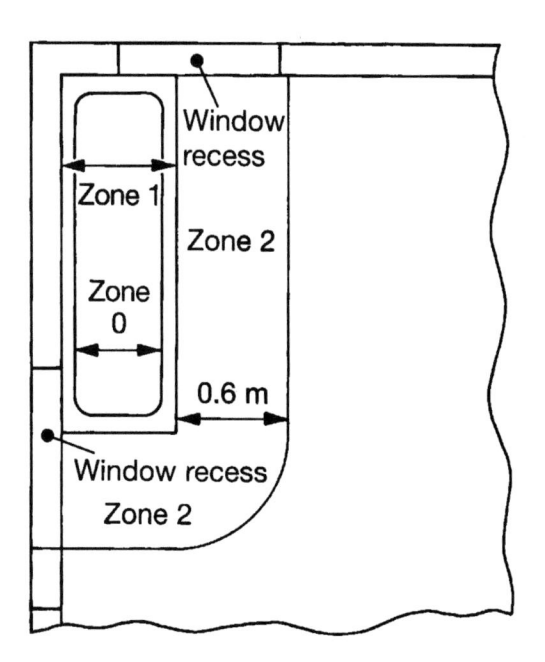

- Window recess
- Zone 1
- Zone 0
- Zone 2
- 0.6 m
- Window recess
- Zone 2

e) Shower basin

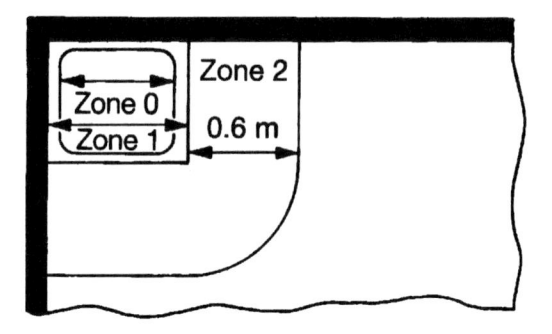

f) Shower basin with permanently fixed partition

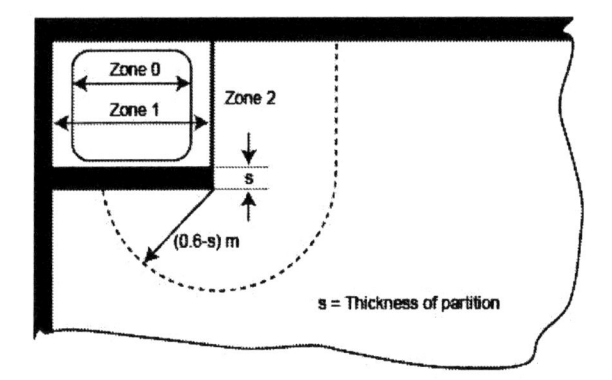

g) Shower without basin

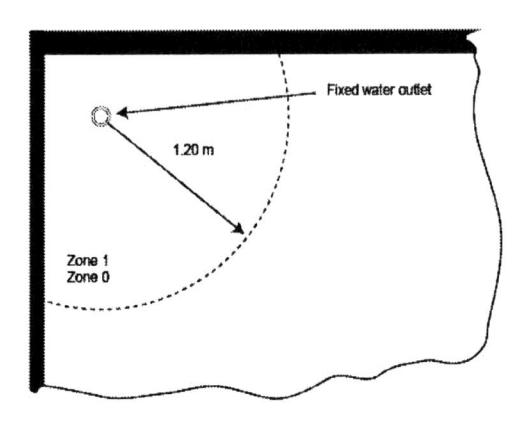

h) Shower without basin but with permanently fixed partition

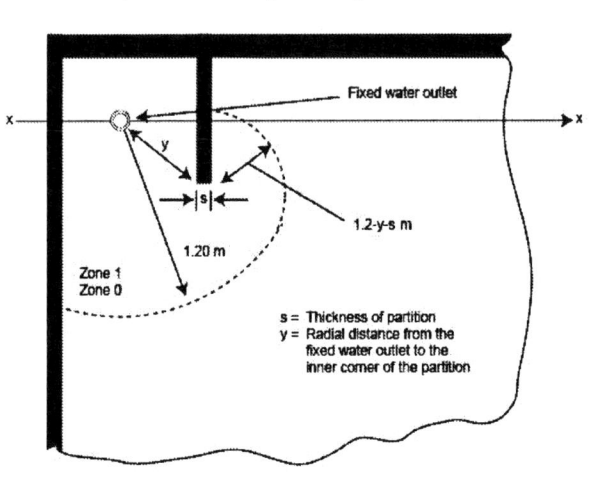

In the diagrams shown on the previous pages, Zone 0 is the interior of the bath tub or shower basin. When showers are installed in a room without a basin the height of Zone 0 is 0.10 m and its horizontal extent is the same as Zone 1.

The area below the bath or shower basin is classed as Zone 1 if access is without the use of a tool. If access is only by the use of a tool then this space is considered to be outside of the zones.

Electric shock protection

Where SELV or PELV is used, basic protection for equipment within zones 0, 1 and 2 shall be provided by either:

- basic insulation complying with Regulation 416.1, or
- barriers or enclosures to IP2X or IPXXB (see Regulation 416.2).

Additional protection

RCDs

Additional protection for all circuits shall be provided by the use of one or more RCDs with characteristics as specified in Regulation 415.1.1 (not exceeding 30 mA).

Supplementary bonding

It will often be possible to dispense with supplementary bonding altogether and the conditions for this will be explained after the supplementary bonding section.

Local supplementary equipotential bonding must be used to connect together the terminals of the protective conductor of circuits supplying Class I and Class II equipment to accessible extraneous-conductive-parts in the location including:

- metal water, gas and other service pipes, metal waste pipes
- metal central heating pipes and air-conditioning systems
- accessible metal structural parts of the building (metal door and window frames etc. are not considered to be extraneous-conductive-parts unless they are connected to structural metal parts of the building).

An example of supplementary bonding in a bathroom with metal pipework is shown below.

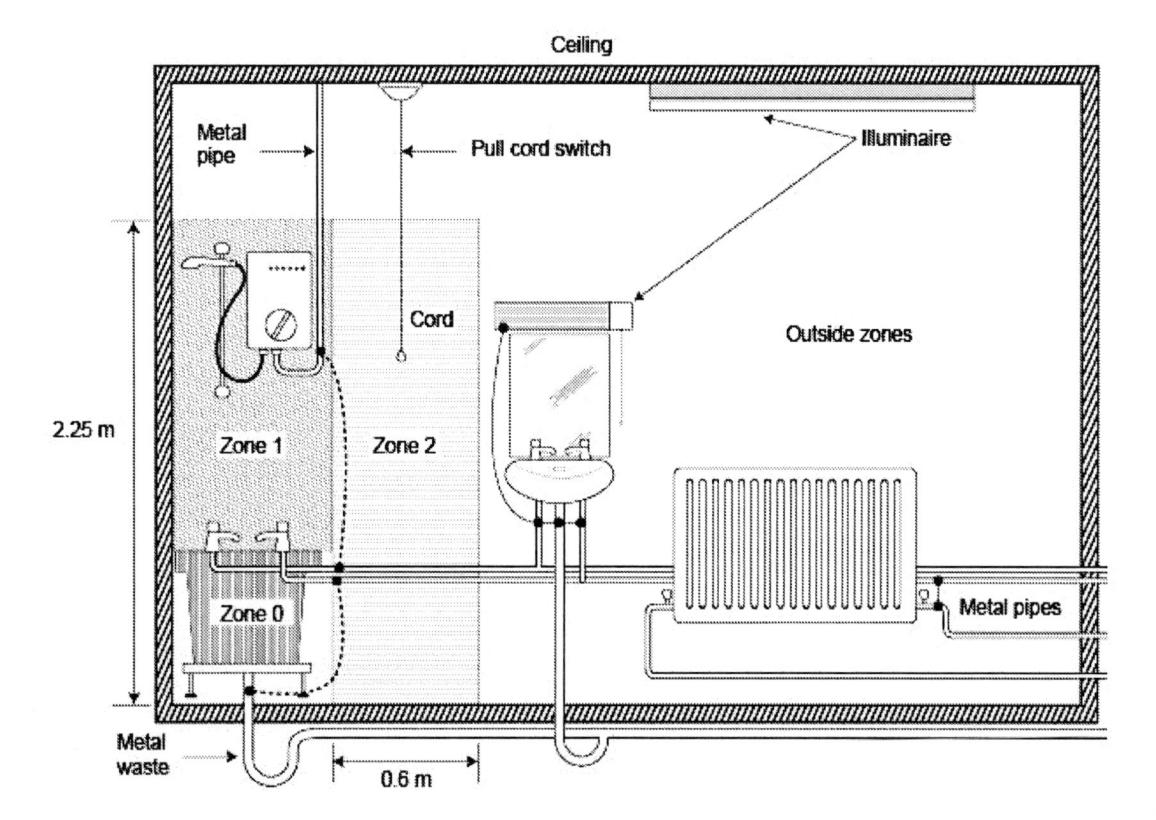

Supplementary bonding may be provided in close proximity to the location, e.g. an airing cupboard adjoining a bathroom or in the roof space above the location. It should preferably be as close to the point of entry of the extraneous-conductive-parts into such locations.

Supplementary bonding may be provided by a:

- conductor
- conductive part of a permanent and reliable nature (e.g. metal pipework with soldered or compression type fittings that are electrically continuous)
- combination of the above.

Plastic pipework

The introduction of plastic pipework and push-fit type fittings for use on metal pipework have had an effect on supplementary bonding requirements, particularly in bathrooms.

For plastic pipe installations there is no need to supplementary bond metal fittings supplied by plastic pipe (e.g. hot and cold taps or radiators supplied by plastic pipes).

The use of plastic pipework can create a safer electrical installation. Installing supplementary bonding to equipment connected to plastic pipework is unnecessary and could reduce levels of electrical safety, rather than increase them.

An example of supplementary bonding in a bathroom with plastic pipework is shown below.

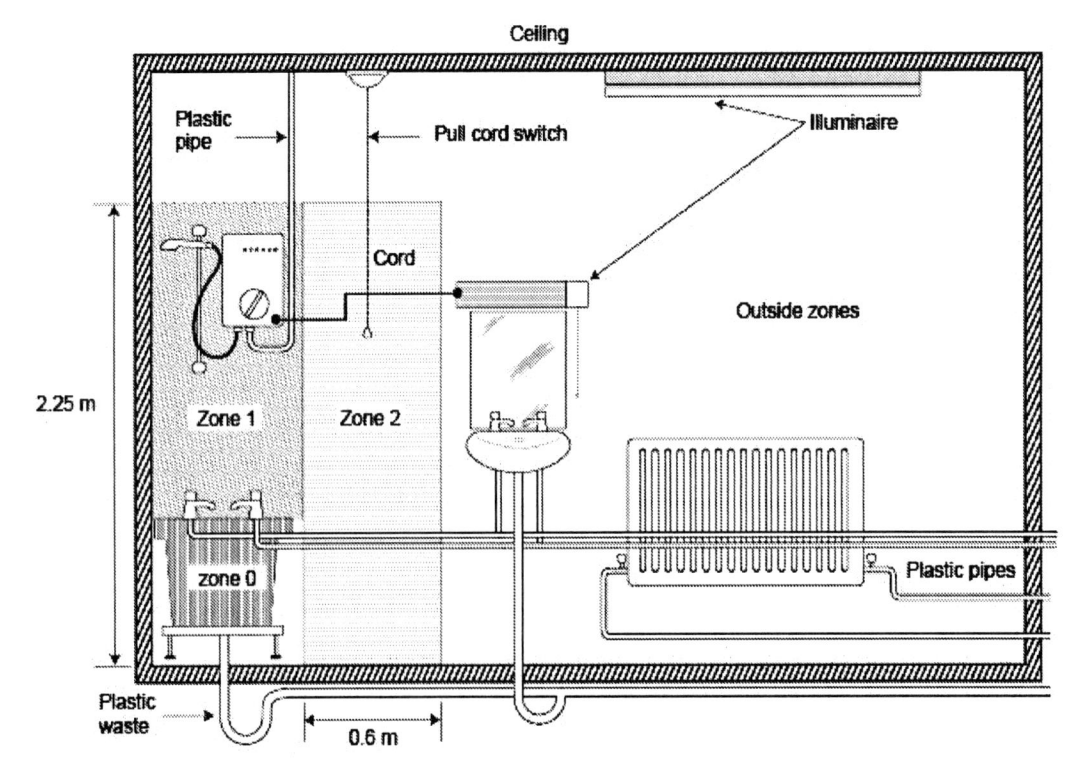

Omission of supplementary bonding

When the location containing a bath or shower is within a building where a protective equipotential bonding system has been installed in accordance with Regulation 411.3.1.2, the **supplementary bonding may be omitted when** all of the following requirements have been met.

- Final circuits of the location comply with Regulation 411.3.2 for automatic disconnection.

- All final circuits within the location have additional protection provided by an RCD in accordance with Regulation 701.411.3.3 (not exceeding 30 mA).

- Extraneous-conductive-parts in the location are effectively connected to the protective equipotential bonding in accordance with Regulation 411.3.1.2.

External influences

Electrical equipment shall have at least the following IP ratings:

Zone 0   IPX7

Zone 1&2 IPX4
          IPX5 (where water jets are used for cleaning purposes)
          This does not apply to shaver supply units to BS EN 61558-2-5 installed in Zone 2 where direct spray from showers is unlikely.

Switchgear

These requirements do not apply to the controls of fixed current-using equipment (suitable for use in a given zone) incorporated in the equipment or to cords of pull switches.

Zone 0   No switchgear or accessories to be installed.

Zone 1   Only switches for SELV (Max 12 V a.c. or 30 V ripple-free d.c.) with the safety supply source being installed outside Zones 0, 1 or 2 to be installed.

Zone 2   No switchgear, accessories with switches or socket-outlets to be installed, except for:

- switches or socket-outlets for SELV circuits (safety supply source must be installed outside Zones 0, 1 or 2)

- shaver supply units to BS EN 61558-2-5.

The following socket-outlets are allowed:

- SELV complying with Section 414

- shaver units to BS EN 6155.2.5.

In addition, 13 A socket-outlets to BS 1362 (the normal socket-outlets for household and similar use) may be installed but must be located at least 3 m horizontally from the boundary of Zone 1.

Current-using equipment

Zone 0   Only fixed and permanently connected SELV current-using equipment specifically designed for that zone to be installed according to manufacturer's instructions.

Zone 1   If it is specifically designed to Zone 1, the following may be installed in accordance with manufacturer's instructions:

- whirlpool baths/units
- electric showers
- shower pumps
- ventilation equipment (extract fans)
- towel rails
- water heaters
- luminaires
- SELV or PELV equipment not exceeding 25 V a.c. or 60 V ripple-free d.c. when the supply source is outside Zones 0, 1 or 2.

Electric floor heating system

Only heating cables relating to product standards or thin sheet flexible heating elements according to relevant equipment standards shall be installed, provided they have a metal sheath, metal enclosure or a fire metallic grid which is connected to the protective conductor of the supply circuit cable. If a protective measure is provided by a SELV supply there is no requirement to connect to the supply circuit protective conductor.

Swimming pools and other basins *(702)*

This section applies to the basins of swimming pools, fountains, paddling pools and surrounding zones. Special requirements may apply to medical pools. Swimming pools which are within the scope of an equipment standard are outside the scope of BS 7671.

Zone classification

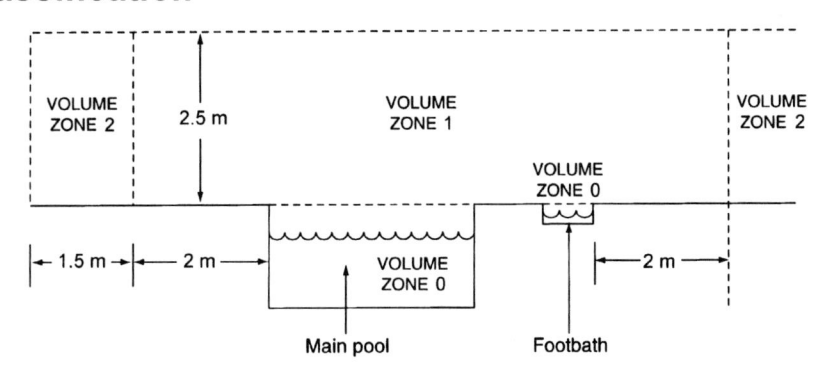

```
VOLUME ZONE 2 | 2.5 m | VOLUME ZONE 1 | VOLUME ZONE 2
                                        VOLUME ZONE 0
|← 1.5 m →|← 2 m →|  VOLUME ZONE 0  |← 2 m →|
                   Main pool          Footbath
```

When diving boards, springboards, starting blocks or a chute exist, then Zone 1 extends 1.5 m from the edge of these items and 2.5 m above the highest surface expected to be occupied by a person when using them.

Note: There is no Zone 2 for fountains.

Application of protective measures against electric shock

In Zone 0 only SELV (maximum voltage 12 V a.c. rms or 30 V ripple-free d.c.) must be used and the safety source must be outside Zones 0, 1 and 2. In Zone 1, only SELV (maximum voltage 25 V a.c. rms or 60 V ripple-free d.c.) can be used with the safety source being outside Zones 0, 1 and 2.

Where electrical equipment is to be used in the interior of basins, which are intended for operation only when people are not inside the basin, one or more of the following methods shall be used:

- SELV with the supply source installed outside Zones 0, 1 and 2. If RCD protection in accordance with Regulation 415.1.1 is provided for the supply circuit the supply source can be installed in Zone 2

- automatic disconnection of the supply using RCDs having characteristics according to Regulation 415.1.1

- electrical separation, in accordance with Section 413 for the supply source, to one item of equipment installed outside Zones 0, 1 and 2. If RCD protection in accordance with Regulation 415.1.1 is provided for the supply source, it can be installed in Zone 2.

Both the socket-outlet supplying such equipment and its control device must have a notice to warn users that the equipment must not be used when persons are inside the basin.

When the supply to a swimming pool is TN-C-S it is recommended that an earth mat or earth electrode be installed and connected to the protective equipotential bonding of the installation. The value should be 20 Ω or less.

The following protective measures must not be used in any zone:

- obstacles
- placing out of reach
- non-conducting location
- earth-free local equipotential bonding.

Additional protection using supplementary bonding

In Zones 0, 1 and 2 any extraneous-conductive-parts shall be connected by supplementary bonding conductors to the protective conductors of exposed-conductive-parts of equipment in the zones. Connection with protective conductors can be achieved in a local distribution board or accessory in the proximity of the location.

Selection and erection of equipment

External influences

Equipment must have at least the following levels of protection:

- Zone 0   IPX8
- Zone 1   IPX4 (IPX5 where water jets are likely to be used for cleaning purposes)
- Zone 2   IPX2 indoor locations
- Zone 2   IPX4 outdoor locations
- Zone 2   IPX5 locations where water jets are likely to be used for cleaning purposes.

Wiring systems

The following applies to surface wiring systems and those embedded in walls, ceilings or floors when their depth is less than 50 mm.

- Metallic sheath or coverings of wiring systems must always be connected to the supplementary equipotential bonding of the installation (Zones 0, 1 and 2).
- Cables should, where possible, be installed in PVC conduit (Zones 0, 1 and 2).
- Wiring systems installed in Zones 0 and 1 shall only supply equipment in those zones.

Additional requirements for fountains

Electrical cables, which supply equipment in Zone 0, shall be installed outside the basin rim as far away as possible and cable runs to equipment shall be kept short. In Zone 1 installed cables shall have mechanical protection to AG2 and, if submersed in water to a depth AD8, cable type H07RN8-F to BS 7919 is suitable. In the case where installation is to depths greater than 10 m, cable manufacturers should be consulted.

Electric pumps shall comply with BS EN 60335-2-41.

Junction box

In Zones 0 or 1 a junction box shall not be installed. If a SELV circuit is used, a junction box may be installed in Zone 1.

Switchgear, control gear and accessories

Switchgear, control gear and socket-outlets must not be installed in Zones 0 and 1.

However, for swimming pools where it is not possible to locate socket-outlets or switches outside of Zone 1, then socket-outlets and switches may be installed providing they have non-conductive cover plates and are:

- more than 1.25 m outside the border of Zone 0
- at least 0.3 m above the floor
- protected by an RCD or electrical separation (the safety isolating transformer being outside Zones 0 and 1).

Other equipment

In Zones 0 and 1, only current-using equipment specifically for swimming pools is allowed to be installed.

Underwater luminaires

Only fixed luminaires complying with BS EN 60958-2-18 shall be used if they are in contact with water.

When underwater lighting is located behind watertight portholes, and they are serviced from behind, they shall comply with the relevant part of BS EN 60598. Careful installation is required to ensure there is no conductive connection between any exposed-conductive-part of the luminaire and conductive parts of the porthole.

Special requirements for electrical equipment in Zone 1

For filtration systems, jet stream pumps or other equipment, designed to be fixed in a swimming pool or other basin, in Zone 1 the equipment shall:

- be within an insulating enclosure Class II or equivalent insulation with mechanical impact protection (AG2) medium severity

- only be accessible via a door or hatch using a key or tool. Opening of the door or hatch shall disconnect all live conductors. Supply cables and the main means of disconnection shall be to Class II or equivalent insulation

- have a supply circuit protected by the three methods specified on page 7/9 for application of protective measures against electric shock.

Where swimming pools have no Zone 2, lighting equipment supplied by other than a SELV source at 12 V a.c. rms or 30 V ripple-free d.c. can be installed on a wall or ceiling in Zone 1, provided that:

- the circuit is protected by automatic disconnection and RCD protection is provided in accordance with Regulation 415.1.1

- equipment must be at least 2 m above the lower limit of Zone 1.

Rooms and cabins containing sauna heaters *(703)*

Scope

This section applies to sauna cabins which have been erected on site and are situated in a room or a particular location, and covers the sauna heating appliance and its location. Prefabricated sauna units complying with an appropriate standard are not covered.

Zone classification

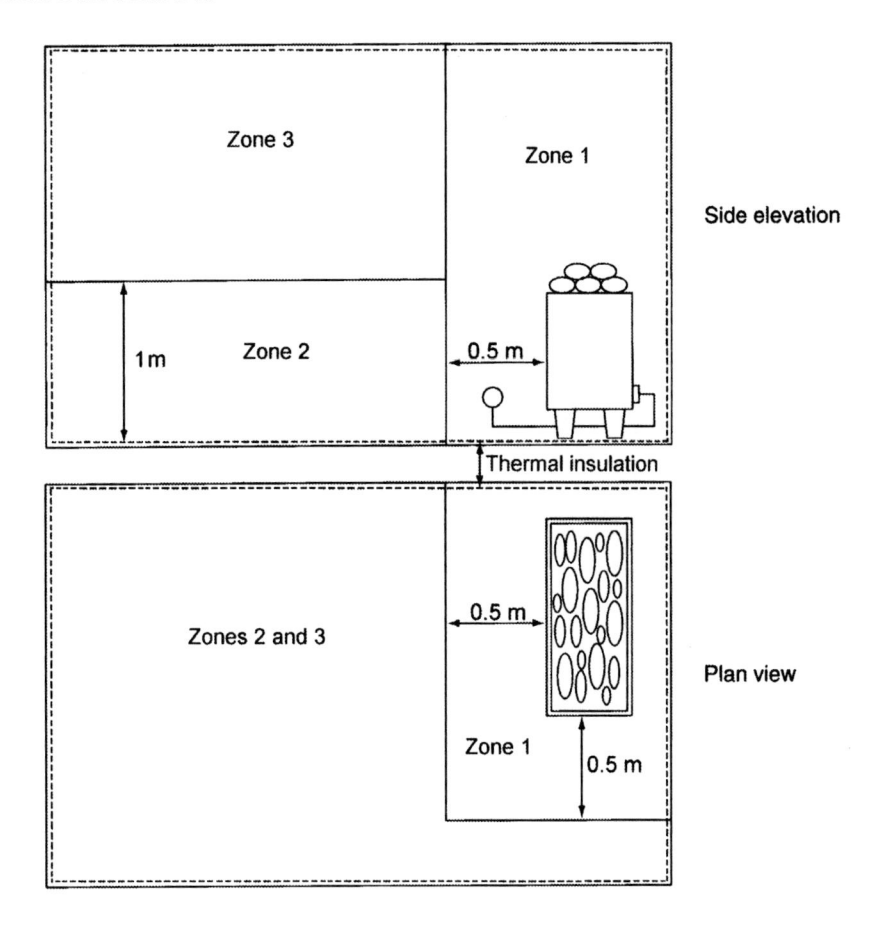

Side elevation

Plan view

Protection against electric shock

When SELV or PELV supply systems are used one of the following may be used:

- basic insulation complying with Regulation 416.1
- barriers or enclosures affording protection to IP2X or IPXXB.

Note: Unless specified by the manufacturer of the sauna RCD protection is not required.

External influences

The equipment shall be protected to at least IPX4, but if water jets are being used for cleaning then a minimum of IPX5 is required.

Zone requirements

Zone 1   Sauna equipment only and equipment directly associated with it.

Zone 2   No special requirement concerning heat-resistance of equipment.

Zone 3   Equipment suitable for an ambient temperature of 125°C and cable insulation and sheaths able to withstand temperatures of 170°C.

Wiring systems

Wiring systems should preferably be installed outside the zones on the cold side of any thermal insulation. If wiring has to be installed in Zones 1 or 3 on the warm side of any thermal insulation, it shall be a heat-resisting type suitable for the temperature as previously stated.

Metal conduits and metallic sheath cables, if used, shall be installed so they are not accessible when the sauna is in normal use.

Switchgear

Any switchgear and control gear which is not built into the sauna, with the exception of thermostats and thermal cut-outs, must be installed outside of the sauna.

Other equipment

Heating appliances used in saunas shall comply with BS EN 60335-2-53 and always be installed to manufacturer's instructions.

Construction and demolition site installations *(704)*

Scope

This section applies to the following types of work:

- new building construction
- repairs, alterations, extension or demolition of existing buildings
- engineering construction
- earthworks
- similar works.

This section does not apply to site offices, cloakrooms, meeting rooms, canteens, restaurants, dormitories and toilets (which are covered under the general requirements of the Regulations), or to installations covered by IEC 60621 series 2 or other installations where equipment similar to that used in surface mining operations is involved.

Protection against electric shock

Socket-outlets and other circuits supplying hand-held equipment up to and including 32 A rating shall be protected by one of the following:

- reduced low voltage (Regulation 411.8)
- RCD (Section 411 and Regulation 415.1.1)
- electrical separation of circuits (Section 413) where each socket-outlet and item of hand-held equipment is supplied from a separate transformer winding or an individual transformer
- SELV or PELV (Section 414).

The preferred practices for the supply of portable hand lamps and tools are:

- reduced low voltage system (as referred to in Technical Data Sheet 7A), for portable hand lamps for general use and portable hand tools and lighting up to a maximum of 2 kW
- SELV system for portable hand lamps, to be used in damp locations or confined spaces.

TN-C-S supply systems shall only supply fixed buildings on construction sites.

Automatic disconnection

For a circuit supplying a socket-outlet(s) exceeding 32 A, an RCD with a residual operating current not exceeding 500 mA shall be provided.

Selection and erection of equipment

All assemblies for electricity distribution on construction and demolition sites shall comply with BS EN 60439-4. Plugs and socket-outlets of 16 A or greater shall comply with BS EN 60309-2.

Wiring systems

All cables shall be adequately protected against mechanical damage, especially those installed across site roads and walkways.

Low temperature 300/500 V thermoplastic (BS 7919) or equivalent flexible cable shall be used for reduced low voltage systems. For applications exceeding this voltage use flexible cable H07RN-F (BS 7919) or equivalent, rated at 450/750 and resistant to water and abrasion.

Switchgear and control gear

Assemblies for construction sites (ACS) need to incorporate devices for isolation and switching of the incoming supply. Isolating devices shall be capable of being secured in the off position, using a padlock or located inside a lockable enclosure.

ACS current-using equipment shall incorporate the following:

- overcurrent protective devices
- fault protection devices
- socket-outlets, if required.

Refer to Technical Data Sheet 7A.

Agricultural and horticultural installations *(705)*

Scope

This section applies to fixed electrical installations both indoors and outdoors in agricultural and horticultural premises, and to other locations belonging to premises of agricultural and horticultural buildings.

Protection against electric shock

Automatic disconnection of supply

The following disconnection devices shall be provided:

- final circuits not exceeding 32 A – 30 mA RCD maximum
- final circuits exceeding 32 A – 100 mA RCD maximum
- all other circuits – 300 mA RCD maximum.

SELV and PELV supplies

Basic protection shall be provided by:

- basic insulation (Regulation 416.1)
- barriers or enclosures affording protection to IPXXB or IP2X (Regulation 416.2).

Supplementary equipotential bonding

All livestock locations, where exposed-conductive-parts and extraneous-conductive-parts can be touched by the livestock, shall be supplementary bonded. This also includes any metal grids which have been laid in floors.

Extraneous-conductive-parts in or on the floor (e.g. metal reinforcement for concrete) and spaced floors (made of prefabricated concrete) shall also be connected to the supplementary equipotential bonding.

Any metal grid shall be erected to protect it against mechanical stresses and corrosion.

TN-C-S supply systems are recommended only where a metal grid has been laid in floors. Refer to the illustration below.

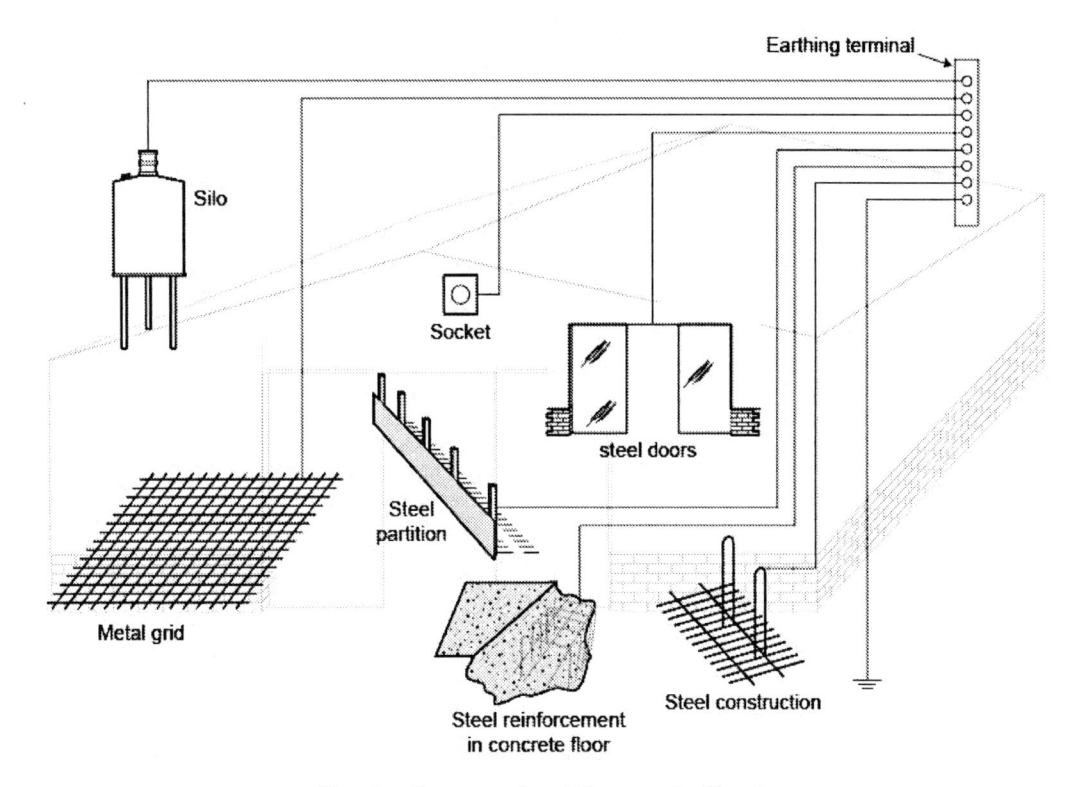

Protection against thermal effects

Electrical appliances which provide heat for breeding or rearing livestock shall comply with BS EN 60335-2-71 and be fixed at an appropriate distance from any combustible material and livestock in order to minimise the risk of burns to livestock and fire. In the case of radiant heaters, the distance shall be greater than 0.5 m or any other distances specified by the manufacturer.

RCDs not exceeding 300 mA shall be installed for fire protection which disconnects all live conductors.

To improve the continuity of the electricity supply, RCDs not protecting socket-outlets should have a time delay or be of the S type.

When a fire risk exists, conductors of circuits supplied at extra-low voltage shall have protection by barriers or enclosures to IPXXD or IP4X or be enclosed in insulating material. For outdoor use, HO7RN-F (BS 7919) cables can be installed to comply with this requirement.

Selection and erection of equipment

External influences

Electrical equipment for both agricultural and horticultural premises shall be IP44 rated or placed in an enclosure to IP44.

Socket-outlets shall be positioned to avoid contact with combustible materials. When installed in locations where the external influences are:

- AD4 presence of water, specifically splashes
- AE3 presence of very small foreign bodies
- AG1 low impact

they shall be provided with suitable protection.

Electrical equipment installed in dairies and cattle sheds shall be suitably protected against corrosive substances (e.g. cattle urine).

Accessibility

Electrical equipment unavoidably accessible to livestock (such as feed and water basins), shall be installed to avoid damage to or by the livestock.

Diagrams

The following information shall be provided for the installation user:

- a plan indicating the location of electrical equipment
- routes of any concealed cables
- a line distribution diagram
- an equipotential bonding diagram showing connection points.

Selection and erection of wiring systems

Wiring systems shall be inaccessible to livestock or be protected against mechanical damage. In agricultural areas around premises, where vehicles and machines operate, suitable cable installation methods shall be used. These include:

- buried in the ground 0.6 m deep with mechanical protection
- in arable or cultivated ground buried 1 m deep
- overhead installation by self-supporting suspension cables at 6 m height.

When conduit systems are installed where livestock is kept, they shall be suitable for classification AF4 and be protected against corrosion to the following, according to BS EN 61386-21:

- indoor use   Class 2 (medium)
- outdoor use   Class 4 (high).

Where wiring systems are exposed to impact and mechanical shock caused by vehicles and mobile machines, they shall be suitable for classification AG3. Conduits, trunking and ducting systems shall provide suitable protection against impact.

Isolation and switching

The electrical installation in either the whole or part of a building shall be provided with a single isolating device (Chapter 53), which isolates all live conductors, including the neutral, for circuits used during harvest time or occasionally.

Isolating devices shall be labelled to show which installations they isolate. Any isolation and switching devices used for emergency switching or stopping shall be inaccessible to livestock. They must be located such that access cannot be impeded by livestock.

Electrical heating appliances shall have a visual indication of the operating position.

Supplementary bonding conductors

The selection of protective bonding conductors shall be made to avoid the following:

- mechanical damage
- corrosion
- electrolytic action.

Suitable examples are:

- hot dip galvanized steel strip (at least 30 mm x 3 mm)
- hot dip round steel (at least 8 mm diameter)
- copper conductor (cross-sectional area at least 4 mm²).

Accessories

In agricultural and horticultural premises the following types of socket-outlets may be installed:

- BS EN 60309-1
- BS EN 60309-2, where interchangeability is required
- BS 1363, BS 546 or BS 196 (20 A maximum).

Luminaires shall protect against dust, solid objects and moisture, e.g. IP54. They shall be suitable for mounting on flammable surfaces and limiting their surface temperature.

Safety services

In the event of a power failure, if the supply for any of the following is not ensured an alternative source of supply shall be provided:

- food
- water
- air
- lighting.

In the case of ventilation and lighting units separate final circuits are required to enable changeover to an alternative source of supply to be effectively achieved.

When an installation requires electrically operated ventilation, one of the following is required:

- a standby electrical source capable of providing an adequate electrical supply, with a notice placed adjacently stating it should be tested periodically
- a monitoring device for the supply and/or temperature, providing a visual or audible warning, which operates independently from the normal supply source.

Conducting locations with restricted movement *(706)*

Scope

This section applies fixed equipment and supplies for mobile equipment in conducting locations where a person's movement is restricted.

A restrictive conductive location is one which consists of a large amount of metal or conductive parts to which a person within that location is likely to come into contact with a substantial part of their body. In such locations, movements are physically constrained and the possibility of preventing contact is limited. An example of such a situation would be a person working inside a large metal boiler, or certain parts of a factory or power station where the floors, walls and equipment are metallic. It is in situations like these that considerable danger could arise to persons using electric hand lamps and tools.

Protection against electric shock and automatic disconnection

For the electrical supply to hand-held tools and mobile equipment, the following methods may be used:

- electrical separation where only one item of equipment is connected to the secondary winding of transformers (Section 413)
- SELV supply in accordance with Section 414.

For the electrical supply to handlamps the following method shall be used:

- SELV supply. A fluorescent luminaire may be used if the SELV circuit which supplies it is via a step-up transformer having windings which are electrically separate.

For the electrical supply to fixed equipment the following methods may be used:

- supplementary bonding. This bonding shall connect exposed-conductive-parts of the equipment and the conductive parts of the location (Section 411)

- Class II equipment or equipment having insulation to an equivalent standard (Section 412). The supply circuit shall be protected by an RCD complying with Regulation 415.1.1

- electrical separation where only one item of equipment is connected to a transformer's secondary winding

- SELV (Section 414)

- PELV. All exposed-conductive-parts and extraneous-conductive-parts within the location shall be equipotentially bonded. The PELV system shall be connected to earth (Section 414).

SELV or PELV supply sources shall be located outside the conductive location.

Electrical installations in caravan/camping parks and similar locations *(708)*

Scope

This section applies to electrical installations in caravans/camping parks and similar locations where facilities are provided for leisure accommodation vehicles, such as caravans, and tents.

The following requirements do not apply to leisure accommodation vehicles or mobile/transportable units themselves. Section 721 covers electrical installations in caravans and motor caravans.

Protection for safety

Protection against electric shock

Only TN-S installations can be used for supplies on site. A TN-C-S system can only be used to supply permanent buildings.

Where a TN-C-S supply has been used for site distribution, the protective conductor must not be connected to the socket-outlets supplying the caravans.

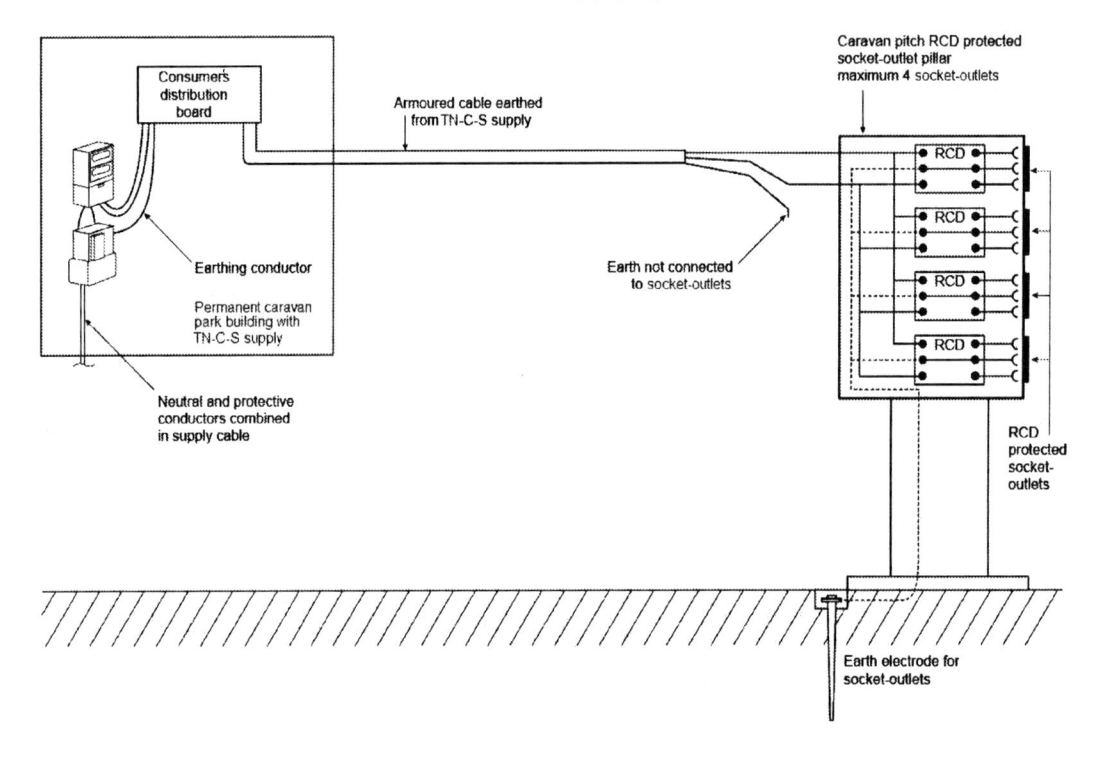

Selection and erection of equipment

External influences

When electrical equipment is installed outside in caravan parks it should comply with the following external influences:

- water   AD4, splashes   IPX4
- foreign bodies   AE2, small objects   IP3X
- mechanical stress AG3, high severity   IK08.

Caravan park wiring systems

The types of wiring systems which are suitable for the distribution circuits for caravans or tent pitches are:

- underground
- overhead.

When underground cables are installed, they shall be at least 0.6 m below ground level, be placed outside caravan pitch areas and any ground where ground anchors or tent pegs are likely to be used. Alternatively, additional mechanical protection is required for the cable to prevent damage.

When overhead conductors are installed, poles and supports shall be located or protected to avoid any damage from vehicles. The installation height is a minimum of 6 m above ground level where vehicles are present and 3.5 m for all other locations.

Switchgear and control gear

Every caravan pitch shall be supplied from a BS EN 60309-2 socket-outlet, which must be of the splashproof type (IPX4) with a minimum current rating of 16 A, individually protected by an RCD, with its own overcurrent protection.

The socket-outlet must be installed between 0.5 m and 1.5 m from the ground measured from the lowest part of the socket-outlet. Socket-outlets must be located within 20 m of the caravan pitch. No more than four socket-outlets shall be grouped together at a single location.

The protective conductors of socket-outlets shall not be connected to the PEN conductor of a TN-C-S system, but shall be connected to an earth electrode complying with Regulation 411.5 for TT systems.

Marinas and similar locations *(709)*

Scope

This section covers circuits which are intended to supply pleasure craft or houseboats in marinas and similar locations. It does not apply to houseboats which are supplied from the public network.

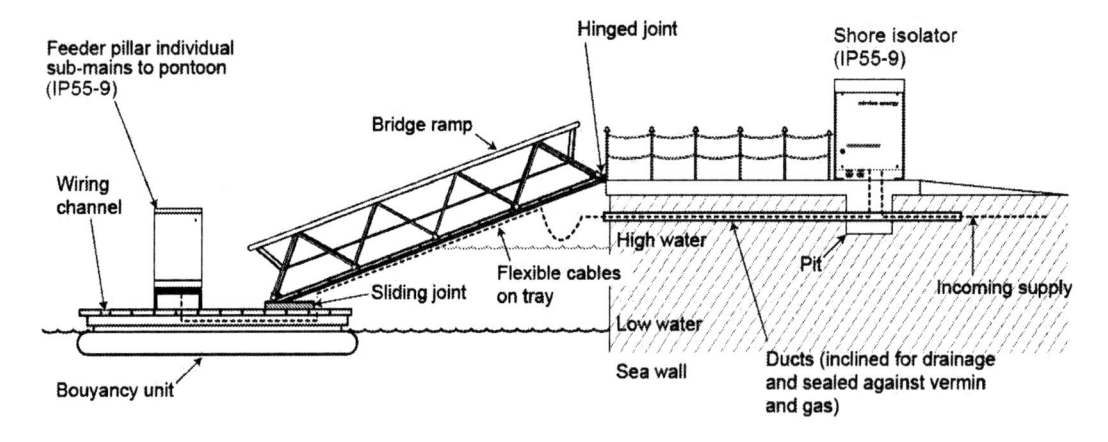

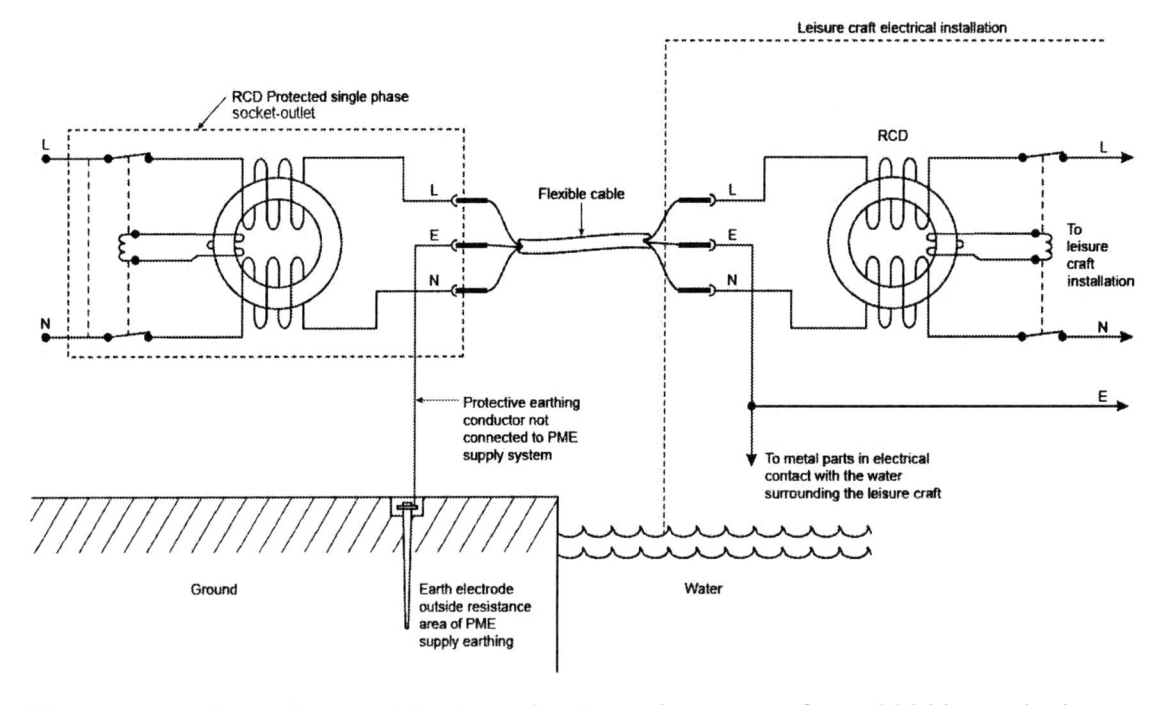

The supply voltages for supplying houseboats or pleasure craft are 230 V a.c. single-phase and 400 V a.c. three-phase supplied from a TN-S or TT system. TN-C-S supply systems shall not be used.

Selection and erection of equipment

Equipment installed on or above a jetty, wharf, pier or pontoon shall be selected to cope with the following external influences:

- splashes (AD4) IPX4
- water jets (AD5) IPX5
- waves (AD6) IPX6
- ingress of small objects (AE2) IP3X
- corrosion or polluting substances (AF2), if hydrocarbons present (AF3)
- impact of medium severity (AG2).

For impact protection, one or more of the following shall be provided for the equipment:

- careful selection of position or location
- local or general mechanical protection
- installation providing a minimum degree of mechanical protection to IK08.

Types of wiring system

The following cables are suitable for distribution circuits:

- underground
- overhead
- PVC covered mineral insulated
- armoured, covered in thermoplastic or elastomeric material
- copper conductor with thermoplastic or elastomeric insulation and sheathed, installed in a suitable wiring system.

The following types of wiring shall not be used:

- cables suspended from or incorporating a support wire
- conduit and trunking with non-sheathed cables
- cables with aluminium conductors
- mineral insulated cables.

Installations shall be installed in such a way that they prevent the build up of water, by having drainage holes and installing equipment on an incline if necessary. The installation of underground cables shall be at a depth to avoid damage (usually a minimum of 0.5 m). Overhead cables shall be insulated and any supporting poles and other supports installed to prevent damage by vehicles. The height of conductors shall be at least 6 m where vehicles pass, otherwise 3.5 m.

Fault protection

Any socket-outlet and any final circuits for supply to houseboats shall be individually protected by an RCD in accordance with Regulation 415.1.1. It must disconnect all poles, including the neutral.

An overcurrent protective device shall protect the fixed supply connection point for supply to each houseboat.

Isolation and switching

Where distribution cabinets are installed they shall be fitted with at least one isolating device, which may isolate up to a maximum of four socket-outlets located in the enclosure.

Other equipment

The types of socket-outlet that can be used are listed below and shall provide protection to IP44 or be installed in an enclosure that does:

- BS EN 60309-1 for loads 63 A and above
- BS EN 60309-2 for loads up to 63 A.

Socket-outlets shall be located as close as possible to the berth being supplied and in a distribution board or separate enclosure.

A socket-outlet may only supply one pleasure craft or houseboat. They shall be installed at least 1 m above the highest water level. In the case of floating pontoons or walkways this can be reduced to 300 mm above the highest water level, provided that suitable precautions are taken to give protection against splashing.

The marine operator shall provide a copy of the instruction on the next page to the pleasure craft operator. A weather protected copy shall be installed adjacent to socket-outlets. The copy illustrated is for a single-phase supply. For three-phase the voltage becomes 400 and the colour red.

Instructions for electricity supply

Berthing instructions for connection to shore supply

This marina provides power for use on your leisure craft with a direct connection to the shore supply which is connected to earth. Unless you have an isolating transformer fitted on board to isolate the electrical system on your craft from the shore supply system, corrosion through electrolysis could damage your craft or surrounding craft.

On arrival

i. Ensure the supply is switched off and disconnect all current-using equipment on the craft, before inserting the craft plug. Connect the flexible cable **firstly** at the pleasure-craft inlet socket and **then** at the marina socket-outlet.

ii. The supply at this berth is 230 V, 50 Hz. The socket-outlet will accommodate a standard marina plus colour blue (technically described as BS EN 60309-2 position 6 h).

iii. For safety reasons, your craft must not be connected to any other socket-outlet than that allocated to you and the internal wiring on your craft must comply with the appropriate standards.

iv. Every effort must be made to prevent the connecting flexible cable from falling into the water if it should become disengaged. For this purpose, securing hooks are provided alongside socket-outlets for anchorage at a loop of tie cord.

v. For safety reasons, only one pleasure-craft connecting cable supplying one pleasure craft may be connected to any one socket-outlet.

vi. The connecting flexible cable must be in one length, without signs of damage and not contain joints or other means to increase its length.

vii. The entry of moisture and salt into the pleasure craft inlet socket may cause a hazard. Examine carefully and clean the plug and socket before connecting the supply.

viii. It is dangerous to attempt repairs or alterations. If any difficulty arises, contact the marina management.

Before leaving

i. Ensure that the supply is switched off and disconnect all current-using equipment on the craft, before the connecting cable is disconnected and any tie cord loops are unhooked.

ii. The connecting flexible cable should be disconnected **firstly** from the marina socket-outlet and **then** from the pleasure-craft inlet socket. Any cover that may be provided to protect the inlet from weather should be securely replaced. The connecting flexible cable should be coiled up and stored in a dry location where it will not be damaged.

Exhibitions, shows and stands *(711)*

Scope

This section covers temporary electrical installations in exhibitions, shows, stands and any mobile or portable displays and equipment.

The supply voltages for such temporary electrical installations are:

- 230/440 V a.c.
- 500 V d.c.

supplied by TN-S or TT supply systems. TN-C-S supply systems shall not be used.

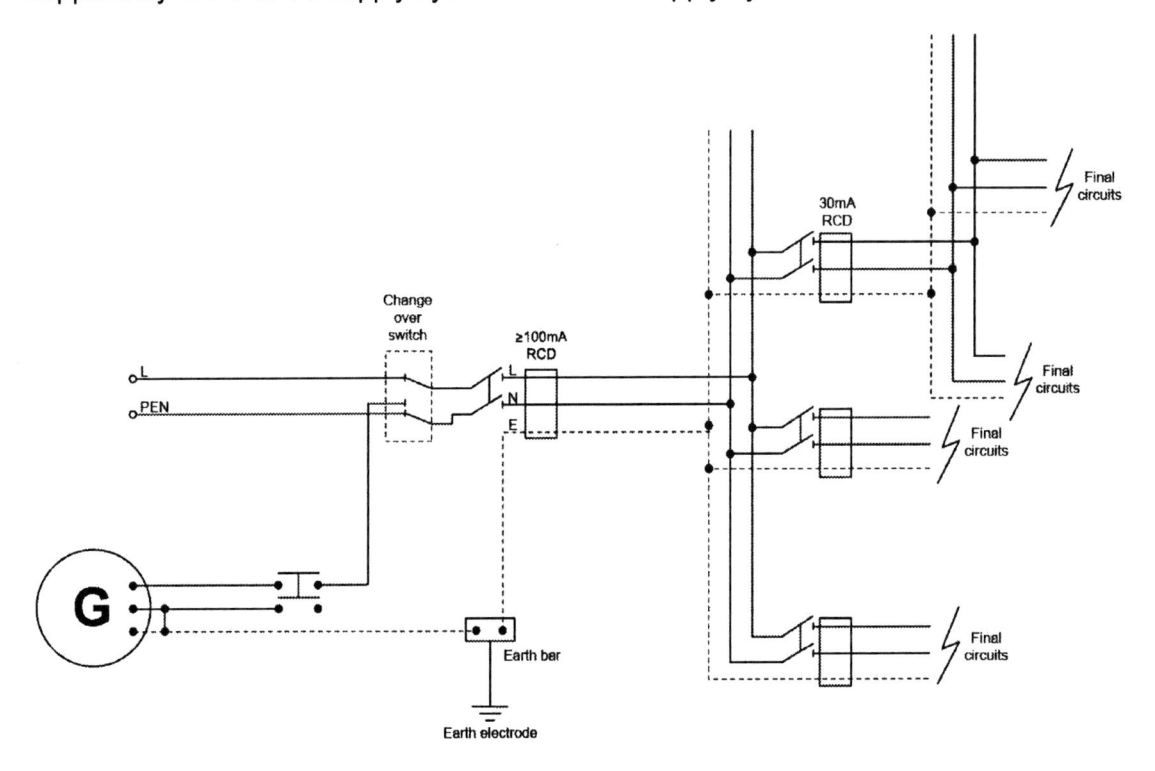

Protection against electric shock

Due to the increased risk of damage to cables, any cable supplying temporary structures shall be protected by an RCD not exceeding 300 mA and with a time delay to BS EN 60947-2, or be an S type to BS-EN-61008-1 or BS EN 61009-1, in order to provide discrimination with final circuits protected by downstream RCDs.

Any structural metal parts which are accessible inside a stand, vehicle, wagon, caravan or container shall be bonded using main protective bonding conductors to the main earthing terminal within the unit.

Types of wiring system

For buildings used for exhibitions which do not have a fire alarm system, the following types of cable shall be used:

- flame retardant low smoke
- single-core or multicore unarmoured cables within a conduit or trunking system providing protection to at least IP4X.

Wiring systems

Cables used for wiring shall be at 1.5 mm² cross-sectional area with copper conductors and thermoplastic or thermosetting insulation. Whenever there is a risk of mechanical damage cable shall be armoured or suitably protected.

Isolation

Each vehicle, stand, unit or separate temporary structure for use by a specific user and each supply distribution circuit for outdoors installations shall be provided with its own isolator, suitably identified and readily accessible.

An emergency switch shall be provided for each separate circuit supplying signs, lamps or exhibits.

ELV transformers and electronic converters

The secondary circuit of transformers and electronic converters shall have a protective device which can be manually reset. The installation of ELV transformers shall be out of arm's reach of the public, adequately ventilated and accessible for testing and maintenance by competent persons.

Additional protection

Every socket-outlet circuit up to 32 A and all final circuits shall be protected by an RCD in accordance with Regulation 415.1.1, with the exception of emergency lighting. SELV or PELV supplies can be used.

Equipment

Lighting equipment or appliances with a high surface temperature shall be guarded and installed to the required standard. If a large number of light fittings or lamps, which generate excessive heat, are to be installed then adequate ventilation shall be provided, especially in ceilings, which should additionally be made of incombustible material. These requirements also relate to showcases and signs.

Any switchgear and control gear not intended to be operated by ordinary persons shall be contained in closed cabinets openable by the use of a key or tool.

Solar photovoltaic (PV) power supply systems *(712)*

Scope

This section covers the electrical installation of PV power supply systems, which only work when connected in parallel with the electricity supply. The solar photovoltaic systems covered need to meet the requirements of the Electricity Safety, Quality and Continuity Regulations 2002. These units are classed as embedded generators.

PV systems

PV modules generate electricity when exposed to daylight. This means they cannot be switched off and therefore special precautions must be taken to ensure that live terminals cannot be readily touched or are not accessible. These terminals will be live during daylight hours and all junction boxes must carry a warning label. PV systems include d.c. wiring which is unfamiliar to most electrical contractors.

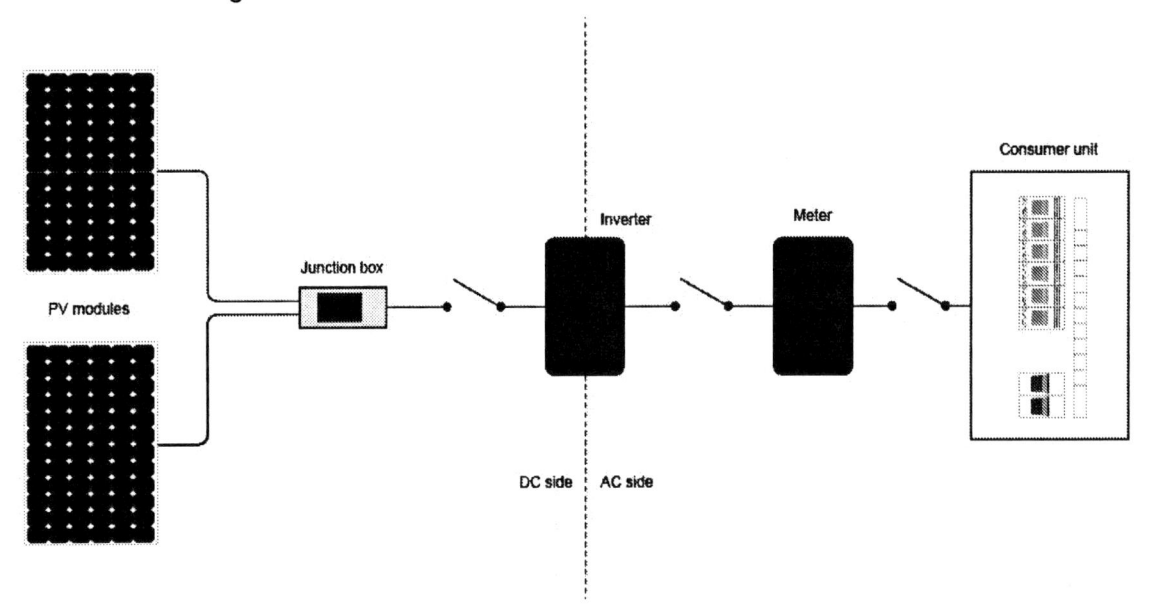

Protection against electric shock

Equipment of the d.c. side of PV equipment is to be considered as energised even when disconnected from the a.c. side and protection by Class II or equivalent insulation is recommended on the d.c. side.

Fault protection

PV supply cables shall be protected against fault currents at the a.c. mains connection point by an overcurrent protective device.

Due to these units being placed outside, often on roof areas, in order to minimise any voltages induced by lightning, the area of wiring loops shall be kept as small as possible.

Standards

All PV modules (array, PV generator junction boxes and switchgear) shall comply with relevant equipment standards. Connection of PV modules in series is allowed up to the maximum operating voltage of the PV modules or the PV inverter, whichever has the lower voltage.

PV modules shall be installed in accordance with manufacturer's instructions for mounting and in such a way that adequate heat dissipation is achieved when maximum solar radiation occurs.

Wiring systems

PV string, PV array cables and PV d.c. cables shall be selected and installed in order to minimise any earth fault and short-circuit risk.

Wiring systems used shall be able to withstand external influences, especially wind, ice, temperature and solar radiation.

Isolating switchgear

When selecting suitable devices for isolation and switching functions between the PV system and the public supply, the PV installation is considered to be the load and the public supply the source.

Switch disconnectors are to be provided for the d.c. sides of PV inverters. Suitable warning labels shall be fitted indicating that parts inside junction boxes, PV generators and PV array boxes may still be live after PV inverter isolation has taken place.

Earthing

When equipotential bonding conductors are installed they shall be run in parallel and in close contact with d.c. and a.c. cables and accessories.

Mobile or transportable units *(717)*

Scope

This section covers electrical installations in mobile or transportable units, provided with a temporary supply by means of a plug and socket-outlet. This section does not apply to the following:

- generators
- marinas
- pleasure craft
- mobile machinery

- caravans
- traction equipment
- electric vehicles.

Examples of the types of units covered by this section are vehicles for technical and facilities purposes which may be used by:

- entertainment industry
- medical services
- advertising
- fire fighting
- workshops
- offices
- transportable catering units.

Protection against electric shock

An RCD shall be provided to ensure automatic disconnection of the supply.

Any accessible conductive parts of a unit, e.g. chassis, shall be connected to earth using finely stranded main protective bonding conductors via the main earthing terminal of the unit. The following cables to BS 6004 are suitable:

- HO5V-K
- HO7V-K.

TN-C-S supply systems shall not be used to supply mobile or transportable units unless the installation is continuously supervised by a skilled or instructed person and the effectiveness of the means of earthing has been confirmed before connection is made.

An IT supply system can be provided by the following means:

- isolating transformer
- low voltage generating set
- transformer providing simple separation to BS EN 61558-1 and the following provisions:
    - insulation monitoring device used without an earth electrode
    - RCD and an earth electrode protecting the transformer in event of failure

Every socket-outlet supplying equipment outside the unit shall be additionally protected by an RCD complying with Regulation 415.1.1 except socket-outlets supplied by PELV, SELV or electrical separation.

Identification

The following information on a notice of durable material shall be displayed adjacent to the supply input point of the unit:

- type of supply that may be connected
- voltage rating
- number of phases and configuration
- on-board earthing arrangement
- maximum power of unit.

Wiring systems

Flexible cable of at least 2.5 mm² shall be used to connect units to the supply source manufactured to HO7RN-F of BS 7919.

The types of internal wiring of units shall be as follows:

- thermoplastic or thermosetting insulated cable to BS 6004, BS 7211, BS 7919 installed in conduit
- the above cables but also sheathed, installed so they are not damaged by sharp edges or abrasion from the unit.

When a gas cylinder storage compartment is fitted to a unit, no wiring systems or equipment shall be installed in it, with the exception of ELV equipment required for any gas controls.

When cables have to run through a gas storage compartment they shall be installed at a height of less than 500 mm above any gas cylinder bases and protected against mechanical damage by being installed in a continuous gas-tight conduit or duct.

Equipment

The connector used to connect the unit to the supply source shall be in accordance with BS EN 60309-2 and the plug shall be on the unit. Appliance inlet and enclosures must be protected to at least IP44.

Electrical installations in caravans and motor caravans (721)

Scope

This section covers electrical installations in caravans and motor caravans where the voltage does not exceed 230/400 V a.c. or 48 V d.c. and where the electrical circuits and equipment are intended for habitation purposes.

Installation requirements

Some of the installation requirements are illustrated below.

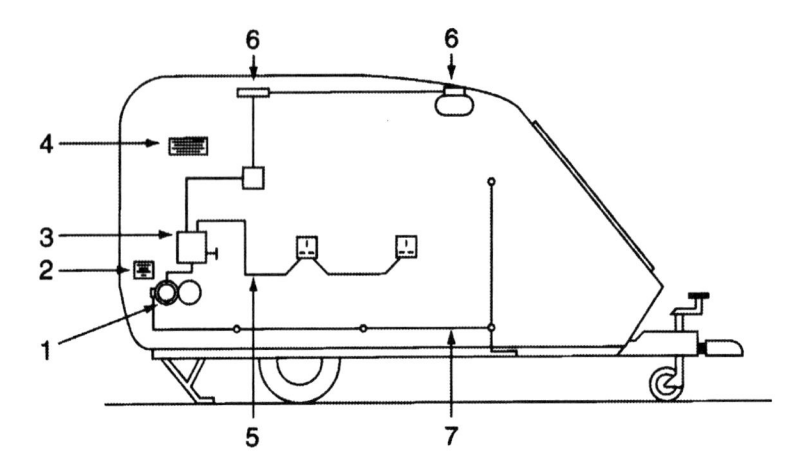

1. Provision of an inlet coupler to BS EN 60309-1 or, if interchangeability required to BS EN 60390-2, mounted no greater than 1.8 m above the ground and readily accessible. The means of connection to the pitch socket-outlet by a flexible cord or cable of maximum length 25 m ± 2 m sized in accordance with Table 721.

2. Inlet coupler marked on the outside indicating the nominal voltage, frequency and rated current of the caravan installation.

3. Protection by automatic disconnection of the supply must include an RCD in accordance with Regulation 415.1.1.

4. Nominal voltage of socket-outlets to be indelibly marked and low voltage socket-outlets shall not be compatible with plugs for socket-outlets supplied at extra-low voltage.

5. Protective conductors installed throughout each circuit within the caravan. Socket-outlets must incorporate an earth contact. Class II equipment may be used.

6. Luminaires, preferably fixed directly to the structure or lining of the caravan. Where a pendant luminaire is installed, some means of securing it to prevent damage during transit must be provided.

7. Structural metal parts which are accessible from within the caravan must be connected to the main earthing terminal within the caravan by means of main protective bonding conductors.

Note: If the caravan structure is fibreglass or other insulating material, the above requirement does not apply to any isolated parts, e.g. fixing brackets, which are not likely to become live in the event of a fault. Metal sheets forming part of the structure of the caravan are not considered to be extraneous-conductive-parts.

- When sheathed cables are installed in inaccessible positions, such as the ceiling or wall, the cable must be supported at intervals of 250 mm for horizontal runs and 400 mm for vertical runs (unless in non-metallic conduit).

- The cross-sectional area of every conductor must not be less than 1.5 mm².

- Accessories that may be subject to moisture to be not less than IP44.
- A notice installed adjacent to the main switch inside the caravan with instructions for electricity supply (see example below).

Instructions for electricity supply

To connect

1. Before connecting the caravan installation to the mains supply, check that:
   - the supply available at the caravan pitch supply point is suitable for the caravan electrical installation and appliances
   - the voltage, frequency and current ratings are suitable
   - the caravan main switch is in the **off** position
   - prior to use, examine the supply flexible cable to ensure there is no visible damage or deterioration.
2. Open the cover to the appliance inlet provided at the caravan supply point, if any, and insert the connector of the supply flexible cable.
3. Raise the cover to the appliance inlet provided on the pitch supply point and insert the plug of the supply cable.

The caravan supply flexible cable must be fully uncoiled to avoid damage by overheating

4. Switch on at the caravan main isolating switch.
5. Check the operation of residual current devices, if any, fitted in the caravan by depressing the test button(s) and reset.

In case of doubt or, if after carrying out the above procedure the supply does not become available, or if the supply fails, consult the caravan park operator, the operator's agent or a qualified electrician.

To disconnect

6. Switch off at the caravan main isolating switch, unplug the cable, first from the caravan pitch supply point and then from the caravan inlet connector.

Period inspection

Preferably not less than once every three years and annually if the caravan is used frequently, the caravan electrical installation and supply cable should be inspected and tested and a report on their condition obtained, as prescribed in BS 7671 Requirements for Electrical Installations published by the Institution of Engineering and Technology and BSI.

Temporary electrical installations for structures, amusement devices and booths at fairgrounds, amusement parks and circuses *(740)*

Scope

This section covers the minimum electrical installation requirements for the safe design, installation and operation of temporary mobile or transportable electrical machines and structures, incorporating electrical equipment. Such machines and structures are intended to be repeatedly installed without safety being compromised.

The supply voltages of temporary electrical installations are:

- 230/440 V a.c.
- 440 V d.c.

TN-C-S supply systems shall not be used. If a suitable alternative is available an IT system shall not be used unless the supply source is d.c.

Regardless of the number of supply sources, the line and neutral conductors from different sources shall not be interconnected after the origin of the temporary installation. The operator's instructions for the electricity supply must be followed.

Protection against electric shock

Every temporary installation shall be protected at the origin by a 300 mA time delay RCD to BS EN 60947-2, or S type to BS EN 61008-1 or BS EN 61009-1.

Additional protection: RCDs

RCD protection in accordance with Regulation 415.1.1 shall be provided for the following final circuits:

- lighting
- socket-outlets up to 32 A
- mobile equipment up to 32 A connected using a flexible cord or cable.

Battery-operated emergency lighting circuits shall be connected via the lighting circuit RCD, except if the following apply:

- circuits are SELV or PELV
- it is protected by electrical separation
- it is placed out of arm's reach if not supplied from socket-outlets.

Additional protection: Supplementary equipotential bonding

Any exposed-conductive-parts and extraneous-conductive-parts in locations intended for animals, where the animals can touch them, shall be bonded together, including any metal grid which has been laid in a floor. Refer to Section 705 (Agricultural and horticultural premises) for further details on equipotential bonding in premises used by livestock.

Selection and erection of equipment

Any switchgear and control gear that will be operated by ordinary persons shall be fitted in cabinets which can only be opened using a key or tool. Any motor not continuously supervised, which is remotely or automatically controlled, shall have a manual reset protective device for protection against excess temperature rises. All electrical equipment shall provide protection to IP44.

Wiring systems

The minimum rated voltage of cables and cords used shall be:

- 450/750 V general
- 300/500 V within amusement devices.

When cables are installed in public areas or where they cross walkways or roads they shall be mechanically protected to external influences (AG2).

Conduit systems shall provide protection against compression and heavy impact. Metallic and composite conduits shall be Class 3 in order to provide protection against corrosion.

Trunking and ducting systems shall comply with BS EN 50085 and be Class 5J in order to provide impact protection.

In areas subject to movement wiring systems need to be of flexible construction, e.g. flexible conduit.

Joints in cables shall be avoided. When this is not possible they should be made using appropriate British Standard approved connectors or the joint must be made inside an enclosure providing protection to at least IP4X or IPXXD.

Isolation and switching

All separate temporary installations for amusement devices and distribution circuits for outdoor installations shall have their own easily accessible isolation device.

Separate circuits shall be provided for any signs, lamps or luminous tubes. The circuits must be controlled by an emergency switch, complying with the following requirements:

- visible
- accessible
- marked to meet local authority requirements.

Other equipment

All luminaires shall be suitably IP rated for the environment, be suitably fixed out of arm's reach (2.5 m), or be provided with suitable guards which can only be removed using a key or tool to prevent injury or ignition of any flammable materials. Extra attention is necessary for lamps in shooting galleries, as they will require protection against projectiles. Floodlights and similar luminaires shall be fixed and positioned to avoid concentration of heat, which could cause a fire by igniting adjacent materials.

In order to provide an adequate number of socket-outlets it is recommended that one be provided for every square metre. The connection point of every amusement device shall be readily accessible and clearly marked to indicate its voltage, current rating and frequency.

Electric dodgems shall be supplied from a circuit providing electrical separation from the mains supply, using either a transformer to BS EN 61558-2-4 or a motor generator, the maximum voltage being 50 V a.c. or 120 V d.c.

When generators supply temporary installations as part of a TN, TT or IT systems, earthing shall comply with Section 542.1 and, except where an IT system is used, the neutral conductor of the generator's star point shall be connected to exposed-conductive-parts of the generator.

Floor and ceiling heating systems *(753)*

Scope

This section covers the installation of direct or thermal storage electric floor and ceiling heating systems.

Fault protection

RCDs not exceeding 30 mA shall be used for automatic disconnection of the supply.

Manufactured heating units without exposed-conductive-parts shall have a conductive covering, such as metal grid with spacings, not more than 30 mm above the heating elements during installation on site. These shall be connected to the installation protective conductor.

In order to avoid unwanted tripping of the RCDs the power rating of the heating system should be limited to:

- 7.5 kW at 230 V
- 13 kW at 400 V.

When a circuit supplies Class II heating equipment additional protection shall be provided by an RCD not exceeding 30 mA.

Selection and erection of equipment

When a floor heating system is installed in a floor where contact with a person's skin or footwear is possible the surface temperature shall be limited to 35°C.

In order to avoid overheating of floor or ceiling heating systems one of the following measures is required to limit the temperature to no more than 80°C in the zone where heating units are installed:

- appropriately designed
- installed in accordance with manufacturer's instructions
- protective devices.

The method of connection of heating units to the electrical installation is preferably by crimped cold tails or other suitable means. Where heating units are installed close to ignitable building structures special measures must be taken to meet the requirements of Chapter 42, such as placing on a metal sheet or in metal conduit or located at least 10 mm in the air from any ignitable structures.

Any heating units installed in ceilings shall provide protection to IPX1 and in floors to IPX7. They shall be installed in accordance with manufacturer's instructions and not cross any expansion joints in a building structure.

There is a requirement to design floor and heating systems such that areas of the floor or ceiling are unheated (heat-free areas). Fixtures to the floor or ceiling can be made in such heat-free areas without inhibiting the emission of heat.

Where the fixing or fitting by screws or nails is required then heat-free areas must be provided. The installer shall inform other contractors that where ceiling and floor heating is installed no fixings are allowed.

Designers or installers of heating systems must provide a plan, fixed adjacent to the distribution board, containing the following details:

- the manufacturer
- number installed
- length/area of unit
- power
- surface power density
- an illustration of the layout
- position and depth
- conductors
- shields
- heated area
- voltage

- resistance when cold
- rating of overcurrent device
- RCD operative current
- insulator resistance
- test voltage used
- leakage capacitance.

On completion information shall be provided by the installer of the heating system to either the owner of the building or the agent, covering:

- a description of the heating system construction
- a location diagram
- information on the control equipment
- heating unit types
- the operating temperature.

The designer/installer shall provide information on the use of the heating system. A copy should be permanently fixed adjacent to the distribution board covering:

- function
- operation of the heating system
- operation of control equipment
- restrictions on placing of furniture and additional floor coverings
- restrictions on furniture height
- where fixings are not allowed.

TECHNICAL DATA SHEET 7A

Electrical supplies on construction and demolition sites

Electricity supply

The supply of electricity on construction sites will normally be provided by one or both of the following:

- a public supply from the local electricity company
- a site generator, where public supply is not practicable or is uneconomic.

Public supply

A public supply of electricity being provided depends on the following:

- written application being made to the local electricity company, as soon as possible at the planning stage
- the name, address, and telephone number of the main contractor and developer, giving the full site address and a location plan
- details of the maximum demand load (in kilowatts) which is likely to be required during construction
- details of the maximum final demand load (in kilowatts) which will be required when the job is complete
- dates when the supply is needed
- a discussion with electricity company staff to determine the necessary precautions to avoid damage or hazards from any existing overhead or underground cables
- the establishment of supply points (where incoming cables will terminate), switchgear, metering equipment and requirements for earthing.

Generators

Generators (even if for stand-by purposes) may be required, and will be powered by petrol or diesel engines. Attention should be given to the siting of such equipment in order to minimise pollution caused by noise and fumes.

- Any private generating plant must be installed in accordance with BS 7375. You are advised to seek advice from the local electricity company.
- If the generator will produce over 55 V a.c. it must be effectively earthed.

- The principle of low voltages and their advantages should be considered further where portable generators are used on site.
- Not all portable generators available for use on site have the 110 V output centre tapped to earth. This is particularly true of generators having dual voltage selectable.
- The metal frame of the generating set should be bonded to the metalwork of the site distribution system, where there is one.

Site distribution

All wiring should conform to BS 7671: Requirements for Electrical Installation, even though much of it will be temporary. Makeshift arrangements cause accidents and must be prohibited.

All switchgear should be freely accessible and capable of being locked in the 'off' position. Regulation 6 of The Electricity at Work Regulations 1989 specifically states that any electrical equipment that may be exposed to adverse or hazardous conditions must be, so far as is reasonably practicable, so constructed or protected that danger is prevented.

Wherever possible a reduced voltage system should be used.

Site offices and other accommodation should be a standard installation to the current BS 7671: Requirements for Electrical Installation.

Site accommodation

Site offices and welfare facilities are the only locations where electrical equipment that runs off a 230 V supply should be in use. The electrical supply panel for such facilities must incorporate a residual current device (RCD) in each circuit. The correct operation of each RCD must be confirmed weekly by operating the 'TEST' button.

The incoming electrical supply to site accommodation must be designed, installed and commissioned.

All portable electrical equipment must be electrical safety (PAT tested) at appropriate intervals as decided by a competent person. This includes:

- common types of office equipment, such as fax machines and photocopiers
- 'kitchen-type equipment', such as kettles, microwave ovens, etc.
- small items, such as chargers for site radios, battery-powered tools, etc.

All units for site use should comply with BS 4363 and installations with BS 7375. Plugs, sockets and couplers must conform to BS EN 60309-2.

BS 4363 recommends use of the following units:

Supply incoming unit (SIU)

- Ratings up to 300 A per phase. These units include main switchgear and metering equipment.

Mains distribution unit (MDU)

- For the control and distribution of electricity on site. 415 V three-phase, 230 V single-phase a.c.

A combined supply incoming and distribution unit (SIDU) may be used in some installations.

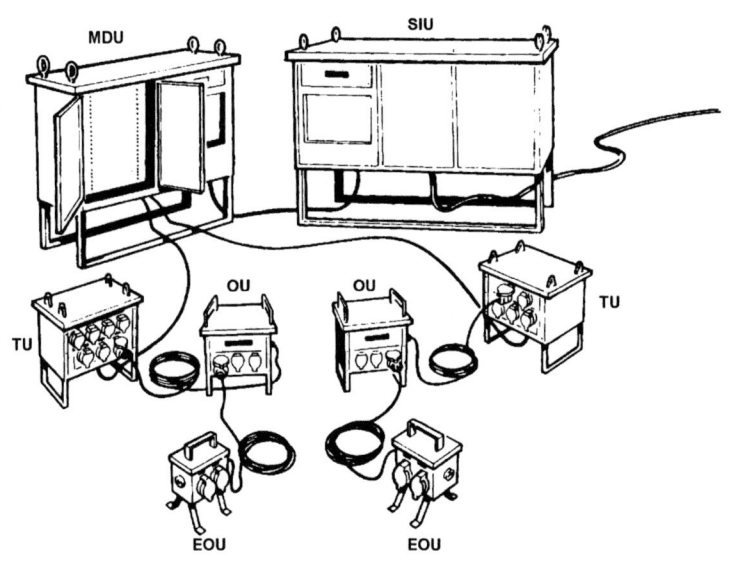

Transformer units (TU)

- TU 1 single-phase 230 V – 110 V.
- TU 3 three-phase 415 V – 110 V.

Transformer units are available with different outlet ratings, i.e. 16, 32 or 60 A. Some units have socket-outlets protected by circuit breakers.

Such transformer units can be used for portable tools and plant, and general floor lighting.

Outlet units (OU)

- 110 V socket-outlet units.
- 16 or 32 A.

Such outlet units can be used for portable tools, floodlighting and extension outlets.

Extension outlet units (EOU)

- 110 V socket-outlet.
- 16 A.

Such units can be used for portable tools, local lighting and hand lamps.

Markings

All supply, distribution and transformer units should be marked with the warning sign shown below from BS 5378 Safety signs and colours.

DANGER 415 V

A supplementary sign with the word **danger**, and indicating the highest voltage likely to be present, should be placed below the warning sign.

Earthing

All metal parts of the distribution systems and fixed appliances not carrying a current must be effectively earthed in accordance with BS 7430 Code of Practice for Earthing, to either:

- the metallic sheath and armouring of the incoming supply cable
- the earthed terminal supplied by the supply authority
- a separate earth electrode system.

Periodic maintenance, inspection and testing is essential.

Monitored earthing systems are recommended for all transportable plant operating at any voltage above 110 V and supplied with flexible cables. In these systems, a very low current circulates continuously in the earthing circuit. If this circuit is broken or interrupted, the supply to the plant is automatically cut off until the earth path is made effective.

Plugs, socket-outlets and couplers

Only components to BS EN 60309-2 should be used. This covers both single and three-phase supplies and is intended to prevent plugs designed for one voltage being connected to sockets of another. This is achieved by different positions of the key-way in plug and socket. Examples are shown below.

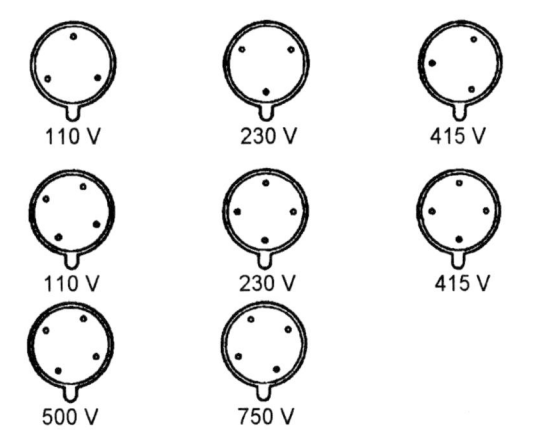

110 V	230 V	415 V
110 V	230 V	415 V
500 V	750 V	

CABLE SELECTION

BS 7671 Part 8

Current-carrying capacities of cables (Refer to Appendix 4 of BS 7671) *(523)*

Installation methods

The rating of a cable depends on its ability to dissipate the heat generated by the current it carries. This, in turn, depends in part on the type of installation. Table 4A1 (Appendix 4) of BS 7671 lists standard methods of installation with examples identified by numbers. These classifications are used in the current-carrying tables.

Example: From Table 4D5A of BS 7671, it can be seen that ordinary PVC insulated and sheathed twin and earth cable installed in an enclosure such as conduit has a current capacity of approximately 75% of what it would support when clipped to a surface.

Size	Installation method			
Conductor csa	Ref method A enclosed in a conduit in an insulated wall	Ref method 102 installed in a stud wall with thermal insulation with cable touching the inner wall surface	Ref method C clipped on surface	Voltage drop Per amp Per metre
2.5 mm² twin with cpc	20 A	21 A	27 A	18 mV/A/m

Factors which affect the ability of a cable to lose heat (other than its physical characteristics) are:

- ambient (surrounding air) temperature
- cable grouping
- thermal insulation.

Ambient temperature

The rate of heat loss from a cable also depends on the difference in temperature between the cable and the surrounding air. A correction must be made to the current-carrying capacity where the cable is to be installed in situations where high or low ambient temperatures may be expected to occur.

Tables 4B1 and 4B2 of BS 7671 give correction factors to be applied to the tabulated current-carrying capacities depending upon the actual ambient temperature of the location in which the cable is to be installed and the type of circuit protective device.

Cable grouping

Cables installed in the same enclosure and all carrying current will get warm. Those near to the edge of the enclosure will be able to transmit heat outwards but will be restricted in losing heat inwards, while cables in the centre may find it difficult to lose heat at all.

The correction or rating factors for this effect are given in Table 4B1 to 4C2 of BS 7671. These relate to touching and spaced cables. 'Spaced' means a clearance between adjacent cables of at least one diameter. Where the horizontal clearances between adjacent cables exceed twice the overall cable diameter, no reduction factor is required.

For a particular circuit the circumstances may change throughout its length, i.e. the ambient temperature or the number of cables bunched together may vary, or there may even be a change in the method of installation. Where this is the case the most onerous figure, i.e. that giving the **lowest** current-carrying capacity, must be used. Alternatively, the correction factors may be applied to the length of run specifically affected, but this will require the size of the cable to be increased for this length.

The correction factors for mineral insulated cables installed on a perforated tray are covered in Table 4C1 of BS 7671. Cables buried in the ground are in Tables 4B3 and 4C2.

Notes:

1) *The factors given in Tables 4B1 to 4C2 of BS 7671 relate to groups of cables of one size.*

2) *If due to known operating condition, a cable is expected to carry a current not more than 30% of its grouped rating, it may be ignored when calculating the rating factor for the rest of the group.*

Thermal insulation

For cables run in a space where thermal insulation is likely to be applied, Regulation 523.6.6 and the derating factor of Table 52.2 shall be applied.

Determination of the size of cables

Having established the design current of a circuit (I_b) and selected the type and size of protective device (I_n), it is necessary to determine the size of cable to be used.

Procedure

Find the tabulated value of current (I_t) by dividing the nominal current of the protective device (I_n) by any applicable correction factor(s), for example:

- ambient temperature (Tables 4B1 and 4B2)
- grouping (Tables 4C1 to 4C2)
- insulation (Table 52.2)
- if the protective device is a BS 3036 fuse, further divide by 0.725.

Then select a suitable cable from the tables in Appendix 4 of BS 7671, after obtaining the reference method from Table 4A2.

The size of the cable selected must be such that its tabulated current-carrying capacity is not less than the value of the protective device, adjusted as above.

In determining cable size, the correction factors are applied as divisors to the rated current of the overload protective device.

Formula

$$I_t \geq \frac{I_n}{\text{Factors applicable}}$$

When the overcurrent device is other than a semi-enclosed fuse to BS 3036, the cable selected has to be such that its tabulated current-carrying capacity for the method of installation chosen is not less than:

$$I_t \geq \frac{I_n}{C_a \times C_g \times C_i} \text{ amperes}$$

Where:

I_t = tabulated value of circuit current

I_n = nominal current or current setting of the protective device

C_a = correction factor for ambient temperature

C_g = correction factor for grouping

C_i = correction factor if cable is surrounded with thermal insulation

Where the overcurrent device is a **semi-enclosed BS 3036 fuse**, a further correction factor has to be applied, which is 0.725.

$$I_t \geq \frac{I_n}{C_a \times C_g \times C_i \times 0.725} \text{ amperes}$$

Example

A 230 V, 30 A single-phase final circuit consists of a 22 m length of run of PVC insulated cable installed in conduit. The circuit has a design current of 26 A. Determine the minimum size of cable which will comply with voltage drop requirements of Appendix 12 Table 12A.

Maximum voltage drop allowed = $\dfrac{230 \times 5}{100}$ = 11.5 V

Actual voltage drop, vd

= $\dfrac{mV/A/m \times \text{design current, } I_b \times \text{length, } L}{1,000}$

The simplest method to adopt is to determine the **maximum** mV/A/m which will comply with the **maximum** voltage drop **allowed**, in this case 11.5 V.

Transposing the above equation:

$mV/A/m \times I_b \times L = vd \times 1,000$

Maximum mV x A/m = $\dfrac{vd \times 1,000}{I_b \times L}$

When we use the maximum volts drop figure of 11.5 V, we obtain the maximum mV/Am.

∴ max mV/A/m = $\dfrac{11.5 \times 1,000}{26 \times 22}$

= $\dfrac{11,500}{26 \times 22}$

= $\dfrac{11,500}{572}$

∴ max mV/A/m = **20.1 mV/A/m**

By referring to Table 4D1A of BS 7671, any cable with 20.1 mV/A/m or less will therefore give an actual voltage drop of 11.5 V or less when installed in accordance with the method to be used. For example, the cable is enclosed in conduit, which is reference method B (see Table 4D1B in Appendix 4 of BS 7671).

Table 4D1B, column 3 shows:

2.5 mm² = 18 mV/A/m (hence voltage drop less than **20.1 V**)

i.e. $\dfrac{18 \times 26 \times 22}{1,000}$ = **10.29 V**

As the actual voltage drop for cable is 10.29 V and the maximum allowable voltage drop is 11.5 V, a 2.5 conductor is satisfactory.

The diagram on the next page is included to illustrate the voltage drop constraint on ring final circuits under different loading conditions.

For simplicity, only half of the ring circuit has been shown.

3 Kw heater   3 Kw heater

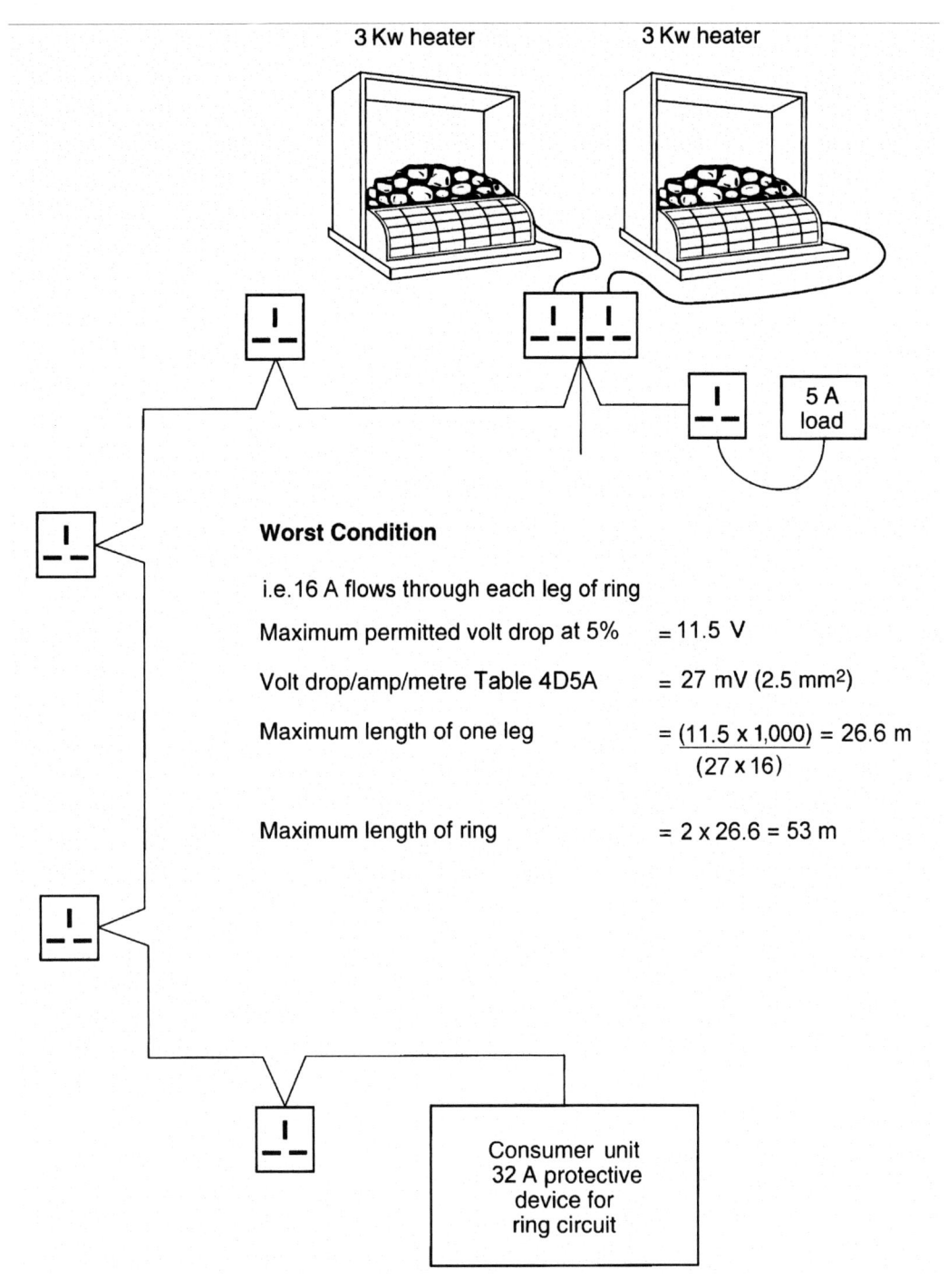

5 A load

Worst Condition

i.e. 16 A flows through each leg of ring

Maximum permitted volt drop at 5%   = 11.5 V

Volt drop/amp/metre Table 4D5A   = 27 mV (2.5 mm^2)

Maximum length of one leg   = $\dfrac{11.5 \times 1{,}000}{27 \times 16}$ = 26.6 m

Maximum length of ring   = 2 × 26.6 = 53 m

Consumer unit
32 A protective
device for
ring circuit

**Voltage drop constraints (32 A domestic ring circuits)
PVC/PVC cable clipped to the surface**

Selecting cables for circuits and checking for compliance with Regulations

Certain tables and formulae used in this section are not part of BS 7671. Reference has been made to the IEE On-site Guide.

When installing a circuit, it is necessary to:

- calculate the design current (I_b)
- select the type and nominal rating of the protective device (I_n)
- determine and apply correction factors to I_n
- select cable from tables in Appendix 4 of BS 7671 (I_z)
- calculate the voltage drop and check for compliance
- check that circuit complies with shock protection
- check that circuit complies with thermal constraints.

Example 1

A radial socket-outlet circuit is protected by a 20 A Type B circuit breaker to BS EN 60898. The circuit is wired using 2.5 mm² PVC insulated/sheathed cable with a 1.5 mm² CPC. The circuit length is 16 m and the cable is clipped directly to the surface of a wall throughout its length. Assuming that no correction factors are applicable and that the value of Z_e is given as 0.5 Ω, determine whether the circuit complies with BS 7671. The nominal voltage (U_o) may be taken as 230 V. Assume an ambient temperature of 20°C.

I_b Design current of circuit = 20 A

I_n Protective device, circuit breaker Type B = 20 A

Step 1

Apply correction factors for grouping and ambient temperature.

None apply.

Step 2

Select a suitable cable (I_z) from Appendix 4, BS 7671.

From Table 4A2 obtain the Installation Reference Method

Clipped direct = **Method C**

From Table 4D5 2.5 mm² cable installed has a current-carrying capacity of **27 A**

Step 3

Calculate voltage drop.

Maximum volt drop allowed = 5% of 230 = 11.5 V

Actual voltage drop = $\dfrac{\text{mV/A/m} \times \text{design current } I_b \times \text{length}}{1{,}000}$

From Table 4D5A mV/A/m for 2.5 mm² cable = 18 mV/A/m

∴ Actual vd = $\dfrac{18 \times 20 \times 16}{1{,}000}$

Voltage drop = **5.76 V**

Step 4

Check for compliance with shock protection.

$Z_s = Z_e + (R_1 + R_2)$

Where:

Z_s = system phase-earth fault loop impedance

Z_e = external impedance (impedance of the supply)

R_1 = resistance of circuit line conductor

R_2 = resistance of circuit protective conductor

Ze = 0.5

$R_1 + R_2$ value from Table 9A of the IEE On-site Guide for a 2.5 mm² line conductor/1.5 mm² protective conductor:

$R_1 + R_2$ = 19.51 mΩ/m

Multiplier for conductor operating temperature from Table 9C of the IEE On-site Guide (for standard protective devices found in Tables 41.1 to 41.4 of BS 7671 and where the cpc is a core in a cable which has 70°C insulation).

Multiplier for conductor operating temperature = 1.2

$R_1 + R_2$ = $\dfrac{19.51 \times 1.2 \times 16}{1{,}000}$ (circuit length in metres) = **0.37 Ω**

Actual Z_s = 0.5 + 0.37 = **0.87 Ω**

From Table 41.3 of BS 7671 maximum Z_s for a 20 A Type B circuit breaker to BS EN 60898 is **2.3 Ω**

Z_s satisfactory as 0.87 Ω < 2.3 Ω

Step 5

Check for compliance with thermal constraints.

Calculate value of fault current $I_f = \dfrac{U_o}{Z_s}$

Where:

I_f = earth fault current

U_o = phase earth voltage

Z_s = system phase-earth loop impedance

$$I_f = \dfrac{230}{0.87}$$

Earth fault current = 264 A

Find value of t when I_f = 264 A from time/current characteristic for 20 A Type B circuit breaker in Appendix 3, Figure 3.4 of BS 7671.

In the absence of manufacturer's data:

t = 0.1 s

Find value of k from Table 54.3 (Table 54.3 is used because the cpc is a core in a cable and the conductor is copper with 70°C insulation):

k = 115

$\therefore$ minimum size of protective conductor $S = \dfrac{\sqrt{I^2 t}}{k}$ mm²

Where:

S = minimum size of protective conductor in mm²

I = maximum earth fault current in amperes

t = time in seconds for the protective device to operate

k = factor for specific protective conductors from Tables 54B-F of BS 7671

$$S = \dfrac{\sqrt{264^2 \times 0.1}}{115}$$

S = 0.73 mm²

Nearest standard size of conductor = 1.0 mm², which would satisfy thermal constraints. 1.5 mm² will definitely satisfy the requirements.

Example 2

A 230 V, 13.5 kW domestic electric cooker installed in a house is to be supplied by a 15 m run of PVC insulated and sheathed cable, installed less than 50 mm deep in a stud wall with thermal insulation on one side of the wall surface. The cooker control unit incorporates a 13 A socket-outlet.

The circuit is to be protected by a 32 A BS EN 60898 Type B circuit breaker and the value of external impedance is 0.85 Ω. Determine the minimum size of cable which may be used. Assume an ambient temperature of 20°C.

Step 1

Determine full load current.

$$I = \frac{W}{V} = \frac{13.500}{230}$$

= 58.7 A

Step 2

Determine design current I_b.

Apply allowance for diversity from Table 1B, IEE On-site Guide.

Diversity for cooker = 10 A + 30% of remaining current + 5 A

Design current I_b = $10 + \frac{(30 \times 48.7)}{100} + 5$

= 10 + 14.6 + 5

= 29.6 A

Step 3

Protective device I_n chosen:

= 32 BS EN 3036 fuse

Step 4

Select cable size.

Select from Table 4D5 of BS 7671 (reference method C)

6 mm² size = 35 A

∴ cable size to be used is **6 mm² phase** with **2.5 mm² cpc**

Step 5

Calculate voltage drop.

$$\text{Maximum vd} = 5\% \text{ of } 230 \text{ V}$$
$$= 11.5 \text{ V}$$

$$\text{Voltage drop} = \frac{\text{mV/A/m} \times I_b \times \text{length}}{1{,}000}$$

From Table 4D5 voltage drop for 6 mm² = 7.3 mV/A/m

$$\therefore \text{actual voltage drop} = \frac{7.3 \times 29.5 \times 15}{1{,}000}$$

= **3.24 V** *(satisfactory)*

Step 6

Check for shock protection constraints.

From Table 41.3(c) of BS 7671, Z_s max = 1.44 Ω

(Table 41.3 used because of socket-outlet in cooker control unit which requires a 0.4 s disconnection time)

Actual:

$$Z_s = Z_e + (R_1 + R_2) \text{ Ω}$$

$$Z_e = 0.85 \text{ Ω}$$

$$(R_1 + R_2) = 10.49 \text{ mΩ/m (from Table 9A, IEE On-site Guide)}$$

(Assuming 6 mm² line conductor/2.5 mm² cpc)

Conductor operating temperature 70°C PVC multiplier = **1.20** (Table 9C, IEE On-site Guide)

Actual:

$$Z_s = 0.85 + \frac{(10.49 \times 1.20 \times 15)}{1{,}000}$$

= **1.04 Ω**

Check:

actual $Z_s \leq Z_s$ max? *(Yes – shock protection satisfied)*

Step 7

Check for thermal constraint.

$$\text{fault current } I_f = \frac{U_o}{Z_s}$$

$$= \frac{230 \text{ V}}{1.04}$$

$$= \mathbf{221 \text{ A}}$$

Value of time for 221 A fault current for a BS EN 60898 Type B circuit breaker from the time current characteristic Figure 3.4 A in Appendix 3 of BS 7671:

t = 0.1 s

Value of k from Table 54.3 of BS 7671 = **115**

Minimum size of protective conductor $S = \frac{\sqrt{I^2 t}}{k}$

$$S = \frac{\sqrt{2,212 \times 0.39}}{115}$$

S = 0.6 mm²

Nearest standard size of protective conductor greater than 0.6 mm² is 1 mm² so 2.5 mm², which is incorporated in the cable, is satisfactory.

Additional protection

Since the cable was installed less than 50 mm deep a 30 mA RCD shall be selected to provide additional shock protection or a 32 A RCBO.

Conventional final circuit design using Table 7.1 of the IEE On-site Guide

Circuit design normally involves extensive use of tables and calculation which can get quite involved. The IEE On-site Guide contains Table 7.1 for conventional circuits which enables the maximum cable run to be established to comply with the BS 7671 without calculation.

In order to use the table safely, the proposed circuit should comply with the following assumptions.

The installation is supplied by one of the following systems:

- TN-C-S with a maximum loop impedance Z_e of 0.35 Ω
- TN-S with a maximum loop impedance Z_e of 0.8 Ω
- TT with RCDs installed as Section 3.6 of the IEE On-site Guide.

The final circuit is to be connected to a consumer unit at the origin of the installation.

The installation method used complies with those specified in Table 4A of BS 7671 for:

- Method 1 sheathed cables clipped or embedded in plaster
- Method 3 cables run in conduit or trunking
- Method 6 PVC sheathed cables in an insulating wall or above a thermally insulating ceiling or single-core PVC insulated cables in conduit in a thermally insulated wall, etc.

Throughout the length of the circuit the temperature does not exceed 30°C.

Any grouping of cables will have to be taken into consideration (see Section 7, IEE On-site Guide).

Protective device characteristics are in accordance with Appendix 3 of BS 7671 with a fault current tripping time for circuit breakers of 0.1 s or less.

A 13 A immersion heater circuit protected by a Type B 16 A circuit breaker wired in 2.5 mm² PVC sheathed cable with a 1.5 mm² cpc is supplied from a 230 V single-phase TN-C-S system. The maximum length of run is 33 m.

Radial circuit

Current rating Amps	Cable CPC size mm²		Protective device	Cable installation method		Maximum length in metres			
				PVC cable	Thermo setting cable	TN-S 0.4 s	TN-S 5 s	TN-C-S 0.4 s	TN-C-S 5 s
16	2.5	1.5	Type B circuit breaker	M6	M6	33	33	33	33

Note: M6 indicates that the installation methods M1, M3 or M6 from Table 4A of BS 7671 may be used.

Circuit design

An 8 kW electric shower is to be installed in the bathroom of a domestic dwelling. The length of proposed cable route is 30 m clipped to the surface supplied from the spare way in a consumer's unit fitted with Type B circuit breakers. The electricity supply is a 230 V TN-C-S system.

Design a suitable installation to comply with the requirements of BS 7671 using Table 7.1 of the IEE On-site Guide to minimise the number of calculations required.

Step 1

Determine full load current (I).

$$I = \frac{watts}{volts} = \frac{8{,}000}{230} = \mathbf{34.8\ A}$$

Step 2

Determine design current.

No diversity allowance for the first two showers installed.

Design current I_b = **34.8 A**

Step 3

Select type and size of protective device.

Since the consumer unit is fitted with Type B circuit breaker, use **Type B 40 A** since this is the next size of circuit breaker > 32 A.

Step 4

Apply correction factors, e.g. for temperature, etc.

None apply.

Step 5

Check circuit complies with shock protection requirements (although the Regulations no longer give exclusive disconnection times for bathrooms and normal Regulations apply, it would be prudent to continue to use 0.4 s due to the hazardous nature of the location).

Using Table 7.1 of the IEE On-site Guide, for a 10 mm² cable with a 4 mm² cpc using a 40 A Type B circuit breaker on a TN-C-S system:

the maximum length of run is **53 m**.

Since the actual cable run is 30 m, which is less than the 53 m maximum, the circuit complies with BS 7671.

Summary

Before installing any conventional final circuit the following must be considered:

- Can the consumer unit or distribution board carry the planned additional load?
- What type of protective device is to be used?
- What type of cable and installation method is to be used?
- What are the ratings of the protective devices available?
- What type of earthing arrangement is being used?

- What is the maximum disconnection time for the circuit; 0.4 s or 5 s?
- What isolation and switching requirements are necessary?
- What labels are required to be fitted?
- Is the earth loop impedance value below the values of 7.1(i) or 7.2.4(ii) of the IEE On-site Guide?
- Is an RCD or RCBO required? (Socket-outlets for equipment outdoors or if the supply system is TT, a 30 mA RCD must be installed. Certain equipment in bath and shower rooms requires RCD protection.) All RCDs or RCBOs to comply with BS 4293, BS 7288, BS EN 61008 or BS EN 61009.

Note: *On all systems where socket-outlets supply equipment outdoors, protection by an RCD with a maximum operating current ($I_{\Delta n}$) of 30 mA is required.*

SIZING CONDUIT AND TRUNKING SYSTEMS   BS 7671
Part 9

Other mechanical stresses

Damage to the sheath or insulation of cables or conductors must be minimised during the installation, use or maintenance of a wiring system.

Where the wiring system is designed to be withdrawable, sufficient access to draw cables in or out must be provided.

In order to comply with the above requirements a method employing a unit system is described in the IEE On-site Guide, Appendix 5, where each cable is allocated a factor. The sum of the cable factors for the cables which are to be run in the same enclosure is then compared with a conduit or trunking factor given in the tables for different sizes of conduit or trunking, in order to determine the size of conduit or trunking required.

	From Appendix 5 of the IEE On-site Guide	
	Cables	**Conduit**
Straight runs of conduit not exceeding 3 m	Table 5A	Table 5B
Straight runs of conduit exceeding 3 m or any length incorporating bends or sets	Table 5C	Table 5D
Trunking	Table 5E	Table 5F

Note: For conduit systems, a bend is classed as 90° and a double set is equivalent to one bend.

Where there are other sizes and types of cable in the trunking, the space factor must not exceed 45%.

Only mechanical considerations have been taken into account in determining the factors in the tables. **The electrical effects of grouping have not**.

As the number of cables increases in a conduit or trunking, the current-carrying capacity of those cables decreases in accordance with the application of the grouping factors (Table 4B1 to 4C2 of BS 7671).

Calculations using Appendix 5 of the IEE On-site Guide

Example 1

A lighting circuit for a village hall requires the installation of a conduit system with a conduit run of 10 m with two right-angle bends. The number of cables required is ten 1.5 mm² PVC insulated. What is the minimum size conduit that should be chosen for the installation?

Step 1

Select correct table for cable factors for conduit runs over 3 m or with bends, **Table 5C, IEE On-site Guide**.

Step 2

Obtain factor from **Table 5C** for 1.5 mm² cable = **22**.

Step 3

Apply factor to number of cables in run = 22 x 10 = **220**.

Step 4

Select correct table for conduit systems runs in excess of 3 m or with bends, **Table 5D, IEE On-site Guide**.

Step 5

Obtain from **Table 5D** a factor for the 10 m run with two bends which is greater than 220. The table gives a factor of **260** for **25 mm** conduit.

Answer

The minimum size conduit is **25 mm**.

Note: For other cables and/or conduit, consult the manufacturer's information.

Space factor

When other sizes and types of cable or trunking are used, the space factor must not exceed 45%. Space factor is the ratio (as a percentage) of the sum of the overall cross-sectional area (csa) of all the cables (including any sheath and insulation) to the internal csa of the enclosure in which they are installed.

In this situation, it is necessary to carry out the following procedure to determine the size of trunking required after the cable size has been decided.

- Consult the cable manufacturer's literature to obtain the overall dimensions of the cable including the insulation.
- Work out the cross-sectional areas of the cables which are to be installed by using the formula:

cross-sectional area = $\frac{\pi \times d^2}{4}$

- Add together the individual cross-sectional areas of the cables concerned and obtain the total cross-sectional areas of the cables.
- Obtain the size of trunking by using the following formula which will allow a 45% space factor:

    let A = the csa of the trunking required

    $A \times \frac{45}{100}$ mm² = total csa of the cables

    then $A = \frac{100}{45} \times$ total csa of cables

- Check the manufacturer's trunking sizes and select one size. Convert to a csa and compare with the calculated value.

Example 2

A steel trunking is to be installed as the wiring system for eight single-phase circuits each having a design current of 35 A.

40 A BS 88 Part 2 fuses will be used as the overcurrent protective devices and PVC insulated copper cables will be installed. Determine the size of trunking required.

Step 1

C_g is the grouping factor and, from Table 4B1 of BS 7671, the value for eight circuits = **0.52**

Minimum current-carrying capacity of cables:

$$I_t = \frac{I_n}{C_g} = \frac{40}{0.52} = \mathbf{77\ A}$$

Using Table 4D1A of BS 7671 and installation method installed in trunking Reference Method B, the minimum size which can be used is 25 mm² cable.

Step 2

For 25 mm² cable, the overall csa is **63.8 mm²** (obtained from the manufacturer's data).

Step 3

Total csa of cables = 16 x 63.8

= **1,021 mm²**

Step 4

Minimum trunking csa $= \dfrac{100}{45} \times$ total csa of cables

$\qquad\qquad\qquad\qquad = \dfrac{100}{45} \times 1{,}021$

$\qquad\qquad\qquad\qquad = \mathbf{2{,}269\ mm^2}$

Step 5

From the manufacturer's data the nearest trunking size greater than **2,269 mm²** is **50 mm x 50 mm (2,500 mm²)**.

PROJECTS

BS 7671
Part 9

Note: Copies of manufacturer's cable data and certain British Standards' data is required in order to complete Assignments C and D.

Assignment A

Installation at Corsby Village Hall

Circuit to cooker control unit not incorporating a 13 A socket-outlet.

Wiring system

- BE heavy gauge steel conduit incorporating four bends enclosing PVC insulated cables.

Information given

- TN-S system, 230 V.
- Circuit design current I_b = 45 A.
- Overcurrent protective device – BS 1361 cartridge fuse.
- Length of run = 14 m.
- No correction factors.
- External earth fault loop impedance Z_e = 0.4 Ω; K value for steel conduit = 47.
- Ambient temperature 20°C.

Required to:

- determine rating of overcurrent protective device, I_n
- select cable size (from tables)
- determine effective current-carrying capacity of cable, I_z
- calculate voltage drop in cable and verify that circuit complies with voltage drop constraint
- determine conduit size
- determine maximum Z_s (from tables)
- calculate actual Z_s and verify that circuit complies with shock protection constraint
- calculate earth fault current, I_f

- determine disconnection time (from time/current characteristics)
- determine minimum csa of circuit protection conductor and verify that circuit complies with thermal constraints
- determine actual csa of conduit size to be used as cpc.

Assignment B

Installation at Corsby Village Hall

Circuit to water heater in the kitchen; heater is 5 kW 230 V instantaneous type.

Wiring system

- Heavy gauge PVC conduit incorporating six bends, with PVC insulated cables.
- TN-S system, 230 V.

Information given

- Overcurrent protection device – BS 88 Part 2 HRC fuse.
- Length of run = 18 m.
- No correction factors.
- $Z_e = 0.4\ \Omega$.
- Ambient temperature 20°C.

Required to:

- calculate design current, I_b
- determine rating of device, I_n
- select line conductor size
- determine current-carrying capacity of cable, I_z
- calculate voltage drop and verify circuit complies with voltage drop constraint
- determine maximum Z_s
- calculate actual Z_s and verify circuit complies with shock protection constraint
- calculate earth fault current, I_f
- determine disconnection time
- determine cpc size and verify that circuit complies with thermal constraint
- determine PVC conduit size.

Assignment C

Installation at Corsby Village Hall

- Circuit to H & V control panel located in the boiler house.
- Isolation device is 15 A S P and N isolator.

Wiring system

- Heavy duty MICC cable with PVC sheath.
- TN-S system, 230 V.

Information given

- Design current, I_b = 11 A.
- Overcurrent device is a BS 88 Part 2 HRC fuse.
- Length of run = 17.5 m.
- Ambient temperature in boiler house = 40°C.
- Z_e = 0.4 Ω.

Required to:

- determine rating of device, I_n
- select cable size (from tables)
- calculate voltage drop and verify circuit complies with voltage drop constraint
- determine maximum Z_s
- calculate actual Z_s
- calculate fault current, I_f
- determine disconnection time.

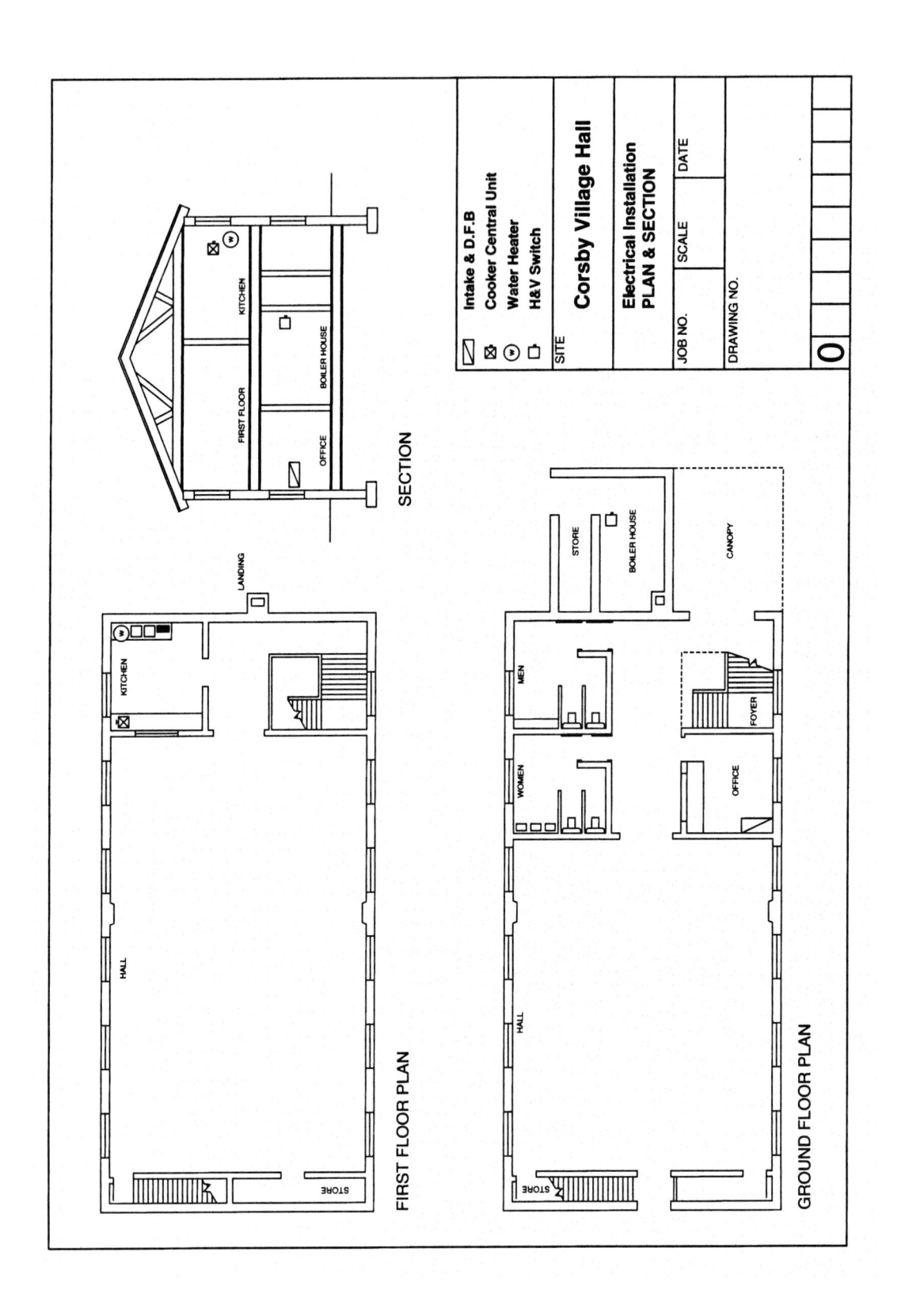

Corsby Village Hall

Electrical Installation PLAN & SECTION

Legend:
- ⌁ Intake & D.F.B
- ⊠ Cooker Central Unit
- ⓦ Water Heater
- ▯ H&V Switch

SECTION — showing KITCHEN, FIRST FLOOR, BOILER HOUSE, OFFICE

FIRST FLOOR PLAN — KITCHEN, LANDING, HALL, STORE

GROUND FLOOR PLAN — STORE, BOILER HOUSE, CANOPY, MEN, WOMEN, FOYER, OFFICE, HALL

E1: BS 7671 (February 2008)

MODEL ANSWERS

BS 7671 Part 9

Assignment A

Information given

Supply – TN-S system	=	230 V
Design current I_b	=	45 A
Protective device	=	BS 1361 cartridge fuse
Length of run	=	14 m
Z_e	=	0.4 Ω
K for steel conduit	=	47 (from Table 54.5 of BS 7671)

No correction factors apply

a) For rating of protective device (I_n)

   From Regulation 433.1.1 $I_n \geq I_b$ = **45 A**

b) For cable size

   From Regulation 433.1.1 $I_z \geq I_n$ = **45 A**

   From Table 4D1A of BS 7671 (reference method B)
   10 mm² may be suitable (57 A, 4.4 mV)

c) Current-carrying capacity of cable (I_z)

   From Regulation 433.1.1 $I_z \geq I_n \geq I_b$

   From Table 4D1B (reference method B) of
   BS 7671 for 10 mm² I_z = **57 A**

d) Voltage drop from Appendix 12 maximum permissible vd = 5%

   $$= \frac{5}{100} \times 230 = 11.5 \text{ V}$$

   Actual voltage drop

   $$= \frac{mV/A/m \times I_b \times \text{length}}{1,000} = \frac{4.4 \times 45 \times 14}{1,000} = \textbf{2.77 V}$$

   Satisfactory as actual voltage drop (2.77 V) < permissible voltage drop 11.5 V

e) Determine conduit size

Length of run = 14 m

Number of bends = 4

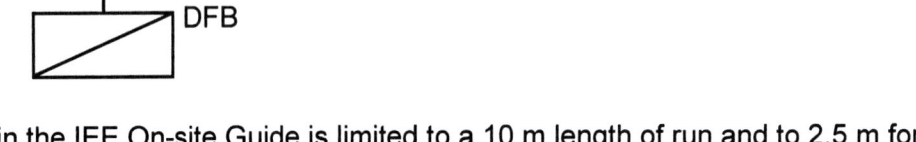

As Table 5D in the IEE On-site Guide is limited to a 10 m length of run and to 2.5 m for runs incorporating four bends, choice is either:

a) halve the run and consider circuit in two sections with intermediate draw in box, or

b) carry out calculation using space factor formula.

Method (a) is preferable since the conduit run repeats itself and the calculations are fewer and easier.

2 x 90° bends at each end, take length = 7 m

From Table 5C, IEE On-site Guide,
cable factor for 10 mm² = 105

for 2 conductors 105 x 2 = 210

From Table 5D, IEE On-Site Guide conduit
size for 7 m run with two bends = 25 mm²

f) For maximum earth fault loop impedance (Z_s)

Table 41.1 of BS 7671 states that disconnection should occur within 0.5 s.

From Table 41B1(b) for fuses to BS 1361

Maximum Z_s = **0.96 Ω**

g) For actual Z_s

$Z_s = Z_e + R_1 + R_2$

$Z_e = 0.4$ Ω

From BS 4568-1:1981 for steel conduit must not exceed $R_2 \leq 0.005$ Ω/m.

From Table 9A, IEE On-site Guide, R_1 = 0.00183 Ω/m for 10 mm² line conductor x 1.04 from Table 9C, IEE On-site Guide = 0.00190 Ω/m.

$$\therefore Z_s = 0.4 + (0.00190 \times 14) + (0.005 \times 14)$$
$$= 0.4 + 0.0266 + 0.07$$
$$= 0.5 \, \Omega$$

Shock constraint satisfactory as actual Z_s (0.5 Ω) < maximum Z_s (0.96 Ω)

h) For earth fault current (I_f)

$$I_f = \frac{U_o}{Z_s} = \frac{230}{0.5} = \textbf{460 A}$$

i) For disconnection time (t)

From Figure 3.1, Appendix 3 of BS 7671, t = **0.3 s**

j) For thermal constraint

From Table 54.5 of BS 7671, K for steel = **47**

$$\therefore \text{cross-sectional area } S = \frac{\sqrt{I^2 t}}{k} = \frac{\sqrt{460^2 \times 0.3}}{47} = \frac{252}{47} = \textbf{5.36 mm}^2$$

k) Actual csa of cpc

From page 5/40 of study notes, csa of 25 mm HG conduit = **131 mm²**

∴ **25 mm² conduit is suitable**

Assignment B

Information given

Load – Water Heater	=	5 kW
Supply – TN-S U_o	=	230 V
Overcurrent protection	=	BS 88-6
Length of run	=	18 m
Z_e	=	0.4 Ω

Wiring system is PVC conduit (6 bends) and PVC insulated cables

No correction factors apply

a) For design current (I_b)

$$I_b = \frac{P}{U_o} = \frac{5 \times 1{,}000}{230} = \mathbf{21.7\ A}$$

b) Rating of protective device (I_n) select 25 A size

Check $I_n \geq I_b$

From Table 41.4 of BS 7671, 25 A BS 88-2.2 fuse is available

∴ $I_n = 25$ A

c) Line conductor size

From Regulation 433.1.1 $I_z \geq I_n = \mathbf{25\ A}$

From Table 4D1A and 4D1B of BS 7671 (reference method B)
4 mm² may be suitable (32 A, 11 mV)

d) Current-carrying capacity of cable (I_z)

From Regulation 433.1.1 $I_z \geq I_n \geq I_b$

From Table 4D1A of BS 7671 (reference method 3)
for 4 mm² $I_z = \mathbf{32\ A}$

e) Voltage drop from Appendix 12 maximum permissible vd = 5%

$$= \frac{5}{100} \times 230 = 11.5\ V$$

Actual voltage drop

$$= \frac{mV/A/m \times I_b \times length}{1{,}000} = \frac{11 \times 21.7 \times 18}{1{,}000} = \mathbf{4.3\ V}$$

Satisfactory as actual voltage drop (4.3 V) < permissible voltage drop 11.5 V

f) For maximum earth fault loop impedance (Z_s)

Table 41.1 of BS 7671 states that disconnection must occur within 5 s

From Table 41.4 of BS 7671 for fuses to BS 88-6

Maximum $Z_s = \mathbf{2.3\ \Omega}$

g) For actual Z_s

$$Z_s = Z_e + R_1 = R_2$$

$$Z_e = 0.4\ \Omega$$

From Table 9A, IEE On-site Guide, $R_1 + R_2$ for 4 mm² line conductor and 1.5 mm² cpc is 0.01671 Ω/m x 1.20 from Table 9C, IEE On-site Guide, = 0.02 Ω/m.

$$\therefore Z_s = 0.4 + (0.02 \times 18)$$
$$= 0.4 + 0.36$$
$$= 0.76\ \Omega$$

Satisfactory as actual Z_s (0.76 Ω) < maximum Z_s (2.3 Ω)

h) For earth fault current, (I_f)

$$I_f = \frac{U_o}{Z_s} = \frac{230}{0.76} = \mathbf{302.6\ A}$$

i) For disconnection time (t)

From Figure 3.3B, Appendix 3 of BS 7671, t < **0.1 s**

j) For thermal constraint

From Table 54.3 of BS 7671, k for PVC insulated copper = **115**

$$\therefore \text{cross-sectional area } S = \frac{\sqrt{I^2 t}}{k} = \frac{\sqrt{302.6^2 \times 0.1}}{115} = \mathbf{0.83\ mm^2}$$

Therefore 1.5 mm² cpc does comply with thermal constraint

k) For PVC conduit size

Consider dividing the circuit into 3 x 6 m lengths incorporating two bends each

From Table 5C, IEE On-site Guide, cable factor for 4 mm²	= 43
From Table 5C, IEE On-site Guide, cable factor for 1.5 mm²	= 22
For 3 cables, (43 x 2) + (22 x 1)	= 108
From Table 5D, IEE On-site Guide, conduit size for 6 m run with 2 bends, conduit factor ≥ 108	= **16 mm**

Note: 16 mm conduit may not be a stock item so it is likely that 20 mm conduit would be used.

Assignment C

Information given

Load – H & V control panel

Supply – TN-S system = 230 V

Isolation – 15 A SP & N isolator

Design current I_b = 11 A

Overcurrent protection = BS 88-6

Length of run = 17.5 m

Ambient temperature = 40°C

Z_e = 0.4 Ω

Wiring system is heavy duty PVC sheathed MICC cable

a) Rating of protective device (I_n)

$I_n \geq I_b$ = **11 A**

From Table 41.4 of BS 7671A, 16 A BS 88-6 fuse is available

∴ I_n = 16 A

b) From Regulation 433.1.1 $I_z \geq I_n$ = **16 A**

From Table 4C1 of BS 7671, for mineral cable (70°C sheath) C_a = 0.8

Correction for ambient temperature (40°C)

$$I_z \geq \frac{I_n}{C_a} = \frac{16}{0.8} = 20 \text{ A}$$

From Table 4G1A and 4G1B of BS 7671 (reference method C)
1 mm² may be suitable (18.5 A, 42 mV)

c) Voltage drop from Regulation 525-01-02 maximum permissible vd = 5%

$$= \frac{5}{100} \times 230 = 11.5 \text{ V}$$

Actual voltage drop

$$= \frac{mV/A/m \times I_b \times length}{1,000} = \frac{42 \times 11 \times 17.5}{1,000} = \mathbf{8.09 \text{ V}}$$

d) Maximum earth fault loop impedance (Z_s)

Table 41.1 of BS 7671 states that disconnection must occur within 5 s

From Table 41.4 of BS 7671 for fuses to BS 88-6

Maximum Z_s = **4.36 Ω**

e) For actual Z_s

$$Z_s = Z_e + R_1 + R_2$$

From manufacturer's data R_1 = 0.0179 Ω/m, R_2 = 0.0019 Ω/m

$$\therefore Z_s = 0.4 + (0.0179 \times 17.5) + (0.0019 \times 17.5)$$

$$= 0.4 + 0.3133 + 0.0333$$

$$= \mathbf{0.7466 \ \Omega}$$

f) For earth current (I_f)

$$I_f = \frac{U_o}{Z_s} = \frac{230}{0.75} = \mathbf{306.6 \ A}$$

g) For disconnection time (t)

From Figure 3.3B, Appendix 3 of BS 7671, t = **< 0.1 s**

MAJOR CHANGES OF BS 7671: 17TH EDITION IEE WIRING REGULATIONS 2008

BS 7671 Part 10

Cover	Red
Issued	January 2008
Implementation	Installations designed after 30 June to comply with BS 7671:2008
Numbering	Has been revised (refer to 'Numbering' on page 1)

Part 1 Scope, object and fundamental principles

Regulation 131.6	Protection against voltage disturbances, electromagnetic influences and electromagnetic emissions generated by installations or equipment. Covered in Chapter 44.
Regulation 132.13	Requirement of documentation for every electrical installation is provided. Covered in Chapter 51, Parts 6 and 7.

Part 2 Definitions

New or revised definitions. Fifteen of these relate to PV solar photovoltaic power supply systems.

Part 3 Assessment of general characteristics

Chapter 35 Safety services	Safety services are often regulated by statutory authorities, whose requirements need to be observed for fire alarms, emergency lighting, fire pumps, rescue service lifts and smoke extract systems. Covered in detail in Part 5, Chapter 56.
Chapter 36 Continuity of service	Assessment of each circuit to be made of any need for continuity of service throughout its intended life, covering earthing, protective devices, multiple power supplies, monitoring equipment and number of circuits. Covered throughout the Regulations.

Part 4 Protection for safety

Chapter 41 Protection against electric shock	Important definition changes from the 16th Edition IEE Regulations (BS 7671:2001) are that protection against: • direct contact becomes basic protection • indirect contact becomes fault protection, which includes overcurrent and short-circuit conditions.

A new category is included called 'additional protection', which covers RCDs and supplementary equipotential bonding.

All socket-outlet circuits up to 20 A are to be protected by 30 mA RCDs used by ordinary persons and mobile equipment not exceeding 32 A for use outdoors.

Earth fault loop impedance tables for 240 V 41B1, 41B2, 41D of the 16th Edition IEE Regulations (BS 7671:2001) have now been replaced by Tables 41.2, 41.3, 41.4 based on 230 V.

New Table 41.5 provides earth fault loop impedance values for RCDs to BS EN 61008-1 and BS EN 61009-1.

FELV is now included as a protective measure.

The above are all referred to throughout the Regulations.

Chapter 42 Protection against thermal effects	Previously Section 48Z of the 16th Edition IEE Regulations (BS 7671:2001). Now Section 422.
Chapter 43 Protection against overcurrent	Conductors in parallel. Previously Section 43 of 16th Edition IEE Regulations (BS 7671:2001). Now Appendix 10.
Chapter 44 Protection against voltage disturbances	This includes a new Section 422 covering protection of low voltage installations against overvoltages due to faults occurring in high voltage systems. Section 443 new Regulations for designers to use risk assessments when designing installations susceptible to overvoltages from lightning strikes.

Part 5 Selection and erection of equipment

Chapter 52 Selection and erection of wiring systems	Busbar trunking and powertrack wiring systems now included. When cables are concealed in walls at depths less than 50 mm 30 mA RCD protection is required. When electrical installation is not under the supervision of a skilled or instructed person 30 mA RCD protection is required. Revised Table 52.2 for cables surrounded by thermal insulation with reduced rating factors than in 16th Edition IEE Regulations (BS 7671:2001).
Chapter 53 Protection, isolation, switching, control and monitoring	Chapter 46, Sections 476 and 537, 16th Edition IEE Regulations (BS 7671:2001), now covered in Chapter 53 with two new sections. 532 devices for protection against risk of fire. 538 monitoring devices.

Chapter 54 Earthing arrangements and protective conductors	Section 607 of the 16th Edition IEE Regulations (BS 7671:2001) covering equipment having high protective conductor currents is now covered in Regulation 543.7.
Chapter 55 Other equipment	New requirements are included covering low voltage generating sets and small-scale embedded generators (SSEGs). Covered in Regulation 551.7.
Section 559 Luminaires and lighting installations	This is a new series of Regulations. Highway power supplies and street furniture, previously Section 611 of the 16th Edition IEE Regulations (BS 7671:2001), is now covered in this section with other outdoor lighting applications.

Part 6 Inspection and testing

This was previously Part 7 in the 16th Edition IEE Regulations (BS 7671:2001).

The minimum value of insulation resistance now given in Table 61, previously Table 71A of the 16th Edition IEE Regulations (BS 7671:2001), are for:

- SELV and PELV tested at 250 V 0.5 π (previously 0.25 π)
- other circuits tested at 500 V 1 π (previously 0.5 π).

Part 7 Special installations or locations

Previously Part 6 of the 16th Edition IEE Regulations (BS 7671:2001).

Section 607 covering high protective conductor currents is now incorporated in Chapter 54.

Section 608 relating to caravans and caravan parks is incorporated into Section 708 electrical installations in caravan/camping parks and similar locations and Section 721 electrical installations in caravans and motor caravans.

Section 611 relating to highway power supplies is incorporated in Section 559.

Other changes to Part 7

Section 701 Locations containing a bath tub or shower basin	Zone 3 has been removed. Circuits must be protected by 30 mA RCD. Supplementary bonding is not required provided that main bonding of other service pipework is effective in accordance with Chapter 41.

	Socket-outlets can be installed in the location greater than 3 from Zone 1.
Section 702 Swimming pools and other basins	Zones A, B and C referred to in the 16th Edition IEE Regulations (BS 7671:2001) are now referred to as Zones 0, 1 and 2.

Basins of fountains now included. |
Section 703 Rooms containing sauna heaters	Zones A, B, C and D referred to in the 16th Edition IEE Regulations (BS 7671:2001) are now referred to as Zones 1, 2 and 3 with some new dimensions given.
Section 704 Construction and demolition site installations	The equation for 25 V maximum fault voltage previously in the 16th Edition IEE Regulations (BS 7671:2001) no longer referred to. Disconnection time of 0.2 s now given.
Section 705 Agricultural and horticultural premises	The equation for 25 v maximum fault voltage previously in the 16th Edition IEE Regulations (BS 7671:2001) no longer referred to. Disconnection time or 0.2 s now given.
Section 706 Conducting locations	Previously Section 606 in the 16th Edition IEE Regulations (BS 7671:2001).
Section 708 Electrical installations in caravan/camping parks and similar locations	New requirement for each socket-outlet to be individually RCD protected.

The following are new sections:

- 709 Marinas and similar locations.
- 711 Exhibitions, shows and stands.
- 712 Solar photovoltaic (PV) power supply systems.
- 717 Mobile or transportable units.
- 721 Electrical installations in caravans and motor caravans.
- 740 Temporary electrical installations for structures, amusement devices and booths at fairgrounds, amusement parks and circuses.
- 753 Floor and ceiling heating systems.

Appendices

The following new appendices are included:

8. Current-carrying capacity and voltage drop for busbar trunking and powertrack systems.
9. Definitions – multiple source d.c. and other systems.
10. Protection against conductors in parallel against overcurrent.
11. Effect of harmonic currents on balanced three-phase systems.
12. Voltage drop in consumers installations.
13. Methods of measuring the insulation resistance/impedance of floors and walls to earth.
14. Measurement of earth fault loop impedance.
15. Ring and radial final circuit arrangements.

APPENDIX 1

Step by step procedure for the selection of a final circuit cable and protective device

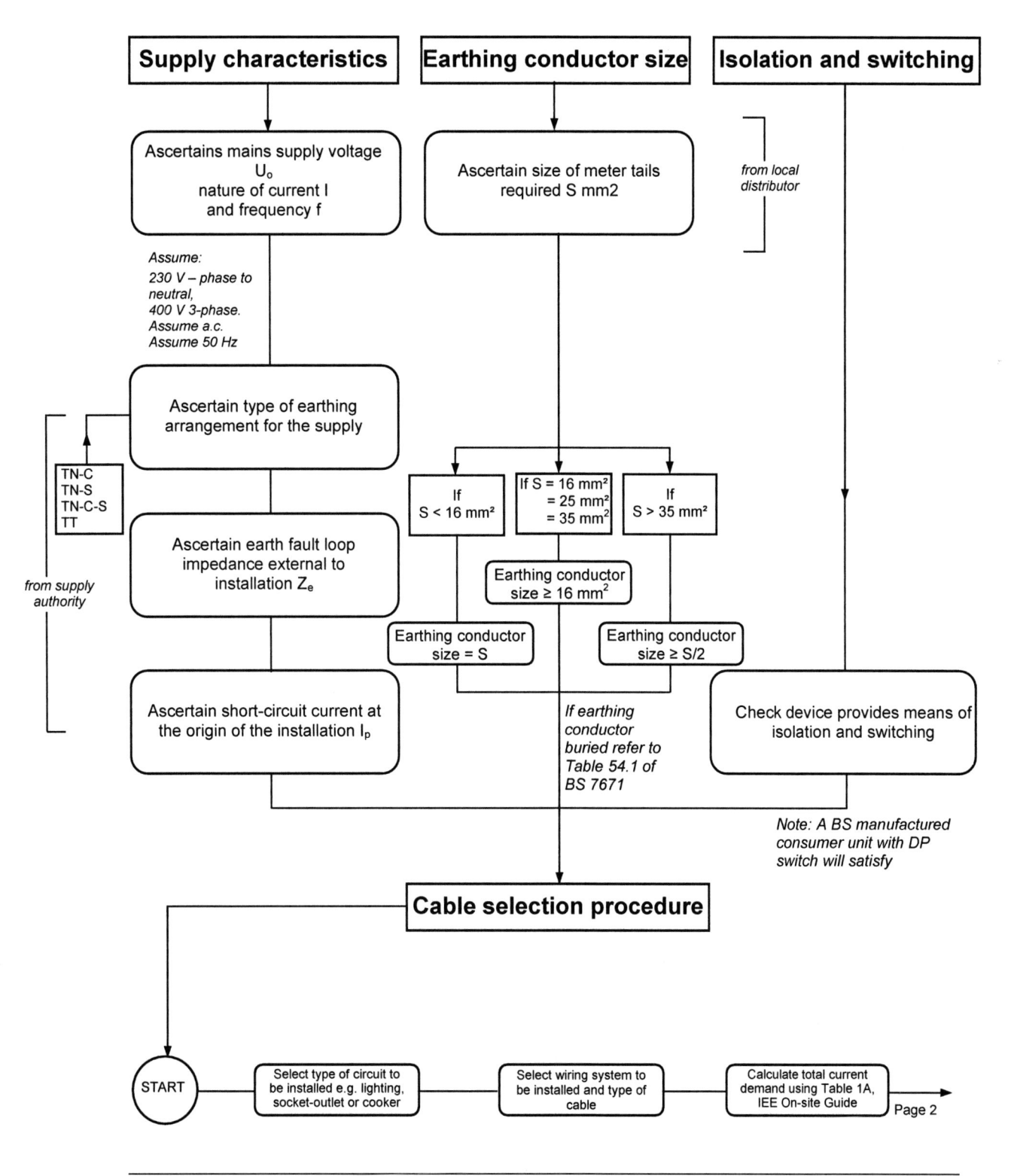

Supply characteristics

Ascertains mains supply voltage U_o nature of current I and frequency f

Assume:
230 V – phase to neutral,
400 V 3-phase.
Assume a.c.
Assume 50 Hz

Ascertain type of earthing arrangement for the supply

- TN-C
- TN-S
- TN-C-S
- TT

from supply authority

Ascertain earth fault loop impedance external to installation Z_e

Ascertain short-circuit current at the origin of the installation I_p

Earthing conductor size

Ascertain size of meter tails required S mm2

- If $S < 16$ mm² → Earthing conductor size = S
- If $S = 16$ mm² = 25 mm² = 35 mm² → Earthing conductor size ≥ 16 mm²
- If $S > 35$ mm² → Earthing conductor size ≥ S/2

If earthing conductor buried refer to Table 54.1 of BS 7671

Isolation and switching

from local distributor

Check device provides means of isolation and switching

Note: A BS manufactured consumer unit with DP switch will satisfy

Cable selection procedure

START → Select type of circuit to be installed e.g. lighting, socket-outlet or cooker → Select wiring system to be installed and type of cable → Calculate total current demand using Table 1A, IEE On-site Guide → Page 2

© Construction Industry Training Board   A/1   E1: BS 7671 (February 2008)

From page 1

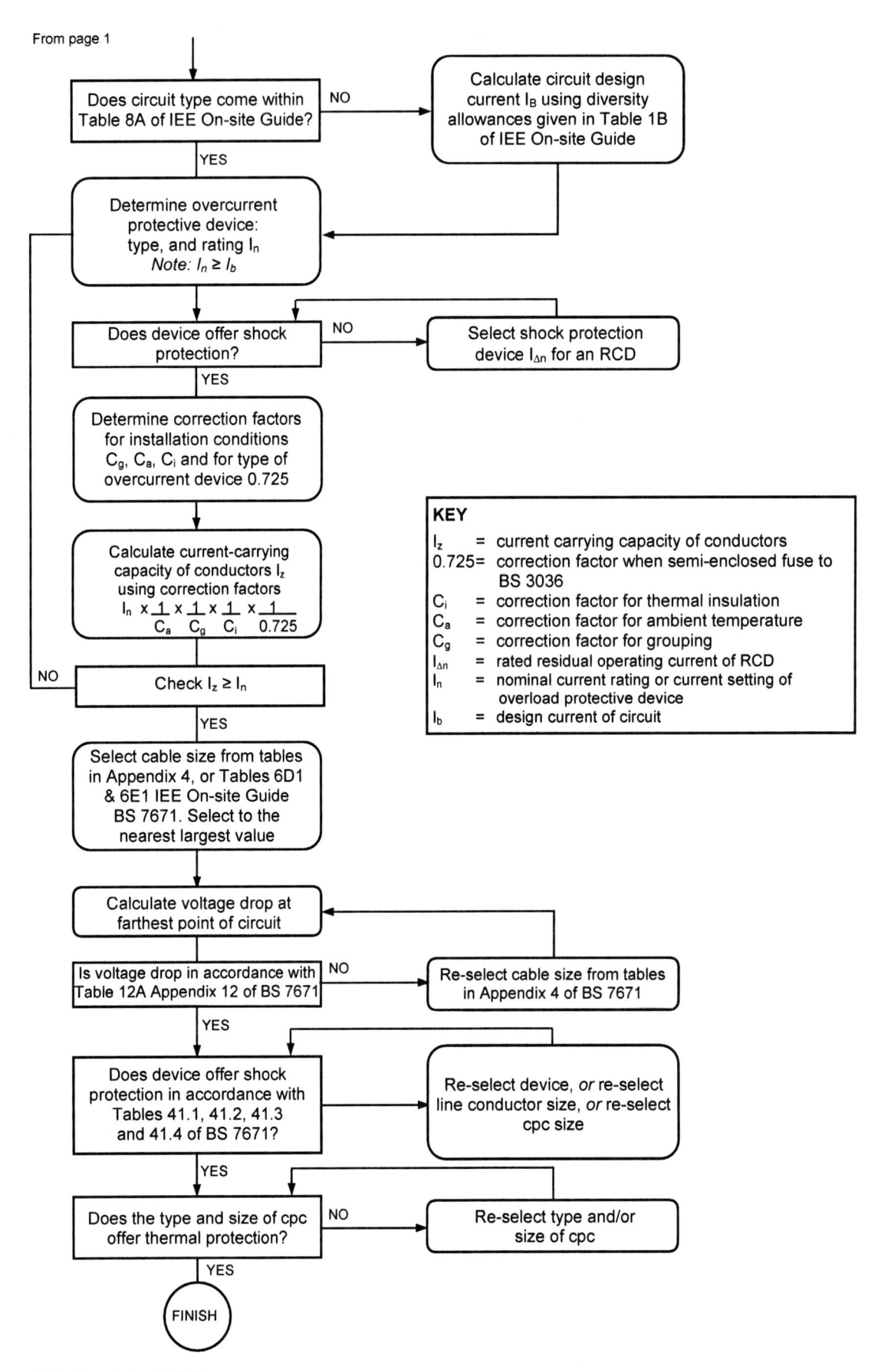

```
                    ↓
        ┌──────────────────────┐   NO    ┌──────────────────────┐
        │ Does circuit type    │────────▶│ Calculate circuit    │
        │ come within Table 8A │         │ design current $I_B$ │
        │ of IEE On-site Guide?│         │ using diversity      │
        └──────────────────────┘         │ allowances given in  │
                  │ YES                  │ Table 1B of IEE      │
                  ▼                      │ On-site Guide        │
        ┌──────────────────────┐         └──────────────────────┘
        │ Determine overcurrent│◀────────────────┘
        │ protective device:   │
        │ type, and rating $I_n$│
        │ Note: $I_n ≥ I_b$    │
        └──────────────────────┘
                  │
                  ▼
        ┌──────────────────────┐  NO   ┌──────────────────────┐
        │ Does device offer    │──────▶│ Select shock         │
        │ shock protection?    │       │ protection device    │
        └──────────────────────┘       │ $I_{Δn}$ for an RCD  │
                  │ YES                └──────────────────────┘
                  ▼
        ┌──────────────────────┐
        │ Determine correction │
        │ factors for install- │
        │ ation conditions     │
        │ $C_g$, $C_a$, $C_i$  │
        │ and for type of      │
        │ overcurrent device   │
        │ 0.725                │
        └──────────────────────┘
                  │
                  ▼
        ┌──────────────────────┐
        │ Calculate current-   │
        │ carrying capacity of │
        │ conductors $I_z$     │
        │ using correction     │
        │ factors              │
        │ $I_n × \frac{1}{C_a} × \frac{1}{C_g} × \frac{1}{C_i} × \frac{1}{0.725}$ │
        └──────────────────────┘
                  │
         NO       ▼
        ┌──────────────────────┐
        │ Check $I_z ≥ I_n$    │
        └──────────────────────┘
                  │ YES
                  ▼
        ┌──────────────────────┐
        │ Select cable size    │
        │ from tables in       │
        │ Appendix 4, or Tables│
        │ 6D1 & 6E1 IEE On-site│
        │ Guide BS 7671.       │
        │ Select to the        │
        │ nearest largest value│
        └──────────────────────┘
                  │
                  ▼
        ┌──────────────────────┐
        │ Calculate voltage    │
        │ drop at farthest     │
        │ point of circuit     │
        └──────────────────────┘
                  │
                  ▼
        ┌──────────────────────┐  NO   ┌──────────────────────┐
        │ Is voltage drop in   │──────▶│ Re-select cable size │
        │ accordance with Table│       │ from tables in       │
        │ 12A Appendix 12 of   │       │ Appendix 4 of BS 7671│
        │ BS 7671              │       └──────────────────────┘
        └──────────────────────┘
                  │ YES
                  ▼
        ┌──────────────────────┐  NO   ┌──────────────────────┐
        │ Does device offer    │──────▶│ Re-select device, or │
        │ shock protection in  │       │ re-select line       │
        │ accordance with      │       │ conductor size, or   │
        │ Tables 41.1, 41.2,   │       │ re-select cpc size   │
        │ 41.3 and 41.4 of     │       └──────────────────────┘
        │ BS 7671?             │
        └──────────────────────┘
                  │ YES
                  ▼
        ┌──────────────────────┐  NO   ┌──────────────────────┐
        │ Does the type and    │──────▶│ Re-select type and/or│
        │ size of cpc offer    │       │ size of cpc          │
        │ thermal protection?  │       └──────────────────────┘
        └──────────────────────┘
                  │ YES
                  ▼
              ( FINISH )
```

KEY

- I_z = current carrying capacity of conductors
- 0.725 = correction factor when semi-enclosed fuse to BS 3036
- C_i = correction factor for thermal insulation
- C_a = correction factor for ambient temperature
- C_g = correction factor for grouping
- $I_{Δn}$ = rated residual operating current of RCD
- I_n = nominal current rating or current setting of overload protective device
- I_b = design current of circuit

* See Regulation 525-01

APPENDIX 2

Cable selection procedure

Quantity or factor	Symbol	Reference	Data calculations	Value
Load current	I	Diversity tables 1A and 1B, IEE On-site Guide, Tables 41.1, 41.2, 41.3, 41.4 or time/current curves. Appendix 3 of BS 7671		
Design current	I_b			
Protective device rating	I_n		Check $I_b \leq I_n$	YES/NO
Cable size				
Grouping factor	C_g	Tables 4C1 to 4C5 of BS 7671		
Ambient temperature	C_a	Tables 4B1 and 4B2 of BS 7671		
Thermal insulating factor	C_i	Table 52.2 of BS 7671		
BS 3036 fuse factor	0.725			
Tabulated current-carrying capacity of a cable	I_t	$I_t \geq \dfrac{I_n}{C_a \times C_g \times C_i \times 0.725}$		
Tabulated current-carrying capacity	I_t	From Tables 4D1A to 4J4A of BS 7671		
Effective current-carrying capacity	I_z	$I_t \geq I_z$	Check $I_b \leq I_n \leq I_z$	YES/NO
Permissible V drop		Safe functioning of equipment or in accordance with Table 12A, Appendix 12 of BS 7671		
Actual V drop		$\dfrac{mV/A/m \times I_b \times length}{1,000}$		
			Check actual voltage drop ≤ permissible voltage drop	YES/NO

Cable selection procedure

Quantity of factor	Symbol	Reference	Data calculations	Value
Shock protection				
Maximum earth fault loop impedance	Max Z_s	Tables 41.1, 41.2, 41.3, 41.4		
External earth fault loop impedance	Z_e	Measure or obtain from electricity company		
cpc size chosen		Select from BS 6004 initially		
Resistance of line conductor	R_1	Table 9A, 9B and 9C IEE On-site Guide, manufacturer's data or BS data		
Resistance of cpc	R_2			
Actual earth fault loop impedance	Z_s	$Z_s = Z_e + (R_1 + R_2)$	Check $Z_s \leq$ Max Z_s	YES/NO
Thermal constraint				
Fault current	I_f	$I_f = \dfrac{U_o}{Z_s}$ (230 V phase to earth)		
Time	t	Read time/current characteristics, Appendix 3, Tables 54.2 to 54.6 of BS 7671		
Factor for cpc	k			
Size cpc	S	$\dfrac{\sqrt{I^2 t} \, mm^2}{k}$	Check actual cpc used is $\geq$ than S above	YES/NO

Cable selection procedure

Quantity or factor	Symbol	Reference	Data calculations	Value
For groups				
Grouping correction factor	C_g	Tables 4C1 to 4C5 of BS 7671		
Tabulated current-carrying capacity required	I_t	$I_t \geq \dfrac{I_n}{C_g \times (0.725)}$		
Alternatively		Cable current-carrying capacity must **not be less than the larger value** of I_t obtained after using **both** of the following formulae provided that circuits of the group are **not** liable to simultaneous overload. $I_t \geq \dfrac{I_b}{C_g}$ and $I_t \geq \sqrt{(1.9)\,I_n^2 + 0.48 \times I_b^2 \,\dfrac{(1-C_g^2)}{C_g^2}}$ (1.9) to be applied only if BS 3036 fuse is used		
		Any correction for ambient temperature or thermal insulation to be applied to I_t after these calculations have been completed i.e. min $I_t \geq \dfrac{I_t}{C_a \times C_i}$		

E1: BS 7671
Index

A

a.c. systems 4/5, 5/13, 5/36, 6/27
access 1/9, 5/2, 5/27
  agricultural installations 7/18
  cables 9/1
  construction sites 7A/2
  control gear 5/65
  earth electrodes 5B/4
  emergency controls 5/31, 5/32, 7/37
  exhibitions, shows and stands 7/29
  isolators 7/29
  joints and connections 5/2, 5/23, 5/42
  main earthing terminals 5/39
  maintenance 1/9, 5/2, 5/27
  outdoor equipment 5/62–3
  safety services 5/65
  socket-outlets 7/38
  swimming pools 7/12
  switchgear 5/65, 7A/2
  temporary installations 7/37, 7/38
  transformers 7/29
  wiring systems 9/1
accessories 3B/1–4, 5/55–6
  agricultural installations 7/19
  caravans 7/34, 7/35
  colour-coding 3C/7
  construction sites 7A/2, 7A/5
  lighting systems 3B/3–4, 5/61
  protective conductors 4/2
  reduced low voltage systems 4/11
  see also cable couplers; lampholders; plugs and socket-outlets; switches
additional protection 4/15–16
  a.c. systems 4/5
  bath and shower rooms 7/5–7
  cables in walls or partitions 5/18
  external influences 5/2
  fire precautions 4/21
  inspection and testing 6/4, 6/27
  major changes in new regulations 10/2
  transformers 7/29
  see also barriers; enclosures; residual current devices; supplementary bonding
additions and alterations 1/6, 1/10, 5/5–7, 5/11, 6/34
agricultural installations 7/16–20
alphanumeric identification, conductors 5/7
alterations *see* additions and alterations
ambient temperature *see* temperatures
amusement parks and devices 7/36–8
animals, protection from 5/19
armoured single-core cables 5/22
assessment of general characteristics, electrical systems 3/1–5, 3A/5
automatic disconnection 4/2–11
  agricultural installations 7/16
  caravans 7/34
  construction sites 7/15
  floor and ceiling heating systems 7/38
  generating sets 5/51, 5/52, 5/53
  inspection and testing 4/6, 6/4, 6/20–1
  IT systems 4/8–10, 6/21
  mobile or transportable units 7/32
  outdoor equipment 5/63
  restrictive conductive locations 7/20–1
  swimming pools 7/12
  TN systems 4/6–7, 6/20
  TN-C systems, residual current devices 4A/2
  see also disconnection times
autotransformers 5/59

B

Band I and Band II voltages 1/3, 5/25–6
bare conductors *see* busbars
barriers
  agricultural installations 7/16, 7/17
  bath and shower rooms 7/4
  electric shock protection 4/12, 4/17, 6/15, 7/4
  electrical separation 4/13
  FELV systems 4/10
  fire precautions 4/23, 7/17
  inspection and testing 6/15
  IP ratings 4/15, 4/17
  outdoor equipment 5/62, 5/63
  reduced low voltage systems 4/11
  safety services 5/65
  saunas 7/13
  SELV systems 4/15
  warning notices 4/17
  see also enclosures; obstacles
basic protection 1/4, 4/16–17
  batteries 5/54
  definition 4/1
  electrical separation 4/13
  FELV systems 4/10
  major changes in new regulations 10/1
  overhead cables 4/17–18
  reduced low voltage systems 4/11
  see also barriers; electrical insulation; enclosures; obstacles; out-of-reach placement
bath and shower rooms 5/58, 7/2–8, 8/12–14, 10/3, 10/4
batteries 3/4, 4/14, 5/54, 5/65
berthing instructions, marinas 7/26–27
birds, protection from 5/19
boats *see* marinas
boilers 5/56–8
  see also water heaters
bonding 1/5, 4/2–4
  additions and alterations 1/6
  boilers and water heaters 5/56, 5/57, 5/58
  cables 5/22
  caravans 7/34
  exhibitions, shows and stands 7/28
  generating sets 5/52, 7A/2
  major changes in new regulations 10/3
  photovoltaic power supply systems 7/31
  plastic pipework 5/48, 7/6
  warning notices 5/10–11
  see also earth-free local equipotential bonding; single point bonding; solid bonding; supplementary bonding

bonding conductors 4/3, 5/47–50, 6/5–8, 7/6
bonding connections, meters 5/48, 5/49
booths, temporary installations 7/36–8
British Standards 2, 5/1
buried cables *see* underground cables
buried earthing conductors 5/38
burns 1/5, 4/23–4
   see also thermal effects
bus shelters, residual current devices 5/63
busbar trunking 4/34, 5/13, 5/24, 10/2, 10/5
busbars 5/3, 5/7

C

cable couplers 3B/3, 5/56
   colour-coding 3C/7
   construction sites 7A/2, 7A/5
   FELV systems 4/10
   reduced low voltage systems 4/11
   see also inlet couplers
cable factors 9/1–4
cable groups 8/2
cable joints *see* joints and connections
cable protection 5/14, 5/18, 7/29, 9/1
cable rating 5/20–2, 8/1–5, 10/5
cable sizing 7/34, 8/2–5
cables
   a.c. circuits 5/13
   access 9/1
   bath and shower rooms 8/12–14
   bonding 5/22
   caravans 7/34
   colour-coding 5/3–7
   construction sites 7/15
   earth monitoring systems 5C/2
   electrical insulation 4/17
   exhibitions, shows and stands 7/29
   fire precautions 4/22, 4/23, 5/24
   fountains 7/11
   identification 5/3–7
   in lift shafts 5/27
   in thermal insulation 5/20–1, 8/2, 10/2
   in walls or partitions 4/5, 5/17–18, 10/2
   joints and connections 5/43, 7/37
   lighting 5/61, 5/62
   major changes in new regulations 10/2
   marinas 7/25
   mobile or transportable units 7/32, 7/33
   over ceilings 5/16–17
   PEN conductors 5/43
   photovoltaic power supply systems 7/31
   reduced low voltage systems 7/15
   safety services 5/65, 5/66
   selection and installation 5/14, 5/24, 8/1–14, A/1–5
   sizing 7/34, 8/2–5
   soil and floor warming systems 5/58–9
   swimming pools 7/10
   temporary installations 7/37
   through wiring of luminaires 5/62
   see also armoured single-core cables; concentric cables; conductors; flexible cables and cords; grouped cables; heating conductors and cables; overhead cables; spaced cables; telecommunications cables; touching cables; underground cables
camping facilities *see* caravan parks
caravan parks 7/22–3, 10/3, 10/4
caravans 7/22, 7/33–5, 10/3
cartridge fuses 4C/2
ceiling heating systems 7/38–40
ceiling roses 3B/4
ceilings, cables over 5/16–17
certification 1/3, 1/4, 6/2, 6/32, 6/33–4, 6B/1
   see also Electrical Installation Certificates
characteristics of electrical systems, assessment 3/1–5, 3A/5
charts 5/7–9, 6/1
   see also documentation
circuit arrangements 3/2–3
circuit breakers 4D/1–6
   earth fault loop impedance testing 6/26
   electrode water heaters and boilers 5/56
   emergency switching 5/31–3
   fault protection 4/25
   fire precautions 4/22
   for mechanical maintenance 5/31
   isolators and 5/29
   location 1/9
   operating characteristics 4E/1
   reduced low voltage systems 4/11
   see also isolators; residual current devices; switches
circuit protective conductors *see* protective conductors
circuits 3C/1–7
   assessment for continuity of service 3/5
   earth-free local equipotential bonding 4/19
   electrical separation 5/13, 5/26, 6/14
   lighting 5/62, 6/23, 7/36, 7/37, 9/2
   plugs and socket-outlets 3C/6–7, 5/45–6
   safety services 5/65–6
   SELV and PELV systems 4/14
   see also cooker circuits; filter circuits; final circuits; radial circuits; ring circuits; telecommunications circuits
circuses 7/36–8
Class II equipment, electrical insulation 4/12
classification
   circuit breakers 4D/2
   electrical systems 3A/2
   emergency evacuation conditions 4/20–1
   external influences 5/2
   safety services 5/64–5
   see also zone classification
clearances
   non-conducting locations 4/18
   out-of-reach placement 4/18
   spotlights and projectors from combustible materials 4/21
   underground cables 5/26
clocks 3B/2, 5/56
co-ordination
   conductors and overload protective devices 4/26–7
   overload and fault current protection 4/34
colour-coding
   alterations and additions 5/11
   cables and conductors 5/3–7

conduits 5/3
earth clamps 5/48
electrical accessories 3C/7
emergency switching devices 5/32
fire-fighters' switches 5/35
plugs and socket-outlets 3B/2, 3C/7
warning notices 5/11
see also identification
combined protective and neutral conductors *see* PEN conductors
combustible materials, fire precautions 4/23
compatibility, electrical equipment 3/3–4
see also electromagnetic influences; mutual detrimental influences
concentric cables, PEN conductors 5/43
condensation 5/27
see also wet conditions
conductors
assessment of general characteristics 3/1
classification of electrical systems 3A/2
co-ordination with overload protective devices 4/26–7
colour-coding 5/3–7
final circuits 3C/1
fire precautions 4/22
identification 1/10, 5/3–7
inspection and testing 6/4, 6/5–13
joints and connections 5/23
outdoor equipment 5/64
protection 9/1
selection and installation 1/8
SELV and PELV systems 4/14
sizing 1/8, 3C/1, 3C/4, 5/22, 5/47, 5/64
through wiring of luminaires 5/62
see also bonding conductors; busbars; cables; earthing conductors; fine wire conductors; heating conductors and cables; line conductors; loaded conductors; mid-point conductors; multiwire conductors; neutral conductors; parallel conductors; PEN conductors; protective conductors; stranded conductors
conduits
agricultural installations 7/19
colour-coding 5/3
drainage 5/15
fault protection omission 4/2
fire precautions 4/23, 5/24
joints and connections 5/42
protective conductors 5/40, 5/41
sizing 9/1–4, 9P1/1–9P1/8
swimming pools 7/10
temporary installations 7/37
see also enclosures
connections *see* joints and connections
construction sites 3B/3, 7/14–16, 7A/1–5, 10/4
consumer unit arrangement, schedule 6/2
consumer units 4B/1–2
continuity of service 3/5, 10/1
control gear
access 5/65
caravan parks 7/23
construction sites 7/16
cooker circuits 3C/5

exhibitions, shows and stands 7/29
functional testing 6/31
identification 5/2
lighting 5/62
safety services 5/65
saunas 7/14
swimming pools 7/11
temporary installations 7/37
see also switchgear
cooker circuits 3C/5
cooking appliances 3B/5–6, 8/9–11, 9P1/1–2, 9P1/1–3
copper tapes, earth electrodes 5B/2
cords *see* flexible cables and cords
corrosion protection 1/7
agricultural installations 7/18, 7/19
earth electrodes 5/38
earthing conductor connections 5B/4
temporary installations 7/37
corrosive environments 5/15
see also wet conditions
couplers *see* cable couplers; inlet couplers; luminaire supporting couplers
creosote, PVC cable separation from 5/15
cross-sectional area *see* sizing
current-carrying capacity *see* cable rating
current demand *see* demand assessment

D

damage prevention *see* protection
damp conditions *see* wet conditions
data processing equipment 5/44
d.c. systems
identification of alterations and additions 5/6–7
insulation monitoring devices 5/35, 5/36
plugs and socket-outlets 3B/2, 5/55
supplementary bonding 6/27
definitions 2/1–4, 10/1
demand assessment 1/7, 3/1, 3B/4, 3B/5–6
see also diversity
demolition sites 7A/1–5, 10/4
diagrams 5/7–9, 6/1, 7/18
see also documentation
direct contact *see* basic protection
discharge lighting 3B/4, 3B/6, 5/30, 5/34
disconnection *see* automatic disconnection; isolation
disconnection devices *see* circuit breakers; emergency controls; isolators; switches
disconnection times
construction and demolition sites 10/4
final circuits 4/5
reduced low voltage systems 4/11
discrimination
overcurrent protection devices 4E/1–2
see also voltage discrimination
diversity 3/1, 3B/5–8
see also demand assessment
documentation
electrical installations 1/9
floor and ceiling heating systems 7/39–40
inspection and testing 6/32

marinas 7/26–7
safety services 5/66
sealing for fire prevention 5/25
see also certification; charts; diagrams; reports
dodgems 7/38
doors, in outdoor equipment 5/62–3
drainage
  conduits 5/15
  marina installations 7/25
drawings *see* charts; diagrams
drives, functional testing 6/31
dual supply, warning notices 5/12
ducting, temporary installations 7/37
duplicate protective conductors 5/45
dust 5/15
duty ratings, circuit breakers 4D/2–3

E

earth clamps 5/48
earth electrode resistance 5B/3, 6/4, 6/18–20
earth electrodes 5/38, 5B/1–4
  caravan parks 7/23
  generating sets 5/51
  inspection and testing 6/5, 6/19–20
  mobile or transportable units 7/32
  swimming pools 7/9
  TT systems 4/7
earth fault loop impedance
  major changes in new regulations 10/2, 10/5
  reduced low voltage systems 4/11
  testing 6/4, 6/21–6
  TT systems 4/8
earth fault protection *see* fault protection
earth-free local equipotential bonding 4/19, 5/11, 7/10
earth mats 7/9
earth monitoring systems 5/42, 5/45, 5C/1–2, 7A/4
earth plates 5B/1
earth rods 5B/2–3
earth terminals, warning notices 5/10–11
earthed equipotential bonding *see* bonding
earthing arrangements 3A/2–5, 4/2, 4/3–4, 5/37–50
  additions and alterations 1/6
  assessment of general characteristics 3/1, 3/2
  caravans 7/34
  construction sites 7A/1, 7A/4
  electrode water heaters and boilers 5/57, 5/58
  generators 7A/1
  insulation monitoring devices 5/35
  major changes in new regulations 10/3
  mobile or transportable units 7/32
  periodic inspection and testing 7A/4
  photovoltaic power supply systems 7/31
  reduced low voltage systems 4/11
  SELV and PELV systems 4/15
  temporary installations 7/38
  see also earth electrodes; earth monitoring systems
earthing conductors 5/10–11, 5/38, 5B/4
earthing terminals or bars 5/39
electric clocks 3B/2, 5/56
electric shavers 3B/2, 5/56, 7/7

electric shock protection 1/4–5, 4/1–19
  agricultural installations 7/16–17
  barriers 4/17
  bath and shower rooms 7/4
  caravans 7/22
  construction sites 7/15
  enclosures 4/12, 4/17, 6/15, 7/4
  exhibitions, shows and stands 7/28
  major changes in new regulations 10/1–2
  mobile or transportable units 7/32
  non-conducting locations 4/18, 6/15
  photovoltaic power supply systems 7/30
  restrictive conductive locations 7/20–21
  saunas 7/13
  swimming pools 7/9–10, 7/12
  temporary installations 7/36–7
  see also additional protection; automatic disconnection; basic protection; bonding; earthing arrangements; electrical separation; extra-low voltages; fault protection; obstacles; out-of-reach placement
electrical accessories *see* accessories
Electrical Installation Certificates 6/33–4
  departures from BS requirements 1/3, 1/4
  electricity supply characteristics 3/2
  inspections 6/2
  prospective fault current values 4/30
electrical insulation 4/12, 4/13, 4/17
  agricultural installations 7/16
  bath and shower rooms 7/4
  batteries 5/54
  connections of earthing conductors to earth electrodes 5B/4
  FELV systems 4/10
  inspection and testing 4/17, 6/4, 6/11–13, 6/14, 6/15
  monitoring 7/32
  monitoring devices 5/35–6
  non-conducting locations 4/18
  outdoor equipment 5/62, 5/63
  PELV systems 4/14
  PEN conductors 5/43
  protective conductors 5/42
  reduced low voltage systems 4/11
  resistance 6/15, 10/3, 10/5
  resistance testing 6/11–13, 6/14, 6/15
  restrictive conductive locations 7/21
  saunas 7/13
  SELV systems 4/14, 4/15, 6/14, 10/3
electrical separation 4/13, 4/19
  circuits 5/13, 5/26, 6/14
  dodgems 7/38
  low voltages 5/26
  restrictive conductive locations 7/20, 7/21
  safety services 5/65
  SELV and PELV systems 4/15, 5/26
  swimming pools 7/9
  telecommunications circuits 5/26
  testing 6/4, 6/14
  warning notices 5/11
  see also barriers; electrical insulation; enclosures
electrical systems
  types of 3A/1–7

see also IT systems; TN systems; TN-C systems; TN-C-S systems; TN-S systems; TT systems; wiring systems
Electricity at Work Regulations 1989 2
Electricity Safety, Quality and Continuity Regulations 2002 2, 3A/5
  information provision 6/1
  overhead lines 4/17
  photovoltaic power supply systems 7/30
electricity supplies
  caravans 7/35
  characteristics 1/6–7, 3/2
  construction sites and demolition sites 7A/1–5
  PELV systems 4/13–14
  protection against interruptions 1/6
  residual current devices 4A/2
  safety services 5/64–5
  SELV systems 4/13–14, 7/16, 7/20, 7/21
  see also batteries; dual supply; emergency power supplies; highway power supplies; photovoltaic (PV) power supply systems; public electricity supply
electrode water heaters and boilers 5/56–8
electromagnetic compatibility 3/4
electromagnetic influences 1/6, 1/9, 4/35–6, 5/12, 5/13
electronic converters 7/29
ELV *see* extra-low voltage
emergency controls 1/9, 5/31–33, 7/29, 7/37
emergency evacuation 4/20–21
emergency lighting, temporary installations 7/36
emergency power supplies 1/7, 3/2, 7/20
  see also generating sets
emergency switching *see* emergency controls
enclosures
  agricultural installations 7/16, 7/17
  as protective conductors 5/13, 5/40, 6/8
  bath and shower rooms 7/4
  cable rating 8/2
  electric shock protection 4/17, 6/15, 7/4
  electrical insulation 4/12
  electrical separation 4/13
  extra-low voltage circuits 4/22
  FELV systems 4/10
  fire precautions 4/20, 4/21, 4/22, 4/23, 7/17
  fixings 4/12
  generating set batteries 5/54
  inspection and testing 6/15
  IP ratings 4/12, 4/15, 4/17
  joints and connections 5/23
  lighting 4/22
  line conductors 5/13
  neutral conductors 5/13
  outdoor equipment 5/62, 5/63
  protection against burns 4/23
  protection of cables from external heat sources 5/14
  reduced low voltage systems 4/11
  saunas 7/13
  SELV systems 4/15
  swimming pools 7/12
  switches 5/30
  temporary installations 7/37
  warning notices 4/17
  see also conduits; trunking
environmental protection 1/7
  see also corrosion protection
equipotential bonding *see* bonding
erection *see* selection and installation
escape routes, fire precautions 4/20–1
exhibitions, shows and stands 7/28–9
expanded polystyrene, PVC cable separation from 5/15
expansion joints, floor and ceiling heating systems 7/39
extension outlet units (EOUs), construction sites 7A/4
exterior installations, fire-fighters' switches 5/34
external earth fault loop impedance, testing 6/24–6
external influences 3/3
  additional protection 5/2
  agricultural installations 7/18
  bath and shower rooms 7/7
  caravan parks 7/23
  classification 5/2
  fire sealing 5/25
  marinas 7/25
  photovoltaic power supply systems 7/31
  saunas 7/13
  swimming pools 7/10
  temporary installations 7/37
  see also environmental protection; non-electrical services
extra-low voltage generating sets 5/50–4
extra-low voltage lamps 5/60
extra-low voltage systems 3A/6
extra-low voltage transformers 7/29
extra-low voltages
  Band I classification 1/3, 5/25, 5/26
  caravan plugs and socket-outlets 7/34
  circuit testing 6/15
  fire precautions 4/22
  line conductor identification 5/6
  range 1/2, 3A/6
  see also FELV (functional extra-low voltage) systems; PELV (protective extra-low voltage) systems; SELV (separated extra-low voltage) systems

F

fairgrounds 7/36–8
fault current, definition 4/24
fault protection 1/5, 4/29–34
  circuit breakers 4/25
  definition 4/1
  disconnection times 4/5
  electrical separation 4/13
  FELV systems 4/10
  floor and ceiling heating systems 7/38
  generating sets 5/51
  major changes in new regulations 10/1
  marinas 7/26
  omissions permitted 4/2
  photovoltaic power supply systems 7/30
  reduced low voltage systems 4/11
  supplementary bonding 4/5

wiring systems near non-electrical services 5/27
see also prospective fault current
fault protection devices 4/32–4, 7/16
see also circuit breakers; fuses
fault voltages 4/35
FELV (functional extra-low voltage) systems 3A/7, 4/10, 10/2
filter circuits 5/44
final circuits 3C/1–5
  arrangement 3/3
  cable selection 8/11–14, A/1–2
  disconnection times 4/5
  electrical separation 5/13
  major changes in new regulations 10/5
  protective device selection A/1–2
  temporary installations 7/36
  see also radial circuits; ring circuits
fine wire conductors, joints and connections 5/23
fire-fighters' switches 5/34–5
fire precautions 1/5, 4/20–3
  agricultural installations 7/17
  barriers 4/23, 7/17
  busbar trunking 5/24
  cables 4/22, 5/24
  conduits 5/24
  enclosures 4/20, 4/22, 7/17
  floor and ceiling heating systems 7/39
  lighting 4/21, 4/22, 4/23, 5/61
  major changes in new regulations 10/2
  outdoor equipment 5/63
  overload protective devices 4/22, 4/27
  residual current devices 4/22, 7/17
  sealing 5/24–5
  soil and floor warming cables 5/59
  wiring systems 4/20–1, 4/22, 5/24–5
  see also safety services; thermal effects
fire propagation 4/23
fire risk zones 1/7, 5/65
fixed equipment and appliances 6/14, 7/8
fixing
  luminaires 5/61
  see also suspended systems
fixings, enclosures 4/12
flammable liquids, escape routes 4/21
flammable materials *see* combustible materials
flexible cables and cords 5/13
  caravans 7/34
  ceiling roses 3B/4
  electrical separation 4/13
  fire precautions 4/23
  lighting 5/61
  mobile or transportable units 7/33
  supplementary bonding conductors 5/50
flexible structures 5/19
floor heating systems 5/58–9, 7/8, 7/38–40
floors 5/16–17, 6/15, 7/17, 10/5
  see also non-conducting locations
flora, protection from 5/19
foreign bodies 5/15
fountains 7/11, 10/4
frequency, inspection and testing 6/33
functional extra-low voltage systems *see* FELV (functional extra-low voltage) systems
functional switching 5/34
functional testing 6/4, 6/30–1
fused spurs 3C/1, 3C/2, 3C/4
fuses 1/9, 4/11, 4C/1–2
  see also HBC fuses

G

gas storage compartments, mobile or transportable units 7/33
generating sets 5/50–4
  bonding 5/52, 7A/2
  major changes in new regulations 10/3
  mobile or transportable units 7/32
  overcurrent protection 4/34
  safety services 3/4
  warning notices 5/12, 7/20
  see also emergency power supplies; low voltage generating sets
generators
  construction sites 7A/1–2
  dodgems 7/38
  neutral conductors 7/38
  SELV and PELV systems 4/13, 4/14
  see also small-scale embedded generators (SSEGs)
grouped cables 5/20, 8/2

H

hand-held tools, conductive locations 7/20
handlamps, conductive locations 7/21
harmonic currents 5/52, 10/5
HBC fuses 4C/2, 4E/1, 4E/2
Health and Safety at Work etc. Act 1974 2
heat-free areas, floor and ceiling heating 7/39
heat hazards *see* burns; fire precautions; temperatures; thermal effects
heat sources, protection from 5/14, 5/27
heating and ventilation control panel, conduit and trunking sizing 9P/3, 9P1/6–7
heating appliances
  agricultural installations 7/17, 7/19
  circuits 3C/5
  fire precautions 4/21, 4/23
  saunas 7/14
  see also boilers; ceiling heating systems; floor heating systems; soil and floor warming cables; surface heating systems; water heaters
heating conductors and cables 5/58–9
heating elements 5/44
high protective conductor currents 5/44–5, 10/3
highway power supplies 5/62–4, 10/3
horticultural installations *see* agricultural installations
hot water appliances *see* boilers; water heaters
houseboats *see* marinas

I

identification 5/2–3
  additions and alterations 5/5–7
  busbars 5/3, 5/7
  cables and conductors 5/3–7

caravan inlet couplers 7/34
earth clamps 5/48
earth electrodes 5B/4
earthing conductors 5/38
emergency controls 5/32, 7/37
fire-fighters' switches 5/34, 5/35
isolators 5/29, 5/30, 7/19, 7/29
mobile or transportable units 7/33
protective devices 5/7
switch wires 5/6
switches for mechanical maintenance 5/31
switchgear 5/2
temporary installations' socket-outlets 7/38
underground cables 5/16
see also colour-coding; warning notices
IEE regulations
major changes 10/1–5
plan and style 1–3
scope, object and principles 1/1–10
immersion heaters, circuits 3C/5
impact, protection against 5/16
index of protection *see* IP (index of protection) ratings
indirect contact *see* fault protection
inlet couplers, caravans 7/34
insects, protection from 5/19
inspection and testing 1/10, 6/1–34
accessibility for 5/2
agricultural installations' standby supplies 7/20
automatic disconnection 4/6, 6/4, 6/20–1
caravans 7/35
construction sites 7A/2, 7A/4
electrical insulation 4/17
major changes in new regulations 10/3
residual current devices 4A/2, 6/26, 6/30–1, 7A/2
sealing 5/25
see also certification; reports
inspection covers, earth electrodes 5B/4
inspection labels 5/10
installation *see* selection and installation
instructed person, definition 2/4
insulated cables *see* cables
insulation *see* electrical insulation; thermal insulation
insulation fault location systems 4/9
insulation monitoring devices (IMDs) 4/9, 4/22, 5/35–36
insulation monitoring systems, overload protection devices 4/28
interlocks, functional testing 6/31
IP (index of protection) ratings
agricultural installations 7/18
barriers 4/15, 4/17
bath and shower rooms 7/4, 7/7
cable protection 5/14
caravan cables 7/35
caravan parks 7/23
code 5A/1
enclosures 4/12, 4/15, 4/17
floor and ceiling heating systems 7/39
marinas 7/25, 7/26
mobile or transportable units 7/33
saunas 7/13

swimming pools 7/10
switchgear 4/21
temporary installations 7/37, 7/38
wiring system sealing 5/24
isolation 1/9, 5/27–30
agricultural installations 7/19
construction sites 7/16
exhibitions, shows and stands 7/29
fire precautions 4/22
generating sets 5/53
marinas 7/26
photovoltaic power supply systems 7/31
temporary installations 7/37
warning notices 5/10
see also disconnection times
isolators 5/30
agricultural installations 7/19
circuit breakers and 5/29
construction sites 7/16
definition 5/28
exhibitions, shows and stands 7/29
identification 5/29, 5/30, 7/19, 7/29
marinas 7/26
PEN conductors 5/43
temporary installations 7/37
IT equipment 5/44
IT systems
automatic disconnection 4/8–10, 6/21
definition 3A/1
earthing arrangements 3A/4, 4/8–9
fire precautions 4/22
generating sets 5/52
insulation fault location systems 4/9
insulation monitoring devices 4/9, 4/22, 5/35, 5/36
insulation monitoring systems 4/28
mobile or transportable units 7/32
overcurrent protection 4/25
overload protection devices 4/28
residual current devices 4/25, 4/28, 5/52
residual current monitoring 5/36
step-up transformers 5/59
temporary installations 7/36, 7/38

J

joints and connections 1/10, 5/23
access 5/2, 5/23, 5/42
cables 5/43, 7/37
conduits 5/42
corrosion protection 5B/4
earthing conductors 5/38, 5B/4
fire precautions 4/22
meters 5/48, 5/49
PEN conductors 5/23, 5/43
protective conductors 5/42
suspended outdoor lighting 5/64
temporary installations 7/37
see also expansion joints
junction boxes 5/13, 7/10, 7/30

K

kitchens
ring circuits 3C/1
see also cooker circuits; cooking appliances

L

labelling *see* identification; warning notices
lampholders 3B/3–4, 4/2, 5/62
legislation 2, 1/3
   see also *Electricity Safety, Quality and Continuity Regulations 2002*
licensed premises 1/3
lift shafts, cables in 5/27
lighting 5/59–61
   accessories 3B/3–4, 5/61, 5/62
   additional protection 4/5
   agricultural installations 7/19
   cables 5/61, 5/62
   conduit sizing 9/2
   control gear 5/62
   enclosures 4/22
   exhibitions, shows and stands 7/29
   fire precautions 4/21, 4/22, 4/23, 5/61
   major changes in new regulations 10/3
   mounting on combustible materials 4/23
   out-of-reach placement 7/38
   overcurrent protection 5/62
   phase earth fault loop impedance testing 6/23
   residual current devices 4/5
   stroboscopic effects 5/62
   swimming pools 7/11, 7/12
   temporary installations 7/36, 7/37, 7/38
   voltage drops 5/22
   wiring systems 5/61
   see also discharge lighting; emergency lighting; outdoor lighting; spotlights; underwater lighting
lightning protection 1/2, 5/37
   bonding 4/2, 4/3
   major changes in new regulations 10/2
   photovoltaic power supply systems 7/30
   see also overvoltage protection
line conductors 1/9, 4/24, 5/6, 5/13, 7/36
live conductors 4/20, 4A/2, 5/23
loaded conductors, number of 5/20
low power supply, safety services 5/65
low-voltage generating sets 5/50–4, 10/3
low-voltage systems 3A/1
   major changes in new regulations 10/2
   overvoltage protection 4/36
   plugs and socket-outlets 3B/1–3, 5/55–6, 7/34
   protection against voltage disturbances and electromagnetic disturbances 4/35
   separation 5/26
low voltages 1/2, 1/3, 3A/1
luminaire supporting couplers 3B/4, 4/10
luminaires *see* lighting

M

main bonding conductors 5/47–9
main earthing terminals 5/39
mains distribution units (MDUs) 7A/3
maintenance 1/9, 3/4, 5/2, 5/27
   see also mechanical maintenance
marinas 7/24–7
marking *see* colour-coding; identification
maximum demand *see* demand
maximum temperatures, PVC insulation 5/20
mechanical maintenance 5/30–1
mechanical protection, marinas 7/25
mechanical stresses, protection from 5/18–19, 5/59
metalwork 4/12, 5/18
meters, bonding connections 5/48, 5/49
mid-point conductors 5/3, 5/7
miniature circuit breakers 4D/1–6
Minor Electrical Installation Works Certificates 6/33, 6/34
mobile equipment 4/5, 7/36
mobile or transportable units 7/31–3
monitoring
   agricultural installations 7/20
   electrical insulation 7/32
   outdoor equipment 5/63
   see also earth monitoring systems; residual current monitoring
monitoring devices 5/35–6, 10/2
   see also insulation monitoring devices
motor caravans *see* caravans
motors 1/9, 4/22, 5/54, 7/37
mould growth, protection from 5/19
multipole switches 5/31, 5/42
multiwire conductors, joints and connections 5/23
mutual detrimental influences 1/9, 5/12
   see also compatibility; compatibility, electrical equipment; non-electrical services

N

NAPIT certificates and reports 6B/1
neutral conductors
   alphanumeric identification 5/7
   classification of electrical systems 3A/2
   colour-coding 5/3
   enclosures 5/13
   generators 7/38
   isolation 5/28
   overcurrent protection 4/25
   protective devices 1/9
   sizing 5/20
   temporary installations 7/36, 7/38
   see also PEN conductors
non-conducting locations 4/18, 6/15, 7/10
   see also floors; walls
non-electrical services 1/9, 3/3–4, 5/27, 5/47–8
   see also mutual detrimental influences
non-fused spurs, ring circuits 3C/2–3
notices *see* warning notices
numbering, IEE regulations 1–2

O

obstacles 4/17, 4/18, 7/10
   see also barriers
oil-cooled transformers, fire precautions 4/20
ordinary person, definition 2/4
out-of-reach placement 4/17–18, 4/21, 7/10, 7/29, 7/38
outdoor equipment 5/62–64
outdoor lighting 5/59–61, 5/62–4, 10/3
outlet units (OUs) 7A/3
overcurrent protection 1/5, 4/24–34
   caravan parks 7/23

electrode boilers and water heaters 5/56
fire precautions 4/22
generating sets 5/52
lampholders 3B/3
lighting circuits 5/62
major changes in new regulations 10/2, 10/5
parallel conductors 10/5
TT systems 4/7–8, 4/24, 4/25
overcurrent protection devices 4/25
construction sites 7/16
final circuits 3C/1
IT systems 4/9
marinas 7/26
operating characteristics 4E/1–2
photovoltaic power supply systems 7/30
reduced low voltage systems 4/11
TT systems 4/7–8
see also circuit breakers
overhead cables 4/2, 4/17, 7/18, 7/23, 7/25
see also suspended systems
overload current, definition 4/24
overload current protection 4/22, 4/25, 4/26–29, 4/34, 5/54
see also circuit breakers; fuses
overload protective devices 4/22, 4/25, 4/26–27, 4/28
see also fuses
overvoltage protection 1/6, 4/36
see also lightning protection

P

parallel conductors 4/29, 4/31–2, 5/21, 10/5
parallel operation, generating sets 5/53
partitions
cables in 4/5, 5/17–18, 10/2
see also walls
passive insulation monitoring devices (IMDs) 5/36
PELV (protective extra-low voltage) systems 3A/7, 4/13–15
agricultural installations 7/16
bath and shower rooms 7/4
electrical separation 4/15, 5/26
generating sets 5/51
identification of line conductors 5/6
major changes in new regulations 10/3
proximity to electrical services 5/26
restrictive conductive locations 7/21
testing 6/4, 6/14–15
PEN conductors 4/22, 5/4, 5/23, 5/28, 5/43
periodic inspection and testing *see* inspection and testing
Periodic Inspection Reports 6/2, 6/33, 6/34
phase earth fault loop impedance, testing 6/21–4
phase sequence, testing 6/4, 6/29
photovoltaic (PV) power supply systems 7/30–1
pipework, bonding 4/2, 5/48, 7/6
placement out of reach *see* out-of-reach placement
plants, protection from 5/19
plastic pipework, bonding 5/48, 7/6
pleasure craft *see* marinas

plugs and socket-outlets 3B/1–3, 5/55–6
access 7/38
agricultural installations 7/18, 7/19
bath and shower rooms 7/7, 10/4
caravan parks 10/4
caravans 7/34
circuits 3C/6–7, 5/45–6
colour-coding 3C/7
construction sites 3B/3, 7/15, 7/16, 7A/2, 7A/5
emergency switching devices 5/31, 5/32
FELV systems 4/10
functional switching 5/34
low-voltage systems 3B/1–3, 5/55–6, 7/34
major changes in new regulations 10/2, 10/4
marinas 7/26
mobile or transportable units 7/32
non-fused spurs 3C/3
PELV systems 4/15
protective conductors 5/45–6
radial circuits 3C/4
reduced low voltage systems 4/11
residual current devices 4/5
ring circuits 3C/1, 3C/2, 3C/3
SELV systems 4/15
swimming pools 7/9, 7/11
switching for mechanical maintenance 5/31
temporary installations 7/36, 7/38
testing 6/10–11, 6/23, 6/26, 6/30–1
voltage discrimination 3C/7
warning notices 7/9
PME (protective multiple earthing) *see* TN-C-S systems
PNB (protective neutral bonding) *see* TN-C-S systems
polarity testing 6/4, 6/5, 6/11, 6/16–17, 6/22
polluting environments 5/15
polystyrene, PVC cable separation from 5/15
portable equipment 4/18, 7/15, 7A/2
portable tools 7/20–1
power demand *see* demand assessment
power frequency fault voltages and stress voltages 4/35
power supplies *see* electricity supplies
powertrack systems 4/34, 5/13, 10/2, 10/5
preservatives, PVC cable separation from 5/15
projectors, fire precautions 4/21
projects, conduit and trunking sizing 9P/1–9P1/8
prospective earth fault currents, generating sets 5/51
prospective fault current 4/29–30, 6/4, 6/27–9
see also fault protection
prospective short circuit current (PSCC), generating sets 5/51
protection
from burns 4/23–4
from electromagnetic influences 1/6, 4/35–6
from heat sources 5/14, 5/27
from impact 5/16, 7/25
from mechanical stresses 5/18–19, 5/59
from radiation 5/19
from seismic effects 5/19
from thermal effects 4/19–24
from voltage disturbances 1/6, 4/35–6
from waves 5/15

from wet conditions 5/27
from wildlife and vegetation 5/19
of conductors 9/1
of earth electrodes 5B/2, 5B/4
of earthing conductor connections 5B/4
of electricity supplies against interruptions 1/6
of low-voltage installations 4/35–6
of or from non-electrical services 1/9, 5/27
see also additional protection; basic protection; cable protection; corrosion protection; electric shock protection; environmental protection; fault protection; fire precautions; IP (index of protection) ratings; lightning protection; mechanical protection; overcurrent protection; overload current protection; overvoltage protection
protective conductors 5/39–50
  alphanumeric identification 5/7
  caravan parks 7/22, 7/23
  caravans 7/34
  Class II equipment 4/12
  classification of electrical systems 3A/2
  colour-coding 5/3–4
  earthing arrangements 4/2
  electrode water heaters and boilers 5/56
  enclosures as 5/13, 5/40, 6/8
  floor and ceiling heating systems 7/38
  generating sets 5/52
  major changes in new regulations 10/3
  residual current devices 4A/2
  SELV systems 4/15
  separated circuits 4/13
  testing 6/4, 6/5–8
  see also bonding; duplicate protective conductors; high protective conductor currents
protective devices 4/25
  co-ordination between conductors and 4/26–7
  generating sets 5/51, 5/52
  identification 5/7
  inspection and testing 6/1, 6/26
  motors 7/37
  parallel conductors 4/32
  positioning 4/27, 4/30–1
  selection and installation 1/8, 1/9, A/1–2
  water heaters and boilers 5/56
  see also circuit breakers; fuses; isolation; overcurrent protection devices; overload protective devices; residual current devices
protective extra-low voltage systems *see* PELV (protective extra-low voltage) systems
protective multiple earthing (PME) *see* TN-C-S systems
protective neutral bonding (PNB) *see* TN-C-S systems
PSCC (prospective short circuit current), generating sets 5/51
public electricity supply, construction sites 7A/1
pumps, fountains 7/11
PV systems *see* photovoltaic (PV) power supply systems
PVC cables, separation from polystyrene and preservatives 5/15

R

radial circuits 3C/4–5
  cable selection 8/6–8, 8/12
  conductors 3C/1
  major changes in new regulations 10/5
  overcurrent devices 3C/1
  protective conductors 5/46
radiation, protection from 5/19
ratings
  circuit breakers 4D/2–3
  see also cable rating; IP (index of protection) ratings
records *see* documentation
reduced low voltage systems 4/10–11, 7/15
reduced voltages, construction sites 7A/2
regulations *see* IEE regulations; legislation
reports 3/2, 6/33–4, 6B/1, 7/35
  see also NAPIT certificates and reports; Periodic Inspection Reports
residual current devices 4/15, 4A/1–2
  a.c. systems 4/5
  agricultural installations 7/16, 7/17
  bath and shower rooms 7/5, 10/3
  cables in walls or partitions 4/5
  caravan parks 10/4
  caravans 7/34
  construction site accommodation 7A/2
  electrode boilers and water heaters 5/57
  exhibitions, shows and stands 7/28
  fire precautions 4/22, 7/17
  floor and ceiling heating systems 7/38
  generating sets 5/52, 5/53
  inspection and testing 6/26, 6/30–1, 7A/2
  IT systems 4/9, 4/25, 4/28, 5/52
  lighting 4/5
  major changes in new regulations 10/2, 10/3, 10/4
  marinas 7/26
  mobile equipment 4/5
  mobile or transportable units 7/32
  notices 5/10
  outdoor equipment 5/63
  overcurrent protection 4/25
  plugs and socket-outlets 4/5
  reduced low voltage systems 4/11
  restrictive conductive locations 7/21
  saunas 7/13
  swimming pools 7/9, 7/11, 7/12
  temporary installations 7/36
  TN systems 4/22, 5/52
  TN-C systems 4A/2
  TN-S systems 5/53
  transformers at exhibitions, shows and stands 7/29
  TT systems 3A/3, 4/7–8, 4/22, 5/52
residual current monitoring 4/9, 5/36
resistance *see* earth electrode resistance; electrical insulation, resistance
restarting
  emergency switching devices 5/32, 5/33
  rotating machine motors 5/54
restrictive conductive locations 4/15, 7/20–1
rewireable fuses 4C/1
ring circuits 3C/1–4, 5/45, 6/4, 6/8–11, 10/5

risk assessments, lightning protection 10/2
rocky soil, earth electrodes 5B/2
rotating machines 5/54–5

S

safety 1/4–6, 5/54
see also electric shock protection; fault protection; fire precautions; overcurrent protection; overvoltage protection; thermal effects
safety services 3/4, 5/64–6
  agricultural installations 7/20
  assessment of supply characteristics 3/2
  electricity supplies 1/7
  major changes in new regulations 10/1
saunas 7/12–14, 10/4
Schedule of Inspection 6/2
schedules 5/8, 6/2
sealing, fire precautions 5/24–5
segregation *see* separation
seismic effects, protection from 5/19
selection and installation
  cables 8/1–14, A/1–5
  conductors 1/8
  electrical equipment 1/10, 3/1–5, 5/1–66
  protective devices 1/8, 1/9, A/1–2
  wiring systems 1/8, 5/13–27
  see also sizing
SELV (separated extra-low voltage) systems 3A/6, 4/13–15
  agricultural installations 7/16
  bath and shower rooms 7/4
  electrical insulation 6/14, 10/3
  electrical separation 4/15, 5/26
  electricity supplies 4/13–14, 7/16, 7/20, 7/21
  generating sets 5/51
  major changes in new regulations 10/3
  restrictive conductive locations 7/20, 7/21
  swimming pools 7/9
  testing 6/4, 6/14
semi-enclosed fuses 4C/1
separated circuits 4/13, 4/19
  see also electrical separation
separation, PVC cables from polystyrene and preservatives 5/15
shavers 3B/2, 5/56, 7/7
short-circuit current *see* fault current
short-circuit protection *see* fault protection
shower rooms *see* bath and shower rooms
shows *see* exhibitions, shows and stands
signs *see* warning notices
single point bonding, cables 5/22
single-phase installations, identification of alterations and additions 5/5
site accommodation, construction sites 7A/2–4
sizing
  bonding conductors 4/3, 5/47
  cables 7/34, 8/2–5
  conductors 1/8, 3C/1, 3C/4, 5/22, 5/47, 5/64
  conduits and trunking 9/1–4, 9P/1–9P1/8
  earthing conductors 4/3, 4/33, 5/38
  neutral conductors 5/20
  protective conductors 5/39–41

skilled person, definition 2/4
small-scale embedded generators (SSEGs) 10/3
  see also photovoltaic (PV) power supply systems
socket-outlets *see* plugs and socket-outlets
soil and floor warming cables 5/58–9
solar photovoltaic power supply systems *see* photovoltaic (PV) power supply systems
solar radiation, protection from 5/19
soldering fluxes 5/15
solid bonding, cables 5/22
solid foreign bodies 5/15
space factors 9/1–4
spaced cables 8/2
special installations 7/1–40, 10/3–4
spotlights, fire precautions 4/21
spurs, ring circuits 3C/2–4, 5/45
standards 5/1
standby systems *see* emergency power supplies; generating sets
stands *see* exhibitions, shows and stands
starters
  rotating machine motors 5/54
  see also restarting
static inverters, generating sets 5/52
stationary batteries, generating sets 5/54
stationary equipment and appliances *see* fixed equipment and appliances
steel enclosures, as protective conductors 6/8
step-up transformers 5/59
stranded conductors, earth electrodes 5B/2
street furniture 4/2, 5/62–4, 10/3
  see also outdoor lighting
stroboscopic effects, lighting 5/62
structural movement 5/19
substations, protection of low-voltage installations 4/35–6
sunlight, protection from 5/19
supplementary bonding 4/16
  a.c. systems 6/27
  agricultural installations 7/16–17, 7/19
  bath and shower rooms 7/5–7, 10/3
  d.c. systems 6/27
  fault protection 4/5
  generating sets 5/52
  major changes in new regulations 10/2, 10/3
  restrictive conductive locations 7/21
  swimming pools 7/10
  temporary installations 7/37
supplementary bonding conductors 5/49–50, 6/5
supply *see* electricity supplies
supply incoming and distribution units (SIDUs) 7A/3
supply incoming units (SIUs) 7A/3
surface heating systems 5/59
surface temperatures
  floor heating systems 7/39
  lighting 4/22
surface voltage gradients, earth electrodes 5B/3
suspended systems
  outdoor lighting 5/64
  see also overhead cables
swimming pools 7/8–12, 10/4

switches and switching 1/9, 5/27–8
  agricultural installations 7/19
  bath and shower rooms 5/58, 7/7
  cooker circuits 3C/5
  enclosures 5/30
  for mechanical maintenance 5/30–1
  identification 5/6, 5/31
  lighting accessories 3B/4, 5/62
  PEN conductors 5/43
  protective conductors 5/42
  swimming pools 7/11
  temporary installations 7/37
  warning notices 5/30
  water heaters and boilers 5/58
  see also circuit breakers; emergency controls; fire-fighters' switches; functional switching; isolation; isolators; multipole switches
switchgear
  access 5/65, 7A/2
  bath and shower rooms 7/7
  construction sites 7/16, 7A/2
  exhibitions, shows and stands 7/29
  fire precautions 4/20, 4/21
  functional testing 6/31
  identification 5/2
  isolation 5/29
  photovoltaic power supply systems 7/31
  safety services 5/65
  saunas 7/14
  swimming pools 7/11
  temporary installations 7/37
  see also circuit breakers; control gear; emergency controls; switches and switching
symbols 5/7–8

T

telecommunications cables, bonding 4/3
telecommunications circuits, segregation 5/26
telecommunications equipment 5/44
telephone kiosks, residual current devices 5/63
temperature control, electrode water heaters and boilers 5/58
temperatures
  automatic disconnection of TN systems 4/6
  cable rating 8/1–2
  cable selection and installation 5/14
  floor and ceiling heating systems 7/39
  protection against burns 4/24
  PVC insulation 5/20
  selection and installation of equipment 1/10
  see also overcurrent protection; thermal effects
temporary installations 7/28–9, 7/36–8
  see also construction sites
terminations 5/23
test instruments
  checking and recalibration 6/34, 6A/1
  earth electrode resistance 6/18, 6/19
  external earth fault loop impedance 6/25
  functional testing 6/30
  insulation resistance 6/12
  phase earth fault loop impedance 6/22
  polarity 6/16
  prospective fault current 6/28
  protective conductors 6/5
  ring circuit conductors 6/9
testing *see* inspection and testing
thermal effects 1/5, 4/19–24, 10/2
  see also fire precautions; temperatures
thermal insulation, cables in 5/20–1, 8/2, 10/2
thermostats, electrode water heaters and boilers 5/58
three-phase systems 4/24, 5/5, 10/5
through wiring, luminaires 5/62
timber treatments, PVC cable separation from 5/15
TN systems
  automatic disconnection 4/6–7, 6/20
  definition 3A/1
  disconnection times 4/5
  fire precautions 4/22
  generating sets 5/51, 5/52
  lampholders 5/62
  overcurrent protection 4/24, 4/25
  residual current devices 4/22, 4A/2, 5/52
  temporary installations 7/38
TN-C systems 3A/1, 3A/4, 4A/2
TN-C-S systems
  agricultural installations 7/17
  bonding conductor sizing 5/47
  caravan parks 7/22, 7/23
  construction sites 7/15
  definition 3A/1
  earthing arrangements 3A/3, 3A/5, 4/4
  exhibitions, shows and stands 7/28
  isolation 5/28
  marinas 7/24
  mobile or transportable units 7/32
  protective multiple earthing (PME) 5D/1–2
  swimming pools 7/9
  temporary installations 7/36
TN-S systems
  bonding conductor sizing 5/47
  caravan parks 7/22
  definition 3A/1
  earthing arrangements 3A/2, 3A/5, 4/3
  exhibitions, shows and stands 7/28
  isolation 5/28
  marinas 7/24
  residual current devices with generating sets 5/53
touching cables 8/2
transformers 5/59
  construction sites 7A/3, 7A/4
  dodgems 7/38
  exhibitions, shows and stands 7/29
  fire precautions 4/20
  mobile or transportable units 7/32
  overcurrent protection 4/34
  SELV and PELV systems 4/13
transportable units 7/31–3
trunking
  as protective conductor 5/40, 5/41
  fire precautions 4/23, 5/24
  sizing 9/1–4, 9P/1–9P1/8
  temporary installations 7/37
  see also busbar trunking; enclosures
TT systems
  automatic disconnection 4/7–8

bonding conductor sizing 5/47
caravan parks 7/23
definition 3A/1
disconnection times 4/5
earth electrode resistance testing 6/19–20
earth electrodes 4/7
earth fault loop impedance 4/8
earthing arrangements 3A/3, 3A/5, 4/4
exhibitions, shows and stands 7/28
fire precautions 4/22
generating sets 5/52
inspection and testing 6/19–20
lampholders 5/62
marinas 7/24
overcurrent protection 4/7–8, 4/24, 4/25
residual current devices 4/7–8, 4/22, 4A/2, 5/52
temporary installations 7/38
two-phase installations 5/5

U

ultraviolet radiation, protection from 5/19
underfloor heating *see* floor heating systems
underground cables 5/16, 5/26, 7/18, 7/23, 7/25
 see also soil and floor warming cables
underground earthing conductors 5/38
undervoltage protection 1/6
underwater lighting 7/11

V

vegetation, protection from 5/19
ventilation 7/20, 7/29
 see also heating and ventilation control panel
verification *see* inspection and testing
vibration 5/18, 5/30, 5/31
voltage bands 1/3, 5/25–6
voltage discrimination, socket-outlets 3C/7
voltage disturbances 1/6, 3/4, 4/35–6, 10/2
 see also overvoltage protection
voltage drops 5/22, 6/4, 6/32, 10/5
voltage gradients, earth electrodes 5B/3
voltage ranges 1/2, 3A/6
voltages
 construction site accommodation 7A/2
 exhibitions, shows and stands 7/28
 marinas 7/24
 temporary installations 7/36, 7/37, 7/38
 warning notices 5/9
 see also extra-low voltages; low voltages; overvoltage protection; power frequency fault voltages and stress voltages

W

walls 4/5, 5/17–18, 6/15, 10/2, 10/5
 see also non-conducting locations
warning notices 1/10, 5/3
 agricultural installation standby supplies 7/20
 barriers and enclosures 4/17
 bonding 5/10–11
 cable core colour interfaces 5/5
 caravans 7/35
 colour-coding of additions and alterations 5/11
 construction site supply, distribution and transformer units 7A/4
 control switches 5/30
 dual supply 5/12
 earth-free local equipotential bonding 4/19
 earth terminals and conductors 5/10–11
 electrical insulation 4/12
 electrical separation 5/11
 fire-fighters' switches 5/34, 5/35
 floor and ceiling heating systems 7/40
 generating sets 5/12, 7/20
 isolation 5/10
 marinas 7/26–7
 mobile or transportable units 7/33
 photovoltaic power supply systems 7/30, 7/31
 plugs and socket-outlets 7/9
 residual current devices 5/10
 safety services 5/66
 voltages 5/9
 see also identification
water heaters 5/56–8, 9P/2, 9P1/3–5
 see also boilers; immersion heaters
waves, protection from 5/15
wet conditions 4/15, 5/14–15, 5/27
 see also corrosive environments
wildlife, protection from 5/19
wire mesh earth electrodes 5B/2
wiring regulations *see* IEE regulations
wiring systems 1/8
 access 9/1
 agricultural installations 7/18–19
 caravan parks 7/23
 construction sites 7/15, 7A/2
 earth-free local equipotential bonding 4/19
 electrical separation 4/13
 escape routes, fire precautions 4/20–1
 exhibitions, shows and stands 7/29
 fault protection near non-electrical services 5/27
 fire precautions 4/20–1, 4/22, 5/24–5
 lighting 5/61
 major changes in new regulations 10/2
 marinas 7/25
 mobile or transportable units 7/33
 photovoltaic power supply systems 7/31
 safety services 5/66
 saunas 7/14
 selection and installation 1/8, 5/13–27
 swimming pools 7/10
 temporary installations 7/37
 see also cables; conductors; conduit; trunking
wood preservatives, PVC cable separation from 5/15

Z

zone classification
 bath and shower rooms 7/2–4, 10/3
 saunas 7/13, 10/4
 swimming pools 7/9, 10/4